African Local Knowledge
& Livestock Health

African Local Knowledge
& Livestock Health
Diseases & Treatments in South Africa

William Beinart
Rhodes Professor of Race Relations
African Studies Centre, University of Oxford

Karen Brown
ESRC Research Fellow
Wellcome Unit for the History of Medicine, University of Oxford

WITS UNIVERSITY PRESS

Published in the UK, USA Europe and Commonwealth excluding Southern Africa by
James Currey
an imprint of
Boydell & Brewer Ltd
PO Box 9, Woodbridge
Suffolk IP12 3DF (GB)
www.jamescurrey.com

and of

Boydell & Brewer Inc.
668 Mt Hope Avenue
Rochester, NY 14620-2731 (US)
www.boydellandbrewer.com

Published in South Africa, Namibia, Lesotho, Zimbabwe and Swaziland in volume form only
by Wits University Press
1 Jan Smuts Avenue
Johannesburg 2001, South Africa
www.witspress.co.za

Rest of the world is an open market between James Currey and Wits University Press.

British Library Cataloguing in Publication Data
A catalogue record for this book is available on request from the British Library

ISBN 978-1-84701-083-4 (James Currey cloth)

Typeset in 11/12.5pt Bembo
by Avocet Typeset, Somerton, Somerset
Printed and bound in the United
States of America

Contents

Contents

Acknowledgements

First of all we would like to thank the many farmers who agreed to be interviewed and provided us with interesting and exciting information about livestock and animal diseases. We will not name them individually here, but they appear in the footnotes, and collectively they made a major contribution to our understanding of local knowledge about veterinary medicine.

We owe a deep debt to our excellent and inspirational research assistants and translators – Barbara Kgari in the North West Province, Sonwabile Mkhanywa in Mbotyi and Tumisang Mohlakoana (alias Gavin) in QwaQwa. Thanks to David Swanepoel for sourcing documents at the Ondersterpoort Veterinary Library and for hosting our trips to Koppies with the help of Sarie Swanepoel. Lenox Mkhanywa and Simphiwe Yaphi assisted with interviews in Mbotyi and Tuffy Kirsten was a generous host and a fund of information. Troth Wells also participated in fieldwork in Mbotyi and took some of the pictures. In Mabeskraal we enjoyed the hospitality of Rebecca Mabe who arranged a number of interviews.

The scope of our research benefited from work carried out in other sites in the Eastern Cape by Vimbai Jenjezwa in the Kat River valley, Andrew Ainslie in Peddie District and Mike Kenyon in Masakhane, near Alice. Andrew Ainslie was joint author of an article emerging from the research. Tim Gibbs trawled the records on dipping in the Transkei. Roger Davies, a vet, visited Mbotyi and provided invaluable insights into animal health.

The following scientists and academics also provided us with information and contacts: Rudolph Bigalke, Maitseo Bolaane, Christo Botha, Michelle Cocks, Anne Digby, Tony Dold, Kobus Eloff, Heloise Heyne, Kassim Kasule, Riana Kleynhans, Andrie Loubser, Ivan Lwanga-Iga, Kate Makgatho, Viola Maphosa, Patrick Masika, Mothupi Molefe, Simon Mosenogi, Busani Moyo, Johan Naude, Kobus Pretorius, Michael Raito, Antonie Snijders, Arthur Spickett, Jonny Steinberg, Sandra Swart, Oriel Thekisoe, Piet van Zyl and Luvuyo Wotshela.

This research would not have been possible without the financial assistance of the Economic and Social Research Council whose readiness to fund research on extra-European themes and local knowledge is greatly appreciated. The Wellcome Unit for the History of Medicine at the University of Oxford managed the grant and provided strong support for Karen Brown. The School of Interdisciplinary Area Studies and African Studies Centre at Oxford provided financial and other assistance for William Beinart, and the Rhodes Chair of Race Relations supported our report-back visit in 2012.

Abbreviations

AHTs	Animal Health Technicians
AJUSA	*Agricultural Journal of the Union of South Africa*
AJCGH	*Agricultural Journal of the Cape of Good Hope*
ANC	African National Congress
ESRC	Economic and Social Research Council
JAH	*Journal of African History*
JDA	*Journal of the Department of Agriculture, Union of South Africa*
JHMAS	*Journal of the History of Medicine and Allied Sciences*
JRAI	*Journal of the Royal Anthropological Institute*
JSAS	*Journal of Southern African Studies*
JSAVA	*Journal of the South African Veterinary Association*
JSAVMA	*Journal of the South African Veterinary Medical Association*
NAJMR	*Natal Agricultural Journal and Mining Record*
NGO	Non-Government Organisation
NWP	North West Province
OJVR	*Onderstepoort Journal of Veterinary Research*
OJVSAI	*Onderstepoort Journal of Veterinary Science and Animal Industry*
TAJ	*Transvaal Agricultural Journal*

Names of Common Diseases

English	Afrikaans	Setswana	Sesotho	isiXhosa
Anthrax	*Miltsiekte*	*Lebete* (spleen); *kwatsi*	*Mbende* (spleen)	*Ubende; inyama makhwenkwe* or *isifo somkhwenkwe*
Blackwater	*Sponsiekte*	*Serotswana; leotwana; ramokutwane; letsogwane; sephatlho*	*Serotswana*	*IsiDiya*
Footrot	*Vrotpootjie*	*Tlhakwana*	*Mokaka*	
Gallsickness	*Galsiekte*	*Gala*	*Nyoko*	*Inyongo*
Gifblaar	*Gifblaar*	*Mohau*		
Heartwater	*Hartwater*	*Heartwater*		*Amanzana; Umkhondo*
Horsesickness	*Perdesiekte*	*Sterf*		*Isimoliya*
Botulism	*Lamsiekte*	*Magetla* (shoulder); *Mokokomalo*		
Lumpy skin disease	*Knopvelsiekte*	*Bolowetse ba letlano; sekgwakgwa*	*Skin en vel*	*Umhlaza; Ungqakaqha*
Redwater	*Rooiwater*	*Omo khibidu; motlapologo khibidu*		*Mbendeni; ihlwili; amanzabomvu*
Scab	*Brandsiekte*	*Lepalo*	*Lekgwakgwa*	*Ibuwa*
Slangkop (snake head)	*Slangkop*	*Sekaname*		
Three-day stiff sickness	*Drie-dae stywesiekte*			*Nonkhwanyana*
Tulp (tulip)	*Tulp*	*Teledimo*	*Tele*	

List of Maps, Photographs & Tables

Maps

Photographs

Table

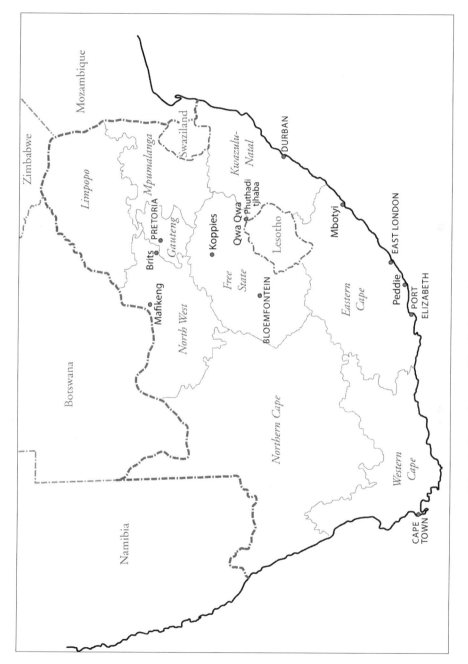

Map 1 General map of South Africa showing the field sites

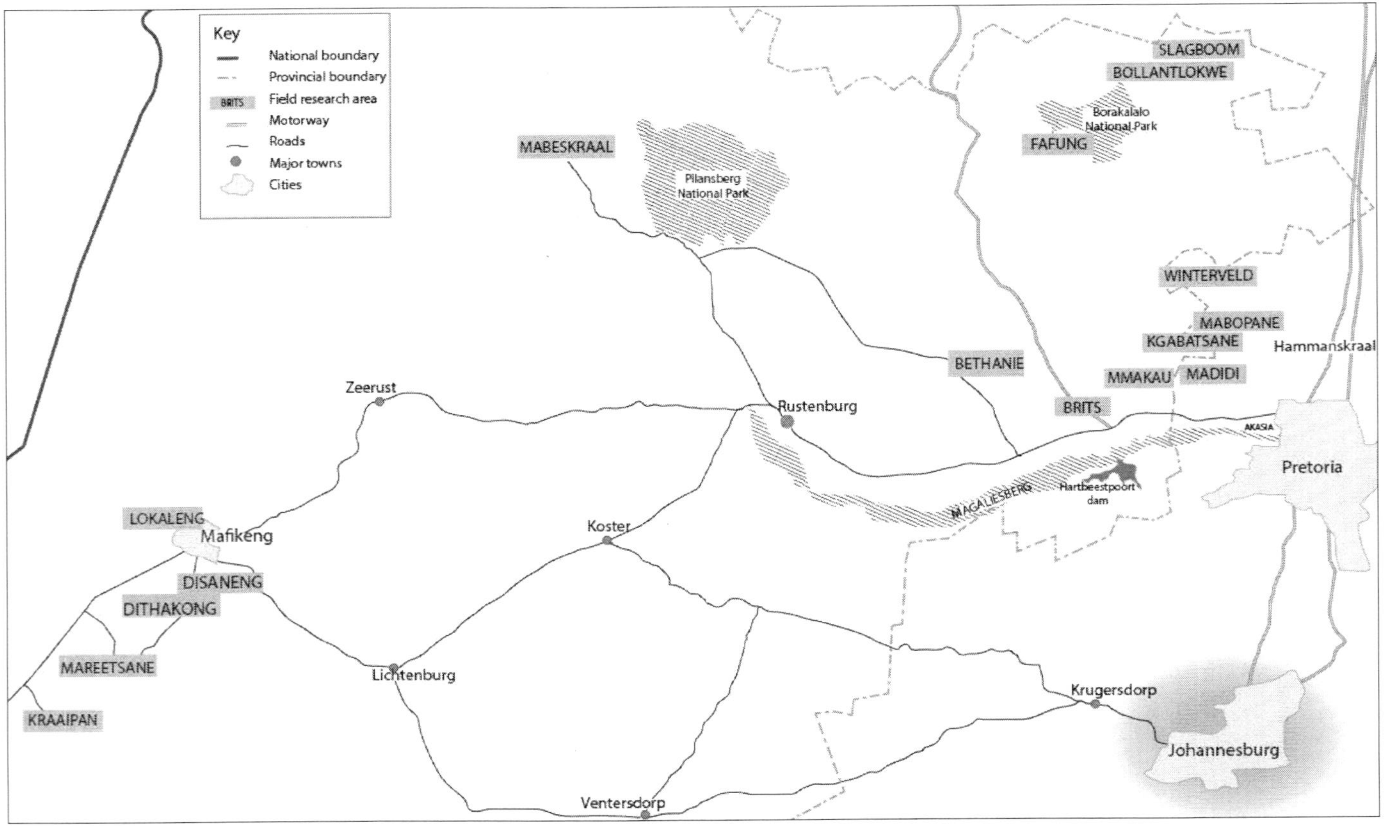

Map 2 North West Province (NWP)

Key

National boundary
Provincial boundary
Field research area
BRITS
Motorway
Roads
Major towns
Cities

Pretoria

Hammanskraal

SLAGBOOM

BOLLANTLOKWE

WINTERVELD

MABOPANE

KGABATSANE

MMAKAU / MADIDI

Borakalalo National Park

FAFUNG

BRITS

BETHANIE

Johannesburg

Krugersdorp

MAGALIESBERG

Hartbeespoort dam

Rustenburg

Pilansberg National Park

MABESKRAAL

Koster

Ventersdorp

Lichtenburg

Zeerust

Mafikeng

LOKALENG

DISANENG

DITHAKONG

MAREETSANE

KRAAIPAN

xiv

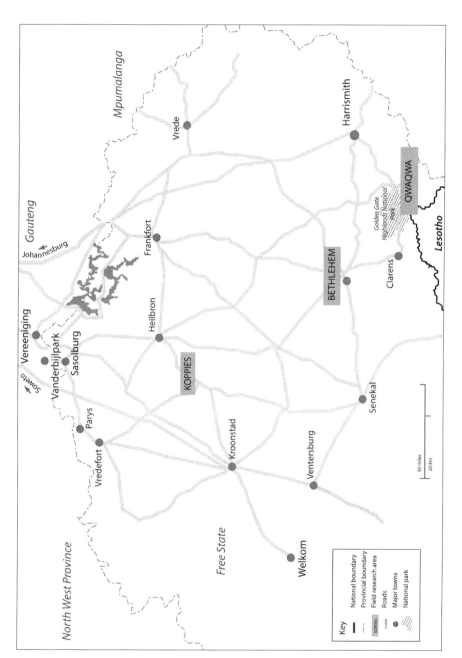

Map 3 Free State

Map 4 Eastern Cape

1

Introduction

African Local Knowledge & Veterinary Pluralism

The road from the bustling town of Lusikisiki to the coastal village of Mbotyi in Mpondoland passes through a large tea plantation then twists down a precipitous escarpment through dense indigenous forest. At any point on this route you are likely to find cattle, horses or goats straying across the way. On the beach cattle often roam the sands and dip their hooves in sea water. Green hills with lush pasture sweep up from the shore line. It is a romantic scene, but there is another side to this bucolic idyll. Many of the cattle have parts of their ears eaten away by ticks which are embedded in every orifice from the eyes to the anus. Even the fattest and healthiest of animals can be assailed by these parasites. Richard Msezwa, a local resident in his sixties, recalled: 'when I was a boy herding you seldom saw ticks in the veld. If you saw a tick you would get excited and call the other herdboys over. Now if you walk through grass your trousers become black with ticks.'[1] All is not well on South Africa's veld.

South Africa is a land not only of politics and gold but also of animals. Its teeming wildlife, once hunted close to extermination, is now conserved and celebrated in renowned reserves. Wildlife was displaced over a long period by livestock in the hands of Africans and then white settlers. Livestock became central to rural society and until recently many South Africans owned or worked with animals. They continue to play an important role in the lives of some black families, both urban and rural. Cattle still have a social and cultural value as well as an intrinsic economic worth. For these reasons farmers have observed the health of their animals and experimented with different ways of preventing and curing diseases. Local veterinary knowledge has constantly evolved, incorporating new ideas about causation and treatment.

---

[1] Richard Msezwa, Mbotyi, 25 February 2008.

1

This book is an exploration of recent African views of animal disease and remedies. It discusses the content of veterinary knowledge in the African-occupied rural areas. Our research revealed a wide interest in disease causation and a search for effective therapeutics. Much has been handed down from previous generations, but contemporary conceptualisations of disease and treatment also result from social interactions between African farmers, various livestock experts and the state. Arguably, this is one of the most important fields of African local knowledge in which millions of people are engaged. Livestock remain a major rural concern and this book is a first attempt at an overarching view of contemporary African veterinary knowledge.

Key contributions

Understanding local knowledge has become a central academic project amongst those interested in Africa and developing countries more generally. This book makes a major contribution at three levels. First, we explore a central body of rural African knowledge in a systematic way, using over 200 interviews. Our core argument is that African approaches to animal health rest largely in environmental explanations. Second, we examine the relationship of local knowledge about livestock disease to scientific knowledge. We discuss the legacy of traditional ideas, and the extent to which African stockowners have incorporated biomedical explanations and treatments. We relate these issues to current debates in history and anthropology about pluralism. We also discuss contestations over knowledge and their implications for effective treatment. Third, we debate changing African approaches to livestock diseases in South Africa during recent decades. African veterinary medical knowledge remains largely unrecorded and this book makes an important contribution in capturing its diversity, as well as its limits, in a number of different localities. Our research is contextualised by a discussion of scientific and biomedical approaches to animal diseases developed in South Africa over the last century. But our main body of material results from interviews conducted with African rural and peri-urban livestock owners.

We record African understandings of disease and the degree of acceptance (and rejection) of biomedical explanations. We also discuss a wide range of treatments from dipping with chemicals to plant remedies and the doctoring of space. Our aim is to understand the significance of and interrelationship between environmental, biomedical and supernatural approaches to livestock diseases. Overall our findings suggest relatively limited penetration of biomedical ideas about germs, or parasites such as ticks, in the explanation of disease. The dominant form of understanding has rested in environmental and nutritional concerns. African livestock owners were also attentive to supernatural threats, but we argue that witchcraft is now less prominent, and more ambient forms of supernatural danger have become prevalent.

Context: livestock diseases in South African history

Some livestock diseases and environmental hazards, such as poisonous plants, were present in pre-colonial South Africa, but the epidemiological picture became far more complicated as a result of colonialism. Livestock gradually colonised the area that became southern Africa over the last 2,000 years, moving down the continent with African migrants, and then adopted by Khoisan people already established in the region. African livestock, especially Nguni-type cattle, short-haired goats and fat-tailed sheep, tended to be small and hardy with some resistance to local diseases and conditions through a long period of exposure. The Dutch East India Company and European settlers introduced new breeds of livestock and other domestic animals (horses and pigs), from the seventeenth century. Incorporation of foreign breeds such as merino sheep, angora goats and Friesland cattle accelerated after Britain assumed control of the Cape of Good Hope during the Napoleonic Wars (1806). Imported animals were generally more susceptible to African infections and they brought new maladies with them. Cattle diseases such as lungsickness (introduced in the 1850s), redwater (1870s), rinderpest (1890s) and East Coast fever (from about 1902) devastated many herds.

For many African people, the arrival of new diseases coincided with the loss of land. Afrikaner settlers gradually displaced Khoisan people in the drier, western half of the Cape. By the early-nineteenth century, settlers dominated the land in these areas and with the help of the new colonial state penetrated into Xhosa territory in the Eastern Cape. Following British annexation, a relatively small section of the Cape's Afrikaner population mobilised themselves for a Great Trek into the interior in the 1830s. Simultaneously, African chiefdoms became more centralised following the rise of the Zulu kingdom in the early decades of the nineteenth century. The trekkers were involved in sequential battles for land with the Sotho, Zulu, Tswana and Pedi states of the interior, carving out significant tracts on the highveld and establishing the two Boer republics. British colonial power followed in their wake, widening the boundaries of the Cape and Natal. The destruction of the Zulu kingdom (1878–1885), signalled the end of independent African political authority and annexations were completed in the 1890s. Finally, British forces defeated the Afrikaner Republics of the Transvaal and Orange Free State during the South African War (1899–1902). This conflict paved the way for the Union of South African as a single country in 1910. By this time, perhaps 80 per cent of South Africa's land had come under white ownership in private tenure.

Africans were defeated politically, and subjected to a white-controlled and increasingly segregationist state. But they did not lose all their land and livestock. Much of the land they retained was in the heart of their old kingdoms and well suited to the mix of arable and pastoral production that had underpinned their economies. While the 1913 Natives Land Act and subsequent legislation curtailed the rights of black people to buy new farms, it also reserved these remaining areas for African occupation, mostly in forms of customary tenure. Many black families were also able to keep livestock as labour tenants on white-owned land. In the 1930s, about 80 per cent of the African population still lived in rural areas and after the massive losses from rinderpest and East Coast fever, some of their herds recovered.

From the late-nineteenth century, veterinary science became an important tool of government policy. The state became highly interventionist in controlling livestock movements, organising a national campaign of compulsory dipping against tick-borne diseases – a major scourge in South Africa – and introducing vaccines. Government policies aimed primarily at protecting white-owned livestock were implemented nationally and also benefited Africans. Black people owned about half the cattle in South Africa in the 1930s. Reports mounted of environmental degradation in the reserves as more livestock were pastured on restricted areas of veld.

Under apartheid (1948–94) the National Party sought to consolidate segregated land-holdings, as well as to define South Africans by ethnicity, creating a number of self-governing African homelands or 'bantustans'.[2] Eager to deter African people from migrating to the cities, and to reverse the threat of soil erosion, the apartheid regime sought to shore up African agriculture by intervening directly in rural settlement and land use. The state imposed betterment, later called rehabilitation – a system of planning devised before 1948 but energetically pursued afterwards. Scattered African settlements were grouped into villages and arable plots were separated into blocks. Grazing lands were fenced into camps (large paddocks) that could be rotated to conserve the veld. Especially in the early years of betterment, cattle were culled through enforced sales. Destocking, designed to diminish overgrazing, was one of the most unpopular aspects of these measures: it helped to foster Africans' distrust of government and to some extent the veterinary scientists who worked for the state, because they saw it as an assault on their cultural and economic wealth. Betterment and rehabilitation triggered rural rebellions in some areas. This massive exercise in social and environmental engineering was not fully implemented and elements of the plans, such as fenced camps, could not generally be enforced or sustained.

During the homeland era, at the height of apartheid from the 1960s to the 1990s, central control over veterinary services declined, and dipping became irregular in many places. After 1994 the African National Congress (ANC) government curtailed national veterinary provision. Due to budgetary constraints and other priorities, less money has been invested in regular dipping or systematic vaccination campaigns. It is anomalous that the state has partially

---

[2] William Beinart, *Twentieth-Century South Africa* (Oxford, Oxford University Press, 2001).

withdrawn from veterinary provision at the very time that more land is becoming available to African livestock owners. Land reform has proceeded slowly in South Africa but African owners are gradually gaining access to larger areas.[3] Most of the redistributed land is used for livestock rather than for crops. Land reform is essentially livestock reform. This trend is likely to continue and the number of African livestock owners will probably increase.

Consequently, and perhaps counter-intuitively with respect to a society that has urbanised rapidly over the last century, animal health is of growing importance and interest to a greater number of people. A small number of wealthier African farmers own hundreds of animals on private farms, but the evidence is not for a major concentration of animals in fewer hands. In general, hundreds of thousands of African livestock owners – possibly over a million – possess relatively small herds and flocks, probably averaging below 10 cattle per family. The great majority of these livestock are managed on land under forms of communal ownership. Disease is a ubiquitous problem, and one of the biggest obstacles all farmers face.

These shifting patterns of ownership will have an impact both on the incidence of disease and the ways in which diseases are controlled. Large-scale commercial farmers, still largely white, have better resources and knowledge to control diseases and treat their animals with biomedicines. Owners of small herds on communal pastures, where they cannot separate their animals, are in a far weaker position. In recent years they have been left largely to their own devices and many are uncertain as to the best treatments, or find it difficult to meet the costs of commercially manufactured medicines.

In this context, an understanding of African livestock owners' approach to disease becomes especially important. How do they conceptualise diseases? How do people prioritise treatments, and how are these affected by economic constraints? To what extent has local knowledge become infused with biomedical concepts of disease causation and control, given the long history of vaccines, quarantines and dipping?

---

[3] Cherryl Walker, *Landmarked: Land Claims and Land Restitution in South Africa* (Athens OH, Ohio University Press and Jacana Media, 2008); Deborah James, *Gaining Ground: 'Rights' and 'Property' in South African Land Reform* (Abingdon, Routledge-Cavendish, 2006); Ben Cousins, 'Livestock Production and Common Property Struggles in South Africa's Agrarian Reform', *Journal of Peasant Studies*, 23, no. 2–3 (1996), 166–208; William. Beinart, 'Strategies of the Poor and Some Problems of Land Reform in the Eastern Cape, South Africa', in Theunis D. de Bruyn and Peter F. Scrogings (eds), *Communal Rangelands in South Africa: a Synthesis of Knowledge* (Alice, University of Fort Hare, 1998), 81–92; Ruth Hall, 'Land Restitution in South Africa: Rights, Development, and the Restrained State', *Canadian Journal of African Studies*, 38, 3 (2004), 654–71; Doreen Atkinson, *Going for Broke: The Fate of Farm Workers in Arid South Africa* (Cape Town, HSRC Press, 2007), 53–68.

Our research

When we conceived this study, with these questions in mind, our aim was to explore a countrywide picture. Our strategy was not primarily anthropological, based on detailed ethnoveterinary research in one village, but rather to compare areas with different political histories, environments, cultural characteristics and disease patterns. At the same time, we knew that a questionnaire survey was very unlikely to reveal the kind of information and knowledge that we sought. Our approach has therefore been to interview a cluster of individual livestock owners in a number of regions so that we could get sufficient depth to understand local diseases and treatments, and sufficient range to discuss our results comparatively.

In a sense, any African livestock-owning settlement in the country was a potential field site. But we did try to seek diversity. Our research, done separately, has taken us along some exciting and absorbing routes. Karen Brown identified North West Province as an interesting area, because in earlier work she had been struck by the range and density of poisonous plants in that region.[4] It was also well known for its nutritional deficiency diseases, resulting in conditions like *lamsiekte* (botulism) and as an expanding tick frontier. Serendipitously, she examined a doctoral student from the area who put her in touch with a well-connected research assistant, Barbara Kgari. Kgari opened up many paths in the province – especially at Mabeskraal, Mafikeng and Garankuwa.[5] These settlements, part of the former Bophuthatswana homeland, are flat highveld lands, with a mixture of savanna and bushveld, becoming progressively drier to the west. This ecological diversity has implications for the distribution of ticks and poisonous plants.

As a contrast, Brown also sought contacts in the former homeland area of QwaQwa, now in the eastern Free State. It is a mountainous zone, lying in the shadow of the Malutis (Drakensberg). Here she was lucky to find Tumisang (Gavin) Mohlakoana, a zoology student at the Free State University in Phuthaditjhaba, to act as research assistant.[6] Brown's final destination was the small northern Free State highveld town of Koppies. It was the one area where we interviewed white commercial farmers as well as African livestock owners from the adjacent township of Kwakwatse. Koppies is a largely Afrikaans-speaking area and at some distance from the nearest apartheid-era homeland. Here the route was also fortuitous in that the librarian at the Onderstepoort Veterinary

---

[4]   Karen Brown, 'Poisonous Plants, Pastoral Knowledge and Perceptions of Environmental Change in South Africa, c. 1880–1940', *Environment and History*, 13, 3 (2007), 307–32.

[5]   The student was Christina Kgari-Masondo.

[6]   Thanks to Oriel Thekisoe, of the Zoology department, for putting us in contact with Gavin.

Photo 1.1 View of Mbotyi, looking down to the sea (WB, 2011)

Photo 1.2 Hut and scenery at Lokaleng near Mafikeng, NWP (KB, October 2009)

Photo 1.3 Heavily populated village of Lusaka in QwaQwa (KB, March 2011)

Institute (Pretoria), David Swanepoel, who has provided outstanding support for this project, owns a farm in the area. This zone provided a strong contrast regarding local understandings of diseases, in the sense that white Afrikaner farmers generally espoused a biomedical approach to animal health.

William Beinart went back to Mpondoland on the east coast of the country, formerly in Transkei, where he had researched on rural history and resistance in earlier years.[7] His route was also fortuitously shaped by contact with a potential assistant, Sonwabile Mkhanywa, in the coastal village of Mbotyi. The area looked particularly promising because it was close to some of the largest indigenous forests in South Africa – likely to provide a wide range of medicinal plants. The east coast is also notorious for its ticks and Mpondoland has a history of traditionalist political resistance. Mkhanywa was an excellent translator and research assistant with a wide knowledge of animals and plants, as well as extensive connections in the village. Mbotyi, situated in an area of high rainfall and dense vegetation, provided an interesting contrast to Karen Brown's research areas.

Additional information on local knowledge in the Eastern Cape came from a trio of other investigators who have helped to supplement the range of our research. Andrew Ainslie wrote his doctorate on livestock management and smallholder livelihoods in the Peddie District of the former Ciskei.[8] He went back to this area with new questions generated by our research in mind. Mike Kenyon interviewed in Masakhane, an area near the heart of the old Christianised mission communities around Alice and Fort Hare. Finally, William Beinart co-supervised a Fort Hare Masters student, Vimbai Jenjezwa, who focused on similar questions in the Kat River valley north of Fort Beaufort.[9] This is a particularly interesting area socially because it has contiguous white-owned commercial farms, coloured smallholders and African communities.

Our material is therefore drawn from nine main sites. Five were in the interior – Mafikeng, Mabeskraal and settlements north of Pretoria such as Garankuwa, in North West Province, as well as Koppies and QwaQwa in the Free State. Four were in the Eastern Cape – Mbotyi and Peddie near the coast and the Kat River valley (mainly Tamboekiesvlei) and Masakhane (near Alice) in the hinterland. Collectively we interviewed about 100 people in North West

[7] William Beinart, *The Political Economy of Pondoland, 1860–1930* (Cambridge, Cambridge University Press, 1982) (the spelling is now generally Mpondoland); William Beinart, 'Ethnic Particularism, Worker Consciousness and Nationalism: the Experience of a South African Migrant, 1930–1960' in S. Marks and S. Trapido (eds), *The Politics of Race, Class and Nationalism in Twentieth Century South Africa* (London, Longman, 1987), 286–309.

[8] Andrew Ainslie, 'Keeping Cattle? The Politics of Value in the Communal Areas of the Eastern Cape Province, South Africa', unpublished PhD thesis, University College, London (2005).

[9] Vimbai Jenjezwa, 'Stock Farmers and the State: A Case Study of Animal Healthcare Practices in Hertzog, Eastern Cape Province, South Africa', unpublished MA, University of Fort Hare (2010).

Province, 40 in QwaQwa, 10 in Koppies and 100 in the Eastern Cape (30 in Mbotyi and Tamboekiesvlei/Hertzog; 20 in Peddie and Masakhane). There is a slight weighting to North West Province sites in some of the chapters, which reflects the range of interviews as a whole. In Mbotyi, William Beinart conducted extended follow up sessions with a few specialists and our Eastern Cape evidence in this book comes largely from this area. In addition we could draw on the published field work of Patrick Masika and his colleagues at Fort Hare for the Eastern Cape and Deon van der Merwe's thesis on the Madikwe District of North West Province.[10] Tony Dold and Michelle Cocks, working in the Eastern Cape, have also highlighted the significance of local plant knowledge in relation to the health and healing of herds.[11]

None of the researchers were fluent speakers of the relevant African languages and all were dependent on interpreters. Most of the interviews with livestock owners, particularly those by Brown and Beinart, were in the vernacular and simultaneously translated into English. Relatively few were recorded so our notes are all in English. All our quotations are the English words of our translators – except in a few cases where the informant spoke English. When appropriate we have noted vernacular words for diseases, ideas and plants. In North West Province the local language is Setswana, in QwaQwa Sesotho and in the Eastern Cape isiXhosa. Interviews with both whites and blacks in Koppies were largely in Afrikaans. We have attempted to think about all of this

[10] P. J. Masika, A. Sonandi and W. van Averbeke, 'Tick Control by Small-Scale Farmers in the Central Eastern Cape Province, South Africa', *Journal of the South African Veterinary Association (JSAVA)* 68, 2 (1997), 45–48; P. J. Masika, W. van Averbeke and A. Sonardi, 'Use of Herbal Remedies by Small-scale Farmers to Treat Livestock Diseases in Central Eastern Cape Province, South Africa', *JSAVA,* 71, 2 (2000), 87–91; O. T. Soyelu and P. J. Masika, 'Traditional Remedies used for the Treatment of Cattle Wounds and Myiasis in Amatola Basin, Eastern Cape Province, South Africa', *Onderstepoort Journal of Veterinary Research (OJVR)*, 76, 4 (2009), 393–7; B. Moyo and P. J. Masika, 'Tick Control Methods used by Resource-limited Farmers and the Effect of Ticks on Cattle in Rural Areas of the Eastern Cape', *Tropical Animal Health Production*, 41 (2009), 517–23; P. J. Masika, A. Sonandi and W. van Averbeke, 'Perceived Causes, Diagnosis and Treatment of Babesiosis and Anaplasmosis in Cattle by Livestock Farmers in Communal Areas of the Central Eastern Cape Province, South Africa', *JSAVA,* 68, 2 (1997), 40–44; D. van der Merwe, G. E. Swan and C. J. Botha, 'Use of Ethnoveterinary Medicinal Plants by Setswana-speaking People in the Madikwe area of the North West Province of South Africa', *JSAVA,* 72, 4 (2001), 189–96; Deon van der Merwe, 'Use of Ethnoveterinary Medicinal Plants in Cattle by Setswana-speaking People in the Madikwe Area of the North West Province', MSc Thesis, University of Pretoria (2000). See also Dowelani Edward Ndivhudzannyi Mabogo, 'The Ethnobotany of the Vhavenda', MSc Thesis, University of Pretoria (1990) and M. Hlatshwayo and P. A. Mbati, 'A Survey of Tick Control Methods used by Resource-poor Farmers in the Qwa-Qwa Area of the Eastern Free State Province, South Africa', *OJVR*, 72, 3 (2005), 245–9.

[11] A. P. Dold and M. L. Cocks, 'Traditional Veterinary Medicine in the Alice District of the Eastern Cape Province, South Africa', *South African Journal of Science,* 97, 9-10 (2001), 375–9; Michelle Cocks, *Wild Resources and Cultural Practices in Rural and Urban Households in South Africa: Implications for Bio-Cultural Diversity Conservation* (Grahamstown, Rhodes University, 2006).

material comparatively and have drawn out similarities and differences in the various localities. Our main poles of comparison are the sites investigated by Brown and Beinart. We use the first person plural throughout the text. Beinart and Brown did not interview together, except for a final report-back trip in November/December 2012. But we thought that the plural would be better than inserting our individual names continuously in the text. It is also true that we did very little as individuals because we were both accompanied by research assistants for the bulk of the time.

Our interviews were held with people who owned livestock. This clearly influenced our understanding of the significance of animals in rural society as a whole. But the logic of this approach resulted from our concern to discover veterinary knowledge and practice, rather than to assess the overall attitudes to livestock. Our interviews, which were designed to find out about local forms of knowledge, tended to concentrate largely on older people; younger men were interviewed in some sites. The majority of our informants were men, as men still dominate the pastoral sector. However, in North West Province and Mbotyi we met a number of women who helped their husbands with the livestock, or else were widowed and had taken on the task of looking after their animals. By and large, women entertained similar ideas to men when it came to understanding diseases.

Although we have a reasonable density of interviews in a number of sites, we have not conducted systematic village level surveys over a long period. This no doubt impacts on the material we deploy. We are largely reliant on people's reports or descriptions of what they thought and did in relation to the identification and treatment of diseases. We did not systematically observe postmortems, diagnostic procedures and treatments in all sites. However, the fieldwork included some observation. Researchers visited kraals and pastures with stockowners and accompanied Animal Health Technicians (AHTs see chapter 2) on their rounds. Some watched dipping, inoculation procedures and animal slaughters. We were shown medicinal plants and in Mbotyi participated in making one of the well-known medicines for gallsickness. In this site we also observed some diagnoses and treatments by a qualified veterinarian.

While we have approached the issues as historians, our book focuses largely on the recent past. At many points, especially in chapters 2 and 4, historical background is sketched because it is essential for context and argument. But we do not try to pursue a systematic discussion of historical changes that might have been better elucidated by life history interviews. Our interviews were present minded – though often with reference to the past as a way of contrasting current ideas and problems. Informants frequently enunciated a binary view of tradition and modernity, even if their practices were more hybrid. For example they would talk of remedies as traditional, and learnt from their ancestors, even when an exotic plant or industrial material was included in their repertoire. Older men complained about the youth for abandoning tradition, and trying to be modern, even though their own lives were deeply immersed in a modern economy. Partly because of the interview strategy, informants also tended to talk about the past in an undifferentiated way, prefixed by such phrases as 'in the old days'. On occasion, we did try to pursue dates, for example in periodising informants' sense of tick resurgence in North West Province (chapter 2) or the demise of milking in Mbotyi (chapter 7). But informants sometimes found it difficult to attach particular years or decades to historical processes and this is reflected in some of our discussions. There are also sections where we look back to anthropological texts of the 1930s in order to explore continuities and contrasts with the present. Again, because of our interview strategy, we are not always able to fill in the complex sequence of change – for example around ideas about supernatural forces. The book would have been different if research had focused on detailed life-stories of fewer informants.

We have named those who imparted information, as well as our intermediaries, because we wish to acknowledge the source of our material. However, we have omitted the names of informants who wished to remain anonymous. Few people that we approached refused to talk to us. Questions about livestock and healing are of great concern to rural stockowners. Some were suspicious about our motives and wondered whether such interviews would bring them any benefits. Nevertheless, many respondents found it interesting to talk about this subject. We encountered a degree of caution about revealing aspects of knowledge, particularly in relation to treatments. Yet many were happy to describe traditional plant remedies in detail. We thought that interviewees may be reluctant to discuss witchcraft and ritual pollution openly, especially because we were not generally in touch with individuals for long periods so that trust could be built up. Yet many were willing to explain ideas and concepts surrounding supernatural causes of disease. We address these in the final chapters as they featured in interviews, and are important in the overall analysis, but we cannot claim to have a detailed ethnographic understanding of contemporary notions of the occult.

In addition to the interviews with livestock owners, we spoke to a number of veterinarians, scientists and AHTs. While we have some grounding in the history of animal diseases in South Africa, and have read scientific publications about them, we do not have specific training or expertise in this area. Specialists alerted us to the significance and epidemiology of particular diseases and helped us to identify them. Veterinarians as well as AHTs were helpful in explaining the policies and everyday operations of the veterinary services. Attendance at stock days revealed something of the interface between experts and African livestock owners. Scientists at Onderstepoort Veterinary Institute provided useful information on such complex areas as tick resistance and evolving methods of dealing with tick-borne diseases, as well as pharmacology and toxicology. There is also a rich record of publications on medicinal plants in South Africa and a number, such as aloe species, have been analysed for their chemical properties.[12] With reference to these, and to specialists, we have tried to identify the diseases and plants that we have encountered through discussion with informants. However, we were not able to specify all of them and in some cases have called them only by their local names in different languages.

Historiography: towards medical and environmental pluralism

Veterinary knowledge is not exactly the same as that about human disease and treatment, but there are many overlapping areas. As the analysis of the history of medicine is better developed, this is where we should begin. Some key texts on the history of colonial medicine emphasised the contestations over know-ledge between colonisers and colonised. David Arnold, for example, argued that science 'delineated the relationship of power and authority between rulers and ruled'.[13] Studies in the environmental history of Africa tended to focus on similarly polarised systems of knowledge.[14] However, since Steven Feierman and John Janzen's collection on health and healing in Africa (1992), the historiography has laid greater stress on medical pluralism. They suggested that African approaches to healing were essentially flexible and drawn from multiple sources.[15] The focus is less on polarities between local and scientific ways of knowing, and more on cultural exchanges and the evolution of pluralistic medical systems.

Two recent books on South Africa exemplify this genre. Anne Digby sees African healing practices as 'profoundly adaptive'. Historically, African patients have tended to be eclectic in their choice of health care and they have based

[12] John Mitchell Watt and Maria Gerdina Breyer-Brandwijk, *The Medicinal and Poisonous Plants of Southern and Eastern Africa* (Edinburgh, Livingstone, 1962); Ben-Erik van Wyk and Nigel Gericke, *People's Plants: A Guide to Useful Plants of Southern Africa* (Pretoria, Briza, 2000); Dold and Cocks, 'Traditional Veterinary Medicine'; O. M. Grace, M. S. J. Simmonds, G. F. Smith and A. E. van Wyk, 'Therapeutic Uses of *Aloe* L. (Asphodelaceae) in Southern Africa', *Journal of Ethnopharmacology*, 119, 3 (2008), 604–14; L. J. McGaw and J. N. Eloff, 'Ethnoveterinary Use of Southern African Plants and Scientific Evaluation of their Medicinal Properties', *Journal of Ethnopharmacology*, 119, 3 (2008), 559–74.

[13] David Arnold (ed.), *Imperial Medicine and Indigenous Societies* (Manchester, Manchester University Press, 1988), 2; Maryinez Lyons, *The Colonial Disease: a Social History of Sleeping Sickness in Northern Zaire, 1900–1940* (Cambridge, Cambridge University Press, 1992).

[14] James Fairhead and Melissa Leach, *Misreading the African Landscape: Society and Ecology in a Forest-savanna Mosaic* (Cambridge, Cambridge University Press, 1996); Roderick Neumann, *Imposing Wilderness: Struggles over Livelihood and Nature Preservation in Africa,* (Berkeley CA, University of California Press, 2002); for an analysis William Beinart, 'African History and Environmental History', *African Affairs*, 99, 395 (2000), 269–302.

[15] Steven Feierman and John Janzen (eds), *The Social Basis of Health and Healing in Africa,* (Berkeley, University of California Press, 1992), 1–25

their decisions on 'availability or perceived utility'.[16] She takes issue with Fanon's contention that biomedicine was generally rejected as a tool of empire because it threatened cultural and social identities. Most Africans in South Africa, she argues, used biomedicine in conjunction with indigenous medicine. Nevertheless, the majority of African people continued to consult traditional healers. Karen Flint's monograph on the development of Zulu medicine during the colonial period shows how the binary between 'traditional' and 'modern' therapeutics is an artificial one. Medicine provided a forum for cultural exchange, and colonialism helped to create a multi-therapeutic society. Despite state support for western medicine, there was no shift towards a biomedical hegemony. Africans negotiated this new medical culture, absorbing elements that they conceived as useful, rather than rejecting it or accepting its epistemology and practices in totality.[17]

Waltraud Ernst has described medical pluralism as 'a variety of medical approaches existing alongside each other, at times in competition and at times in collaboration with or complementary to each other'.[18] Instead of looking at medicine simply as a reflection of power relations, or a source of tension between rulers and subjects, it is also essential to consider interactions and collaborations. The development of medical traditions weds the local with the global and generates hybridised medical knowledges. Historical accounts that look for syncrecy, Ernst suggests, rather than only rejection or opposition, can provide new insights into how medical knowledge has grown. The role of intermediaries is also important in this regard. Anne Digby and Helen Sweet, for instance, have shown that black nurses in South Africa played an important role in explaining western medicine to black patients. But they also altered biomedical explanations to fit in with their own personal and cultural world views, providing new medical conceptualisations and practices.[19] Cross-racial rural interactions – illustrated by Sandra Swart and Charles van Onselen in the former Transvaal – also produced distinctive local medical practices.[20]

In most contexts, medical knowledge of different people incorporates a range of ideas and curative traditions. Significant minorities in western societies dispute aspects of biomedical science, or what they call allopathic treatment,

[16] Anne Digby, *Diversity and Division in Medicine: Health Care in South Africa from the 1800s* (Oxford, Peter Lang, 2006), 32–33.

[17] Karen Flint, *Healing Traditions: African Medicine, Cultural Exchange and Competition in South Africa, 1820–1948*, (Athens OH, Ohio University Press, 2008).

[18] Waltraud Ernst, 'Plural Medicine, Tradition and Modernity. Historical and Contemporary Perspectives: Views from Below and from Above', in Waltraud Ernst (ed.), *Plural Medicine, Tradition and Modernity, 1800–2000* (London, Routledge, 2002), 1–18, quotation from page 4.

[19] Anne Digby and Helen Sweet, 'Nurses as Culture Brokers in Twentieth-century South Africa', in Ernst, *Plural Medicine, Tradition and Modernity,* 113–29.

[20] Sandra Swart, '"Bushveld Magic" and "Miracle Doctors" – an Exploration of Eugène Marais and C. Louis Leipoldt's Experiences in the Waterberg, South Africa, c.1906–1917', *Journal of African History (JAH)*, 45, 2 (2004), 237–55; Charles van Onselen, "Race and Class in the South

and seek complementary therapies. Although the majority of our material evinces examples of pluralism, we also encountered explicit rejection of biomedicine. Historically, this is evident from early-twentieth-century resistance to dipping. A current example emerged in an interview with Zulu Mokoena from Lusaka in QwaQwa. He expressed a disdain for western medicines and claimed: 'I only use Sotho medicine because it is safe and effective. I want to show respect for my ancestors and my culture.'[21] The fact that there were African people who did not openly embrace western medicine, and continued to adhere to some older elements of local knowledge and practice, emphasised the limits of colonial or western biopower as a cultural or political force.[22]

Most of the published history on African veterinary ideas has focused on East African pastoralists, such as the Maasai, who clearly developed a good knowledge of the disease environment.[23] They knew that ticks were dangerous, that tsetse flies brought death, that certain plants were toxic and that wildlife could be the host for livestock infections. They practised transhumance to optimise grazing and avoid seasonal tick and tsetse belts. They may have understood that some exposure to infection was important to boost immunity and they used various medicinal herbs to treat diseases. Many plants had emetic or purgative properties to flush out infection.

Until recently, historians of South Africa examined veterinary issues in African societies largely at a tangent – as part of a discussion of rural resistance. Since Charles van Onselen wrote about conflicts over rinderpest controls in the 1890s, opposition to veterinary regulations have been a significant subtheme in the historiography of hidden struggles and rural protest.[24] Mordechai Tamarkin

(contd) African Countryside: Cultural Osmosis and Social Relations in the Sharecropping Economy of the South-Western Transvaal, 1900–1950", *The American Historical Review*, 95, 1 (1990), 99–123; C. van Onselen, *The Seed is Mine: the Life of Kas Maine, a South African Sharecropper 1894–1985* (Oxford, James Currey, 1996).

21 Zulu Mokoena, Lusaka, 2 March 2011.

22 Lyn Schumaker, Diana Jeater and Tracy Luedke, 'Introduction. Histories of Healing: Past and Present Medical Practices in Africa and the Diaspora', *Journal of Southern African Studies*, 33, 4 (2007), 707–14.

23 Richard Pankhurst, 'The History and Traditional Treatment of Rabies in Ethiopia', *Medical History*, 14, 4 (1970), 378–89; Helge Kjekshus, *Ecology Control and Economic Development in East African History: the Case of Tanganyika 1850–1950* (London, James Currey, 1996); James Giblin, 'Trypanosomiasis Control in African History: an Evaded Issue?', *JAH*, 31, 1 (1990), 59–80; Richard Waller, 'Tsetse fly in Western Narok, Kenya', *JAH*, 31, 1 (1990), 81–101; Richard Waller and Kathy Homewood, 'Elders and Experts: Contesting Veterinary Knowledge in a Pastoral Community', in Andrew Cunningham and Bridie Andrews (eds) *Western Medicine as Contested Knowledge* (Manchester, Manchester University Press, 1997), 69–93; Lotte Hughes, '"They Give Me Fever': East Coast Fever and Other Environmental Impacts of the Maasai Moves', in Karen Brown and Daniel Gilfoyle (eds) *Healing the Herds:, Disease, Livestock Economies, and the Globalization of Veterinary Medicine* (Athens OH, Ohio University Press, 2010), 146–62.

24 Charles Van Onselen, 'Reactions to Rinderpest in Southern Africa 1896–97', *JAH*, 13, 3 (1972), 473–88; Beinart, *Political Economy of Pondoland*; C. Bundy, '"We don't want your rain, we

has shown that late-nineteenth-century Afrikaner farmers in the Cape strongly resisted scientific prescriptions and state regulations to deal with sheep scab. The Cape government attempted to introduce the first comprehensive dipping programme to kill the mites that caused this malady. Tamarkin argued that aside from the economic disadvantages of government legislation, frontier farmers had their own ways of dealing with animal disease.[25]

Our route into this project, which explores veterinary ideas beyond their role in resistance, has a long genesis. William Beinart became concerned, in the 1980s, to analyse betterment which had provoked so much conflict in the former African reserves or homelands. He argued that this intervention drew on scientific concerns about animal health and conservation. He focused on the formation of environmental thinking in South Africa, and found that veterinary scientists had been significant in early formulations of degradation and conservation.[26] Karen Brown was one of a group of doctoral students at Oxford who produced detailed studies of the relationship between science and public policy in different colonial contexts, particularly in relation to medical and environmental initiatives. The development of veterinary services in South Africa proved to be an important element in the formation of the Department of Agriculture and the colonial state more generally.[27] This paved the way for broader studies on the evolution of scientific knowledge and institutions in South Africa. Onderstepoort Veterinary Institute, established near Pretoria in 1908, was at the heart of internationally important research that helped to transform the understanding of many diseases as well as treatments. With Daniel Gilfoyle, Brown illuminated the research of Onderstepoort's scientists and their international networks.[28] These approaches attempted to move beyond the

---

[contd] won't dip": Popular Opposition, Collaboration and Social Control in the Anti-Dipping Movement 1908–16' in W. Beinart and C. Bundy, *Hidden Struggles in Rural South Africa* (James Currey and Ravan Press, London and Johannesburg, 1987), 191–221; P. Phoofolo, 'Epidemics and Revolutions: the Rinderpest Epidemic in late Nineteenth-Century Southern Africa', *Past and Present,* 138 (1993), 112–43.

[25] Mordechai Tamarkin, *Volk and Flock: Ecology, Identity and Politics among Cape Afrikaners in the Late Nineteenth Century* (Pretoria, University of South Africa Press, 2009).

[26] William Beinart, 'Soil Erosion, Conservationism and Ideas about Development: a Southern African Exploration, 1900–1960', *JSAS,* 11, 1 (1984), 52–83; William Beinart, 'Vets, Viruses and Environmentalism: the Cape in the 1870s and 1880s', *Paideuma,* 43 (1997), 227–52; William Beinart, *The Rise of Conservation in South Africa: Settlers, Livestock and the Environment, 1770–1950* (Oxford, Oxford University Press, 2003).

[27] Daniel Gilfoyle, 'Veterinary Science and Public Policy at the Cape 1877–1910", unpublished DPhil thesis, University of Oxford (2002); Karen Brown, 'Progressivism, Agriculture and Conservation in the Cape Colony circa 1902–1908', unpublished DPhil thesis, University of Oxford (2002); Daniel Gilfoyle, 'Veterinary Research and the African Rinderpest Epizootic: the Cape Colony 1896–1898', *JSAS,* 29, 1 (2003), 133–54.

[28] Daniel Gilfoyle, 'Veterinary Immunology as Colonial Science: Method and Quantification in the Investigation of Horsesickness in South Africa, c. 1905–1945', *Journal of the History of*

critique of science as simply a 'tool of empire' or a perk for the benefit of white farmers.[29]

While we argued for the importance of veterinary medicine and related policy, in South African history, this research nevertheless focused largely on science and the state. There were many indications as to how African communities conceived of these issues and we suggested some interpenetration of scientific and local ideas.[30] We knew, for example, that early settlers quickly learnt from Khoisan pastoralists about the value of trekking between sweetveld and sourveld.[31] Moving livestock seasonally (or transhumance) mitigated the effect of diseases like *lamsiekte*, which made animals lame and could be fatal. We knew that the early veterinarians also learnt about diseases from both white and black farmers. Zulu livestock owners influenced scientific knowledge about *nagana*, the animal disease caused by tsetse flies, and its relationship with the wider environment. The Zulu word *nagana* was even adopted into scientific usage. African farmers also had an intimate knowledge of the veld and were able to point veterinary scientists in the direction of some toxic flora.[32]

This project grew from our increasing awareness of the importance of local knowledge both for the large number of African livestock owners and in the early formation of veterinary and environmental science. Specific research on

[(contd)] *Medicine and Allied Sciences (JHMAS)*, 61, 1 (2006), 26–65; Daniel Gilfoyle, 'Anthrax in South Africa: Economics, Experiment and the Mass Vaccination of Animals, c. 1900–1945', *Medical History*, 50, 4 (2006), 465–90; Karen Brown, 'Tropical Medicine and Animal Diseases: Onderstepoort and the Development of Veterinary Science in South Africa 1908–1950', *JSAS*, 31, 3 (2005), 513–29; Karen Brown, 'From Ubombo to Mkhuzi: Disease, Colonial Science and the Control of *Nagana* (Livestock Trypanosomosis) in Zululand, South Africa, c1894–1953', *JHMAS*, 63, 3 (2008), 285–322; Karen Brown, 'Veterinary Entomology, Colonial Science and the Challenge of Tick-borne Diseases in South Africa during the late Nineteenth and early Twentieth Centuries,' *Parassitologia*, 50 (2008), 305–19.

[29] D. R. Headrick, *The Tools of Empire: Technology and European Imperialism in the Nineteenth Century* (Oxford, Oxford University Press, 1981); J. Krikler, *Revolution from Above Rebellion from Below The Agrarian Transvaal at the Turn of the Century* (Oxford, Clarendon Press, 1993), 77–83; S. Schirmer, 'African Strategies and Ideologies in a White Farming District: Lydenburg 1930–1970', *JSAS*, 21, 3 (1995), 509–27; S. Milton, The Transvaal Beef Frontier: Environment, Markets and the Ideology of Development, 1902–42' in T. Griffiths and L. Robin (eds), *Ecology and Empire: Environmental History of Settler Societies* (Edinburgh, Keele University Press 1997), 199–212.

[30] William Beinart, Karen Brown and Daniel Gilfoyle, 'Experts and Expertise in Colonial Africa Reassessed: Science and the Interpenetration of Knowledge', *African Affairs*, 108, 432 (2009), 413–33.

[31] William Beinart, 'Transhumance, Animal Diseases and Environment in the Cape, South Africa', *South African Historical Journal*, 58 (2007), 17–41.

[32] Brown, 'From Ubombo to Mkhuzi'; Brown, 'Poisonous Plants'; see also Daniel Gilfoyle, 'The Heartwater Mystery: Veterinary and Popular Ideas about Tick-Borne Animal Diseases at the Cape, c. 1877–1910', *Kronos*, 29 (2003), 139–60; Shirley Brooks 'Changing Nature: A Critical Historical Geography of the Umfolozi and Hluhluwe Game Reserve, Zululand 1887–1947', unpublished PhD thesis, Queens University, Kingston (2001).

African ideas seemed essential for a more complete view of veterinary practices in South Africa. We also felt that analysis of African responses to veterinary and related interventions needed to be embedded in a better understanding of indigenous ideas and how these had changed. This implied that more sustained research was required on the interaction between biomedical and local ideas. Our project reflects medical and environmental historiography that has examined the exchange and interpenetration of scientific and local knowledge.

Local knowledge and its limits

Our book builds upon the environmental and medical historiography by adding a veterinary perspective to writings on local knowledge systems. We use the term local, rather than indigenous or customary or traditional, because we want to emphasise pluralism in African veterinary knowledge and treatments. Paul Richards popularised the idea of indigenous knowledge in West African agriculture. He was arguing for adaptability and innovation in a context of social change.[33] Although the idea of indigenous knowledge is sometimes used as a synonym for local with such fluid meanings, we think that term is more appropriate for a deep ethnographic approach to societies or communities which have been less thoroughly incorporated into national or global networks. The communities in which we interviewed, or at least individuals within them, had been highly mobile. Both nineteenth- and twentieth-century South African history was characterised by population movements, the migrant labour system, urbanisation and forced removals under apartheid, all of which contributed to creating local cultural diversity. This is obvious in peri-urban areas such as Garankuwa on the northern boundaries of Pretoria, and home to hundreds of thousands of people. But it also applied to QwaQwa, where there had been massive in-migration from the white-owned farms of the Free State, or even Mbotyi, which had attracted settlers from other parts of Transkei, white holiday-makers, as well as families with an inter-racial or coloured background.

Our informants were not articulating an unadulterated indigenous knowledge; notions of indigeneity are problematic in the context of communities so thoroughly incorporated into a wider national society. There are many other terms for such local knowledge, such as creole or métis, in academic analyses which also stress hybridity and pluralism.[34] However, these have such strong American roots that they do not seem appropriate for South Africa. Nor do

---

[33] Paul Richards, *Indigenous Agricultural Revolution: Ecology and Food Production in West Africa* (London, Hutchinson, 1985).

[34] Carolyn Allen, 'Creole Then and Now: The Problem Of Definition', *Caribbean Quarterly*, 44, 1–2 (1998), 33–49, for discussion of 'creole'; James Scott, *Seeing Like State: How Certain Schemes to Improve the Human Condition Have Failed* (New Haven, Yale University Press, 1998), 311ff., for discussion of 'métis'.

these concepts sufficiently emphasise locality. The term vernacular does offer this sense of place and Beinart has used 'Cape vernacular' to discuss the early settler-controlled colonial pastoral economy that incorporated Khoisan and African ideas. This notion, however, works better for Cape settler society in the eighteenth and nineteenth centuries.[35] In using the term local knowledge, we wish to capture many of these elements, including plural practices, hybrid understandings of disease and treatment as well as the significance of locality and environmental influences.

Similarly, we have avoided the term 'ethnoveterinary' (which is now quite widely used in parallel with ethnobotany and similar coinages), because it might seem to indicate the practices of a particular ethnic group, or a body of knowledge distinct from the many influences of biomedicine. In our research, we needed to find ways of understanding those global impacts, but also recognising that biomedicine has not attained hegemony because many African people have maintained a mixture of beliefs about the causes and treatment of livestock diseases. Again, local knowledge seemed the most trenchant term by which to capture this intellectual space.

One of our most striking findings is that even farmers who used vaccines and modern drugs did not necessarily accept the aetiological explanations that might be expected to inform the administration of these therapeutics. Despite the fact that most of those interviewed were enthusiastic protagonists of dipping, relatively few African livestock owners accepted that ticks transmitted disease. They saw ticks as troublesome but not as vectors of specific infections such as gallsickness and heartwater. Rather they saw the latter as a consequence of seasonal environmental changes, or occasionally attributed them to supernatural forces if their animals fell sick under certain suspicious circumstances. Farmers who invested in the latest vaccines, dewormers and pour-on acaricides to deal with ticks, might also ban women from handling livestock because they adhered to beliefs that women could undermine the fecundity of their animals, or bring misfortune to the cattle kraal.

The term local knowledge is valuable because it encompasses such medical pluralism, rather than suggesting exclusively indigenous or traditional beliefs. Moreover, our research shows that such knowledge was generated and transmitted in specific localities. Local veterinary knowledge, even within one country, has been shaped by particular environmental influences and resources and by specific histories. Such ways of knowing are very largely orally transmitted and they are not generally publicised in written and other media. By and large local knowledge has not been systematically recorded and where it has, publication and dissemination of findings has remained in the academic or expert sphere. Rural small-scale farmers, the practitioners of local veterinary medicine, seldom have access to this material. These are fragmented systems of

---

[35] Beinart, *Rise of Conservation*, 45–47.

knowledge, unevenly handed down and, we suspect, partly forgotten. They do not generally aim to universalise and may be received by individuals incompletely and in different ways. For this reason also, our interviews suggest that the content remains distinctive in different localities.

Another striking finding of our research is that even within the same settlement individuals interpret and treat animal diseases in different ways. The interplay of biomedical and older forms of knowledge amongst African farmers has resulted in a plethora of different ideas about the most effective means of disease control. Fredrik Barth reminds us that people within a single group may participate differentially in multiple social knowledges and that 'branches' of knowledge are distributed differentially across groups of people.[36] This certainly applies to our context, in that individual African livestock owners within the same communities can hold diverse views. Such diversity may also be related to an increasing devolution of responsibility and practice. Since the relaxation in compulsory state programmes there has been growing latitude for innovation and improvisation.

The idea of the local is often offset against global influences. One strand in African studies emphasises the penetration of global forces in the continent as a critique of views of African exceptionalism or isolation. In this sense, also, rural African societies can be seen as participants in the modern world, fully connected to global forces, through a long connection with transatlantic trade, colonialism, capitalism, education and new social media. The impact of such processes is very evident and ubiquitous in rural South Africa: many have been to school and away to work; everyone is a consumer and some have mobile phones. However, 'to say that people live lives that are structured by the modern capitalist world system', James Ferguson argues, 'or that they inhabit a social landscape shaped by modernist projects does not imply that they enjoy conditions of life that they themselves would recognise as modern'.[37] Many rural people have very partial access to the fruits of globalisation and modernity. Older livestock owners we interviewed, for example, were largely excluded from formal schooling systems and had limited opportunity to learn about biomedicines or the general concepts that underpin scientific knowledge. This is a legacy of racial power structures in South Africa. Even now, as cellphones penetrate to the most distant rural settlements, the digital divide effectively separates most African smallholders. But exclusion cannot be the only explanation for their location on the peripheries of global circuits of knowledge. Paul Zeleza emphasises African intellectuals' capacity to rethink the unwanted forces of globalisation and Toyin Falola explores the *Power of African Cultures*.[38] These processes have long been evident in South

---

[36] Fredrik Barth, 'An Anthropology of Knowledge', *Current Anthropology*, 43, 1 (2002), 1–18.

[37] James Ferguson, *Global Shadows: Africa in the Neoliberal World Order* (Durham NC, Duke University Press, 2006), 168.

[38] Paul Tiyambe Zeleza, *Rethinking Africa's Globalization*, Volume 1: *The Intellectual Challenges* (Trenton NJ, Africa World Press, 2003); Toyin Falola, *The Power of African Cultures* (Rochester NY, University of Rochester Press, 2003).

Africa at a grassroots level.[39] As we have noted, some informants remain strongly committed to inherited forms of local knowledge, both in respect of explanations of disease and in relation to the efficacy of local medicines. The idea of the local is important for us in this respect also – that through a complex process of marginalisation and resistance, people's absorption of new scientific knowledge is very partial. They have neither been fully incorporated in global knowledge systems, nor have they wished to be overwhelmed by them.

Terms such as hybridity, pluralism, bricolage and mélange are useful vehicles for avoiding simple dichotomies between scientific and indigenous knowledge. Hybridity has, in some senses, been a celebratory term in anti-colonial and post-colonial thinking, recognising that identities, cultures and ideas are multivalent and dynamic. It has been deployed as a challenge to the essentialist conceptions of culture, or tribe, associated with colonial thinking. But hybridity as a concept also has it limits for our purposes. It may assume purity of systems of knowledge as a precursor to pluralism, and perhaps suggests a specific or static outcome – like a mule or a plant variety with a name. But in our study, both scientific knowledge and local knowledge were evolving at the same time as they were imbricating locally. Arguably, most knowledge is plural or hybrid, especially in contexts of rapid social change such as South Africa experienced over the last century. If this is the case, the terms lose specific meaning.

A further problem concerns the balance between different forms of knowledge in any particular context. Do the ideas of hybridity and pluralism suggest equilibrium between local and scientific bodies of knowledge? Or are we exploring a long-term change from the former to the latter? Equally, how useful are these ideas in exploring the dominant elements in local knowledge systems, when they vary so greatly between different places, and amongst individuals?

Rural people have tended to use an increasing variety of medicines, including home-mixed dipping solutions, local plants, and broad-spectrum antibiotics such as Terramycin (oxytetracycline). Some interviewees asserted the cultural importance not only of livestock ownership, but also of their ways of managing and treating animals. This could imply commitment to older patterns of transhumance, use of medicinal herbs or beliefs in the effectiveness of magical treatment of kraals as well as animals. In Mbotyi, every informant suggested they had recourse to plant medicines for some purposes, and other researchers have suggested about 70 per cent of livestock owners did so.[40] Even amongst those who used plants, however, there may have been different rationales. An AHT in QwaQwa was convinced that a sizeable minority, perhaps 40 per cent, of livestock owners in his area pursued herbal and other strategies specifically as an expression of their Sotho identity, rather than simply because they

---

[39] For the Eastern Cape, Philip and Iona Mayer, *Townsmen or Tribesmen: Conservatism and the Process of Urbanization in a South African City, 2nd edn* (Cape Town, Oxford University Press, 1971), and the debates they triggered.

[40] Masika, et al., 'Use of Herbal Remedies'.

worked.[41] More generally, we encountered people who preferred biomedicines because they seemed to be more efficient, fast-acting and convenient to use. Gathering and preparing plants could be time consuming and they were difficult to store. In Hertzog, for example, in the Eastern Cape, all the respondents said that they procured western drugs when they could afford them.[42] The extent to which farmers used modern drugs could be an indicator of levels of poverty, as well as attitudes to disease.

A consequence of the fragmented and to some degree increasingly individualistic patterns of local knowledge is that there are not necessarily coherent understandings of or even solutions for many of the problems that arise in rural communities, in such old-established areas of concern as livestock. 'The local skills, and capabilities, most of which have some empirical grounding' can be suitable for some problems but less so in the case of new diseases or addressing environmental or veterinary issues of a regional scale, such as an epizootic.[43] Moreover, as we will illustrate (chapter 5), transmission of local knowledge is by no means an open and effective process.

Gaps in knowledge can also lead to uncertainty. Susan Reynolds Whyte offers a more general analysis of uncertainty in respect of medical understanding in rural Uganda.[44] She argues that responses to disease and misfortune, as well as the search for treatment, are pragmatic and most people accept that they will be uncertain in their outcome. Our interviews did not indicate a sense of fatalism about treatment and many were convinced that their remedies were effective. However, the idea of uncertainty is also valuable for our analysis. We do not deploy it to indicate a general state of mind. Informants were operating in a particular historical context, where some were unconfident about older methods and insufficiently convinced of new forms of diagnosis and disease control. Even when they did accept the latter, they could seldom afford the most effective drugs and acaricides – which exacerbated their sense of insecurity. Some accepted that their strategies were sub-optimal and, particularly with respect to tick control, admitted to being relatively powerless.

The authorities that disseminate knowledge are another factor influencing uncertainty and trust: whether transmission is through literature and formal education on the one hand, or experience and verbal communication, on the other.[45] Some of those we interviewed had basic literacy – often in vernacu-

[41] Mothupi Molefe, Phuthaditjhaba, 14 October 2010.
[42] Jenjezwa, 'Stock Farmers and the State', 114.
[43] Christopher Antweiler, 'Local knowledge and Local Knowing: An Anthropological Analysis of Contested "Cultural Products" in the Context of Development', *Anthropos,* 93 (1998), 469–94.
[44] Susan Reynolds Whyte, *Questioning Misfortune: The Pragmatics of Uncertainty in Eastern Uganda* (Cambridge, Cambridge University Press, 1997).
[45] P. Sillitoe, 'Trust in Development: Some Implications of Knowing in Indigenous Knowledge', *Journal of the Royal Anthropological Institute,* 16, 1 (2010), 12–30.

lars rather than English – but relied very largely on shared experience, visual evidence, and oral instruction. They tended to trust family traditions or the ideas of local specialists with regard to disease causation. Use of drugs was more eclectic and they were certainly not closed to information from government officers, agents for drug companies, and retailers. However, they did not necessarily accept biomedical explanations and treatments on the authority of a government officer or written text. To some degree, they applied their own ways of testing efficacy. This could mean that they looked for results other than those sought by scientific specialists. Trust and word of mouth influenced approaches to biomedicines as well as local knowledge.

It is not possible for us to say how effective local medicines have been. While some of the relevant plants have been analysed in laboratories for their medicinal properties (chapters 5 and 6), little research has been done on their application in rural contexts or the outcome of such treatments. From the perspective of the informants, efficacy largely depended on what farmers had identified as the key ailment and what they expected the treatment to achieve. One person's idea of efficacy was not necessarily translatable to others. For example, informants put great emphasis on the laxative quality of medicines. Their judgement as to the curative properties of a plant depended partly on the rapidity and scale of excretion. A secondary element might be the ability of a medicine to relieve the symptoms that farmers had initially observed as a sign of sickness, such as weakness, or eruptions of warts or lesions on the skin. The type of disease strongly affected perceptions of the success of local medical practice.

Understanding the causes and symptoms of livestock diseases

The anthropologist Monica Hunter suggested that there was little differentiation between natural and supernatural explanations of disease in 1930s Mpondoland. She also defended her focus on the supernatural, arguing: 'Much space has here been devoted to witchcraft and magic, but it is commensurate with the part they play in Pondo life. The belief in them permeates the whole of life.'[46] Recent witchcraft studies tend to emphasise its dynamic and contemporary features and also insist on its pervasive presence in people's lives. Authors such as Isak Niehaus and Adam Ashforth suggest that discourses around witchcraft increase or become more intense in periods of rapid social change. Accusations of witchcraft can also be an expression of social insecurity.[47]

[46] Monica Hunter, *Reaction to Conquest: Effects of Contact with Europeans on the Pondo of South Africa* (London, International Institute of African Languages and Cultures, 1936), 319.

[47] Isak Niehaus, *Witchcraft, Power and Politics: Exploring the Occult in the South African Lowveld* (Cape Town, David Philip, 2001); Isak Niehaus, 'Witches and Zombies of the South African Lowveld: Discourse, Accusations and Subjective Reality', *Journal of the Royal Anthropological*

Our overall impression from interviews is that natural explanations were more strongly differentiated from supernatural by the early twenty-first century and that in relation to animal health, at least, natural and environmental causes were overall the most significant. Diseases were seldom attributed to witchcraft itself, although informants linked a number of maladies to ideas of pollution or more ambient supernatural forces. It is possible that conceptions of human health differ from those about animal health. This could be a result of observation by stockowners, both of their animals in the veld and at post-mortems, which are not carried on human bodies. Edward Green argues more generally that naturalistic ideas of disease are more widespread in Africa than often acknowledged, and offers an 'indigenous contagion theory' as a means of analysing them. While we do not agree entirely with his dividing line between the natural and supernatural (chapters 7 and 8), our evidence supports some of his general conclusions.[48]

Environmental and dietary explanations underpinned many of our informants' interpretations of disease and health (chapter 3). In all the areas we visited, livestock owners were attentive to the quality of the veld. Historically, seasonal transhumance has been an important strategy for livestock owners (chapter 4). We found that such practices had largely been curtailed and only continued on a limited scale in Mbotyi and QwaQwa. But transhumance reflected a strong environmental conceptualisation of diseases that linked health with nutritious grasses and seasonal changes. Gallsickness, prevalent throughout the eastern half of South Africa, was probably the most common disease interviewees identified and they generally linked it to the state of the veld.

In all our sites, informants thought that a good appetite was an indicator of good health (chapter 2). Whatever the disease, sick animals often demonstrated an unwillingness to graze and a lethargic disposition (chapter 3). Farmers often carried out post-mortems while butchering the meat for human consumption (chapter 3). In Mbotyi this task sometimes fell to village specialists whom locals believed had a greater understanding of the anatomy of animals and also more experience in identifying the signs of disease. It was striking that despite the many different names for diseases, most African livestock owners within and across our interview sites, recognised the same sorts of symptoms (chapter 3).

Respondents did not generally attribute disease to witchcraft, but under certain conditions some invoked malevolent supernatural forces (chapter 7). This seemed more likely either if animals died suddenly in the kraal, or else if only one owner experienced losses. If diseases appeared with the seasons, or livestock from many kraals contracted the same malady, then it was unlikely to be witchcraft. As a precaution against witchcraft, there was a widespread use of protec-

(contd) *Institute*, 11, 2 (2005), 191–210; Adam Ashforth, *Witchcraft, Violence and Democracy in South Africa* (Chicago IL, Chicago University Press, 2005).

[48] Edward Green, *Indigenous Theories of Contagious Disease* (Walnut Creek CA, Alta Mira Press, 1999).

tive medicines (*Imithi* in isiXhosa, from the name for trees, but commonly called *muthi*). Sometimes farmers believed that their animals had been bewitched because their protective *muthi* had failed. This might result from an evil-doer deploying stronger *muthi*, or it might be a consequence of 'ritual pollution'.[49] This term, used in ethnographies, is uneasy because it applies very largely to a gendered view of misfortune and seems to reinforce patriarchal controls over space and property. A number of our Tswana informants explained that menstruation and widowhood could be detrimental to livestock. In QwaQwa the majority of our informants refused to allow women into the kraal, or to handle cattle, because their mere presence could destroy the protective *muthi*. Some informants in North West Province associated this form of infection with the concept of *mohato*. Women of childbearing age and widows, who carried the shades of their dead husbands, could cast misfortune and infertility on the cattle kraal, if they approached the livestock. A similar term, *umkhondo*, was used in Mbotyi. This, as in the case of *mohato*, derives from the word for footprint but in Mpondoland it was associated with natural conditions or with the deposition of disease in the dew by a metaphorical snake, rather than with women. Neither of these terms appears in the older ethnographies and both may be relatively recent concepts that combine older ideas about witchcraft and pollution with notions of infection. *Umkhondo* differed from witchcraft in that resulting ill health amongst animals was not attributed to the agency of a particular malevolent individual but to more ambient supernatural influences (chapter 7).

By far the most common way of looking after animals was to dose them with solutions made from plants. Farmers collected plants from nearby forests or the veld and sometimes cultivated a few species around the homestead. Barks, roots and leaves from different plants all featured in the *muthi*s that we heard about; the desired effect was often of a purgative nature (chapter 5). Even those who thought biomedicines were more effective, often preferred plants for particularly purposes – for example in dealing with birthing problems such as retained placentas and for cleansing the uterine tract after a miscarriage. Scientific medicine also failed to address the issue of witchcraft, *umkhondo* and *mohato* or to protect the kraal (chapters 7 and 8).

[49]   H. Stayt, *The Bavenda* (London, Oxford University Press, 1931), 38; E. J. Krige and J. D. Krige, *The Realm of a Rain-Queen: A Study of the Pattern of Lovedu Society* (London, Oxford University Press, 1943), 43; Hunter, *Reaction to Conquest*; Harriet Ngubane, *Body and Mind in Zulu Medicine: An Ethnography of Health and Disease in Nyuswa-Zulu Thought and Practice* (London, Academic Press, 1977), 24–29, 77–99; W. D. Hammond Tooke: *Bhaca society: a People of the Transkeian Uplands South Africa* (Cape Town, Oxford University Press, 1962), 21–24; W. D. Hammond-Tooke, *Boundaries and Belief: The Structure of a Sotho Worldview* (Johannesburg, Witwatersrand University Press, 1981), 124–30; *Rituals and Medicine: Indigenous Healing in South Africa* (Johannesburg, Donker, 1989), 93–99; Mary Douglas, *Purity and Danger: An Analysis of Concepts of Pollution and Taboo,* (London, Routledge, 2002), 155; Berthold Pauw, 'Widows and Ritual Danger in Sotho and Tswana Communities', *African Studies,* 49, 2 (1990), 75–99.

Livestock in African hands

Henry Ramafoko from Mabeskraal told us that 'without cattle we are nothing; cattle are part of our nature'.[50] He regarded cattle as integral to Tswana identity and to notions of personhood. Like many people who lived in this village in North West Province, he had spent much of his life as a migrant worker. When he retired from a cement factory in Thabazimbi he decided to invest in cattle, partly in the hope that it would provide some form of financial security for his old age, but primarily because he wished to make his own statement about his identity as a Tswana man. Cattle ownership ensured he maintained a stake in the land and reflected his status as a successful elder. In many families men rather than women still have responsibility for looking after the livestock, although this has changed in some communities. Molelekwa Mabe, also from Mabeskraal, stated that 'Africans are born to herd cattle', but he was concerned that the younger generation, who had moved to the cities, were forfeiting this sense of identity.[51] He thought that urbanisation meant that young people were disavowing traditional knowledge in the belief that this was 'not modern'. Mabe feared that Tswana lore, which he felt had scarcely been recorded, would disappear, and with it important aspects of Tswana culture, as well as some useful medical cures.

Lott Motaung, a retired teacher from Bollantlokwe (North West Province), articulated a long-held belief amongst African owners: it was better to invest in cattle than in banks because they multiplied and the returns were much higher than the interest rates that small savers could hope for.[52] He told us:

> People still settle loans with cattle, and pay traditional healers with cows. Diviners acquire their bones (toala) from cows, especially the most important – the moremogolo extracted from the hind legs. This is vital for the art of prophecy. It's true that bridewealth in cattle is being replaced by cash, but slaughter cattle are still central to feasts and celebrations, binding families together. Cows are useful as they provide draught and milk. Dung is still used to polish floors and skins made into traditional wedding skirts and drums.

As Motaung's cousin, Evelyn, put it 'people may be poor, but with cattle they are rich'.[53]

By no means all of those interviewed, in all areas, shared the idea that livestock were so integral to African culture and identity. Yet they continued to

[50] Henry Ramafoko, Mabeskraal, 2 February 2010.
[51] Molekwa Mabe, Mabeskraal, 1 February 2010.
[52] Lott Motuang, Bollantlokwe, 10 November 2009; Flora Hajdu, *Local Worlds: Rural Livelihood Strategies.in Eastern Cape, South Africa* (Linköping, Linköping University, 2006).
[53] Evelyn Motuang, Bollantlokwe, 10 November 2009.

have economic value as well as cultural significance. In many villages that we visited, livestock were ubiquitous. Social differentiation is shaped primarily by education and income. But wealthier rural families sometimes invested in cattle so that ownership, in terms of both species and numbers, still to some degree reflected the division between richer and poorer residents. Those who did not own livestock would on occasion purchase animals for ceremonial slaughter and consumption. The ability to slaughter for the ancestors, throw a good funeral feast or host a spectacular wedding celebration enhanced one's social status in the community. To varying degrees, the rhythms of rural life are still shaped by handling animals.

Although detailed figures for livestock ownership were published up to about the 1960s, there are no recent national figures for smallholder ownership of livestock, or ownership by race. Furthermore, figures about the differentiation of holdings within African communities are limited to a few local village surveys. Most of our interview sites were in the former homeland areas of South Africa, but we do not have any reliable data about the long-term changes within the former homelands as a whole. National figures suggest that livestock ownership has remained relatively stable since 1996.[54] These statistics are unlikely to be precise because the state no longer collects such data effectively in the communal areas. Although this gap was recognised by the Eastern Cape veterinary services, it is not a priority at a national level.[55] Enumeration at dipping tanks and on inoculation days gave the most accurate figures in earlier decades, but neither practice is now sufficiently widespread to form the basis of national statistics.

Longer-term data exist for livestock in some communal areas, such as Lusikisiki in Mpondoland (now part of O. R. Tambo), in which our Mbotyi research site is located. Cattle numbers are the best recorded and the area of the old magisterial district – very largely under African occupation – has remained stable. The statistics indicate that the absolute number peaked in the 1930s when they averaged around 110,000 annually, declining to about 90,000 in the 1960s.[56] According to one series they increased a little in subsequent years and averaged a little over 90,000 in the 1990s.[57] A figure for 2007 showed about 70,000 cattle but it is likely that counting has been less thorough.[58] The indi-

---

[54] www.daff.gov.za/publications/publications.asp?category=Statistical+information consulted in September 2011 has figures from 1996 to 2010.

[55] Discussion with Dr Ivan Lwanga-Iga, Dohne, 26 March 2012.

[56] Figures collected from annual returns in the publications of the Transkeian Territories General Council and subsequent bodies.

[57] Thembela Kepe, 'The Dynamics of Cattle Production and Government Intervention in Communal Areas of Lusikisiki District' in Andrew Ainslie (ed.), *Cattle Ownership and Production in the Communal Areas of the Eastern Cape, South Africa* (Programme for Land and Agrarian Studies, University of Western Cape, Research Report 10, 2002), 65.

[58] Figures obtained from the Veterinary Section, Department of Agriculture, Lusikisiki.

cations are that, within a fixed African-occupied district, cattle numbers declined gradually from a peak in the 1930s, with some periods of stability, and are now about 60–70 per cent of that level. Numbers in the smaller, dryer Xhalanga district in the western Transkei appear to have remained stable at about 25,000–35,000 from 1920 to 2000.[59] It should not be assumed that numbers were higher in the distant past. Lungsickness, redwater, rinderpest and East Coast fever decimated the herds in the late-nineteenth and early-twentieth centuries. It is likely that the figures recorded in the 1930s were an all-time high because systematic dipping and disease control had increased survival rates. At that time, all of the professionals and officials dealing with the African reserves were convinced that they housed too many animals and that the result was rapid degradation of the pastures.

Because of the gradual decline in absolute numbers of livestock and the increase in both overall population and the number of individual households, the proportion of families with livestock in Lusikisiki has declined slowly since the 1930s.[60] Transkeian figures for that period show about 60 per cent of households owned cattle throughout the region – though this was almost certainly higher in Lusikisiki because of its high rainfall and rich pasturelands. Lusikisiki veterinary office recorded 163 owners in Mbotyi in 2007, about 40 per cent of the c. 400 households. Flora Hajdu's household survey in Cutwini village, close to Mbotyi, broadly confirms such figures. 43 per cent of households owned cattle and others owned goats and sheep. Over 60 per cent had chickens.[61] A greater proportion of people kept livestock at some time. Such percentages are almost certainly higher than in most other districts in Transkei and in most other former homelands, few of which have equally favourable environmental conditions.

Amongst those we interviewed, cattle ownership varied between one animal and a few hundred. We heard of the biggest herds in North West Province. We did not attempt village livestock surveys. Yet we can get some picture of average ownership from local dipping records that survived in a few old registers in a derelict hut near the Mbotyi dipping tank.[62] In 1954 average ownership stood at 13.45 animals. The largest herd numbered 55 and the smallest, one. Per capita ownership declined in later years, so that the average was 9.6 in 1976, 9.2 in 1992 and 9.6 in 2007. The latter figure was higher than the Lusikisiki district

[59]   Andrew Ainslie (ed.), *Cattle Ownership and Production in the Communal Areas of the Eastern Cape, South Africa* (Programme for Land and Agrarian Studies, University of Western Cape, Research Report 10, 2002), 43.

[60]   W. Beinart, 'Transkeian Smallholders and Agrarian Reform', *Journal of Contemporary African Studies*, 11, 2 (1992), 178–99 for a more detailed discussion of livestock statistics.

[61]   Hajdu, *Local Worlds*, 150–51.

[62]   The records in this hut were scattered on the floor, in very poor condition, and had clearly not been used for many years. Four dipping registers for 1954, 1971–72, 1976 and 1996 were cleaned and sufficiently intact to read. They are now stored at the Mbotyi River Lodge.

average of 7.[63] Although per capita ownership has declined a little since the 1950s, the figures suggest a remarkably persistent drive to accumulate livestock. Amongst interviewees, the average holding in Peddie was about 10 head of cattle and in Mbotyi about 12. Cattle holdings of interviewees in North West Province and QwaQwa were not so systematically recorded but showed a similar range – though the biggest herds we encountered of more than 100 were mostly in these areas.

Livestock ownership is not just a feature of the rural areas. Some urban residents kept animals in the towns and dense settlements where we carried out research such as Mmabatho/Mafikeng and Garankuwa. Wealthier residents had cattle and even the poorest kept backyard chickens. Overall, our informants expressed a preference for cattle but some also kept sheep and goats, depending on the composition of the pastures and the dangers of stock theft. In North West Province respondents who could not afford cattle tended to invest in goats as a source of meat and food for ceremonies. Some deliberately kept white goats because traditional healers favoured them for their healing and initiation rituals.[64] Goats were also important for purification rites, such as those performed by widows at the end of their year of mourning.[65]

Unlike noisy goats, sheep are particularly easy to steal and in some areas like QwaQwa, where larceny was a particular problem, sheep were not common. Ecological conditions in Mbotyi discouraged sheep farming. In the western part of North West Province, around Mafikeng and Mareetsane, donkeys were particularly numerous and kept for draught. In remote areas, where people could not afford tractors and cars, they relied on donkeys to collect water from streams and fuelwood from the veld. Africans adopted horses widely in the nineteenth century, often initially for military purposes, and the Sotho bred their own distinctive ponies.[66] Horses were not common in North West Province, possibly due to African horsesickness and shortage of good grazing. Donkeys, by contrast, according to many informants, 'never got sick'.[67] Despite the difficult disease environment, horses were popular in Mbotyi (chapter 3).

Rural surveys show a decline in the proportion of livelihoods generated through agriculture.[68] Many no longer cultivate crops. In Mabeskraal cultiva-

---

[63] Figures from the Veterinary department Lusikisiki District, 2007. We could not find later figures for Mbotyi.

[64] Elizabeth Bokaba, Brits, 9 November 2009.

[65] Michael Mlangeni, Mabopane, 24 October 2009.

[66] Sandra Swart, *Riding High: Horses, Humans and History in South Africa* (Johannesburg, Wits University Press, 2010).

[67] Abraham Meno, Mareetsane, 4 November 2009; Aaron Aobeng, Mareetsane, 4 November 2009; Joseph Keagile, Lokaleng, 3 November 2009; Ernest Medupe, Mafikeng, 10 February 2010; Modisaotsine Taikobong, Magogwe, 11 February 2010; Kgomotso Goapele, Mantsa, 12 February 2010.

[68] For Mpondoland, Hajdu, *Local Worlds*; for North West Province, Elizabeth Francis, *Making a*

tion is generally limited to tilling small vegetable gardens in the back yard, and some may plant an assortment of medicinal herbs. Even in Mbotyi, where rainfall is high, most villagers have abandoned their main arable fields and grow relatively small patches of maize in gardens around their homesteads (chapter 4). Whereas crops do not bring a great return, we have suggested that livestock are still a significant commitment and are surprisingly valuable in cash terms in relation to other produce or possessions. In Mabeskraal people sell meat on the streets and dodge the abattoir, which might discard carcasses on account of diseases or contamination with tapeworms. In Mbotyi, migrant earnings and government grants were the most important sources of income, although there was more local employment (at the hotel and holiday homes) than in most Transkeian villages. The nearby Magwa tea plantation also provided employment. But livestock were a central element in village commerce. In 2008 one head of cattle sold for R4,000–6,000 locally; by 2012 prices increased to R6,000–9,000. In comparison the annual pension in 2008 only amounted to about R900 a month, or R10,800 per year; the minimum wage was about the same. A goat could bring up to R1,000.

Motives for livestock accumulation have been debated in academic literature.[69] Here, as in the case of veterinary and medical ideas, we found diversity in views and practices. As noted above, many older men still see them as a store of wealth; they have social as well as economic value and act as a safety net. Many people perceived an added incentive of rapid increase, although in fact it is equally possible to lose livestock. We asked stockowners in Mbotyi whether they sold livestock to purchase medicines. Tata Alfred Banjela gave a typical traditionalist answer:

> For us – we prefer cattle, we don't want money. We would rather have more cattle than money. More money is more trouble. Money gets used, we just drink it away. Cattle we keep. We can't take a cow into the bar and buy drink with it. Even when you sell, and get money, you are not happy when you see your animal leave you.[70]

Cattle usage has changed in Mbotyi in recent years and they are less multipurpose animals than they used to be. Cows are seldom milked (chapter 7) and the need for draught, and hence ox teams, has declined. Yet they are no less

(cont) *Living: Changing Livelihoods in Rural South Africa* (London, Routledge, 2000); for the former Ciskei, Paul Hebinck and Peter C. Lent, *Livelihoods and Landscapes : the People of Guquka and Koloni and their Resources* (Leiden, Brill, 2007) .

[69] James Ferguson, *The Anti-Politics Machine: 'Development', Depoliticization and Bureaucratic Power in Lesotho* (Cambridge, Cambridge University Press, 1990); Andrew Ainslie, 'Farming Cattle, Cultivating Relationships: Cattle Ownership and Cultural Politics in Peddie District, Eastern Cape', *Social Dynamics*, 31, 1 (2005) 129–56; Ainslie, *Cattle Ownership and Production*; Hajdu, *Local Worlds*.

[70] Tata Alfred Banjela, Mbotyi, 20 December 2011.

valuable, especially for slaughters in a range of ceremonies from funerals and weddings to celebrations for the completion of buildings. Slaughters are often designed to connect the family to the ancestors, for example when children come of age. Cattle are now, perhaps increasingly, part of a cultural economy that is perhaps linked with grassroots enactment of specifically African identity. Thus despite Banjela's common refrain, people did sell or slaughter cattle, and especially goats, in Mbotyi. A few people bred goats specifically for sale. It is impossible to estimate the take-off rate each year because there is no central market. Animals were slaughtered every weekend. Sometimes they were purchased for this purpose. Commercial exchanges take place on a regular basis; transactions and prices were widely discussed. The capital held in cattle alone, at about R6,000 a head, in 2008, was probably over R6 million and goats another R1 million. The turnover from trade and slaughter, with a possible take-off rate of 15–20 per cent, may well be over R1 million a year, and prices increased markedly by 2012 to around R8,000 a head. Chickens, which abound, sold for R30–60 each. Livestock sales, exchanges and slaughter for consumption at ceremonies almost certainly constituted the largest segment of the local agricultural economy.

The mountainous terrain of Sesotho-speaking QwaQwa is densely populated and largely unsuitable for crops. Many people seek work in the main town of Phuthaditjhaba or migrate to other parts of the Free State and Gauteng to find work in the mines and cities. This was one of apartheid's notorious dumping grounds and 'commuter' settlements. There is little grazing land around Phuthaditjhaba and the municipal golf course doubles up as a pasture. Nevertheless, many in the outlying villages aspire to own livestock in the hope that animals will bring them financial security in old age. Like those interviewed in North West Province, people regarded livestock as a safer investment than banks and cattle owners saw their animals as an affirmation of Sotho identity. All the men interviewed recollected their experiences of herding as children and felt this gave them an appreciation of Sotho culture and a love of cattle. Despite ecological challenges, some farmers who had other sources of income hoped one day to run larger herds. Here as elsewhere, livestock owners placed a premium on keeping animals healthy.

Conclusion

Livestock were still important for a significant segment of the rural population in the villages where we interviewed. In one of many post-apartheid ironies, state provision of veterinary services has ebbed at the very moment that opportunities are increasing for African smallholders. Most are left to find their own treatments and remedies. The interviews pointed strongly to plural ideas about the causation and treatment of disease. We were struck that, despite over a century of veterinary interventions in rural South Africa, many black livestock owners had a very limited concept of biomedical causes. Such knowledge had not been transmitted with the dip. Except for those with high levels of education, they tended to attribute diseases to aetiologies that would not generally be accepted in contemporary scientific terms.

With regard to treatments, there was more openness to, and adoption of, biomedicines. Dipping was widely accepted, even if its specific role in controlling diseases was not always appreciated. Many livestock owners were prepared to mix biomedical treatments with other methods, including infusions from plants. Our overall impression from a range of research sites pointed to a growing interest in biomedicines, but an incapacity on the part of poorer black livestock owners to afford them. In this sense the balance within pluralistic approaches appeared to be shifting unevenly towards biomedical ideas. The gaps in knowledge and uncertainty about the best treatments were striking. We judge this not only in relation to efficacy but also the perceptions of livestock owners themselves. While confidence and practice varied, we were struck in our interviews by the diversity of strategies.

2

Ticks, Tick-borne Diseases
& the Limits of Local Knowledge

'The ticks suck the blood from animals and they die; there is no particular disease that causes death but they suck the blood'. (Nongede
Mkhanywa, Mbotyi)[1]

Introduction

South Africa is a hotbed of ticks. Anyone walking through the veld in rural
areas knows their dangers. The Afrikaans word *bosluise* (bush lice) perhaps
conveys their threat more vividly. For humans, they transmit tick bite fever, an
uncomfortable and dangerous disease. They are even more hazardous for livestock. Over a century ago, scientists discovered that ticks are responsible for
transmitting some of the country's most infectious diseases such as the ubiquitous gallsickness (anaplasmosis), East Coast fever (theilerosis), and redwater
(babesiosis) that infect cattle, as well as heartwater (ehrlichiosis) which affects all
ruminants. Despite a century of dipping to eradicate these diseases, ticks remain
a scourge for many livestock owners, black and white, in South Africa.

Delving into the history of tick control is essential because it has preoccu-
pied the state veterinary services and farmers alike. This is an exciting field of
interdisciplinary research, traversing history, anthropology, entomology and
epidemiology. Discussion of dipping as a means of control also opens up fascinating differences in the understanding of diseases and their treatment. We need
to unravel the changing patterns of knowledge, and contestations over
explaining the many tick-borne maladies that beset livestock in South Africa.

[1] Nongede Mkhanywa, Mbotyi, 23 February 2008.

A central part of our story must be the innovative and sustained research that produced the scientific understanding of these diseases, and the rationale for dipping as the primary mode of prevention. Yet scientific knowledge was not static. Dipping, as we will show, worked at its optimum for about half a century. Then evidence gradually emerged of the limits of its efficacy as some tick species developed resistance to acaricides. Scientists were responsive to uncertainties in their knowledge in this area and, over the long term, tried to evolve changing and multiple strategies of disease prevention.

In the early years of state dipping, African people often opposed this imposition, which cut deep into their patterns of livestock management. In part they were contesting state power to control and alter fundamental aspects of rural life. In part, they offered a different understanding of disease transmission and prevention. We do not have a detailed record of how they thought about diseases when dipping was introduced in the early-twentieth century. But our interviews have provided a view of African veterinary knowledge in more recent times. One of the most striking findings of our research is that most of the black stockowners we interviewed did not directly connect ticks with specific diseases. While they were convinced that ticks were harmful, and most knew the symptoms of the diseases that scientists associate with ticks, they explained these diseases in different ways. Yet African knowledge was also changing. Despite an undercurrent of resistance to dipping through the twentieth century, many came to accept its benefits and the practice became embedded in the rhythms of rural life. 'When we were growing up', Nongede Mkhanywa, in his sixties, recalled 'we were dipping often; dipping every Monday, every week, at that time'.[2] Many of the older men we interviewed looked back with a degree of nostalgia to an earlier era when the state organised dipping. Knowledge about ticks, related diseases, and dipping is plastic and mutable in all of these different constituencies. However, we will argue that there remains a significant gap between scientists and commercial farmers on the one hand, and the majority of African smallholders on the other. This impacts on the efficacy of disease control.

Tick-borne diseases do not always kill the livestock that contract them; they are now largely endemic rather than epizootic in South Africa. But they do weaken the overall health of animals, suppress yields of milk and meat and cause significant losses, both from disease and surface lesions. This chapter explores knowledge about tick-borne diseases and their treatment. The first section gives an overview of scientific approaches, the second focuses on state policy and the final part explores African ideas about ticks. We argue that in recent decades tick control has been constrained both by state neglect and the limits of African knowledge.

[2] Nongede Mkhanywa, Mbotyi, 25 February 2009.

Scientific understandings of ticks and efforts at control

Approaches to acarid (tick) management and disease control have changed during the course of the twentieth and twenty-first centuries. Ticks first achieved government veterinary and political attention during the Cape Colony's 1877 *Commission on Diseases of Cattle and Sheep*. A major spur for the enquiry was to investigate the support for, and feasibility of, scab eradication in sheep. This generally non-lethal condition, caused by a mite, greatly reduced the quality and quantity of wool produced, which was then the Colony's major export. Some commercial farmers at the Cape understood that parasites such as the scab acari could transmit disease – although this was not generally accepted amongst white livestock owners. A few Cape farmers also complained to the commissioners that ticks were adversely affecting their livestock. John Webb, from the district of Albany (around Grahamstown) linked the arrival of the bont tick (multi-coloured or variegated; *Ambylomma hebraeum*) with heartwater fatalities in his goats.[3] He believed that this was a new disease in the area and suggested that the boundaries of tick infestation could change. Webb also associated gallsickness – long a scourge in the region – with ticks. This was an early example of local knowledge informing the beginnings of scientific enquiry. William Branford, the first state veterinary surgeon, who served on the Commission, did not immediately respond to this intelligence, but his successor Duncan Hutcheon was gradually persuaded of the link. Research into tick-borne diseases started in earnest with the arrival of the Cape's first official entomologist, Charles Lounsbury, from the United States, in 1894.

In the previous year, American researchers had demonstrated that Texas Fever, characterised by red urine and known as redwater in South Africa, was spread by the blue tick (*Boophilus decoloratus*). The 1890s was an exciting period in the development of parasitology as scientists pinpointed mosquitoes and tsetse flies as vectors of human and animal diseases. Lounsbury threw himself into investigation of the ticks that plagued the Cape countryside.[4] Working with state veterinarians and farmers in the Eastern Cape, where ticks were a particular problem, he demonstrated that a number of diseases were tick-borne and he was able to identify the specific vectors. Experiments linked redwater and gallsickness to the blue tick and, as Webb had suggested, heartwater with the bont tick.

---

[3] Cape Parliamentary Papers, *Commission on Diseases in Cattle and Sheep in this Colony*, G3-1877, 108–11: evidence, John Webb; William Beinart, *The Rise of Conservation in South Africa: Settlers, Livestock and the Environment, 1770–1950* (Oxford, Oxford University Press, 2003), chapter 4.

[4] Karen Brown, 'Political Entomology: the Insectile Challenge to Agricultural Development in the Cape Colony, 1895–1910', *Journal of Southern African Studies*, 29, 2 (2003), 529–49.

East Coast fever reached South Africa in 1902 with cattle imported via the Mozambique coast.[5] By 1904 Lounsbury's experiments revealed that it was carried by the brown ear tick (*Rhipicephalus appendiculatus*).

Lounsbury mapped out the life-cycle of the different species and classified ticks according to how long each species spent on a specific host. Blue ticks went through all three stages (larvae, nymph and adult) of their life-cycle on a single host and relied on blood-meals to progress to the next phase. They lived on an animal for three weeks or more. Eventually the female dropped from the host to lay her eggs in the grass and if infective, passed on redwater or gallsickness to the eggs, so the offspring could infect future hosts. The bont and brown ear were three-host ticks. They too needed blood for each of their transformations, but dropped to the ground to metamorphose at every stage. They did not pass diseases onto their eggs. But if they picked up heartwater or East Coast fever from an infected animal during an earlier part of their life-cycle, they could transmit it to another beast during a more mature stage. Brown ear and bont ticks spent less than a week on a single animal in each of their life phases. The length of time particular ticks spent on a host had implications for dipping programmes. Immersion in a dip tank every three weeks could be sufficient to kill the blue ticks, but more frequent dipping, at least once a week, was necessary to destroy the brown ear and bont ticks.[6] Parasitologists were beginning to understand the complex interaction between the vector, parasite and host in the transmission of key diseases. Lounsbury was at the cutting edge of his field internationally.

The Cape government introduced compulsory dipping to eradicate scab from sheep in 1894, gradually overcoming widespread Afrikaner opposition.[7] Most white farmers became attuned to this strategy and when it was presented as a possible prophylaxis against tick-borne diseases Natal's state veterinarian, Herbert Watkins-Pitchford, published a safe and effective arsenical formula in 1909 which he unimaginatively named 'Laboratory Dip'. Farmers added this chemical to water in a plunge dip tank. The aim was to ensure total immersion by digging a deep trench or plunge and lining it with concrete. Herders drove the animals from a crush pen or kraal down a fenced race which separated them into single file. The race dropped down an incline five to six feet into the tank,

[5] P. Cranefield, *Science and Empire: East Coast Fever in Rhodesia and the Transvaal* (Cambridge, Cambridge University Press, 1991).

[6] Charles Lounsbury, 'Tick Heartwater Experiments', *Agricultural Journal of the Cape of Good Hope (AJCGH)* 16, 11 (1900), 682–7; 'Ticks and African Coast Fever', *Transvaal Agricultural Journal (TAJ)* 2, 5 (1903), 4–13; 'Transmission of African Coast Fever', *Agricultural Journal of the Cape of Good Hope (AJCGH)*, 24, 4 (1904), 428–32; 'Ticks and East Coast Fever', *AJCGH*, 28 (5), 1906, 634–43. See also Lounsbury's discussions in the Report of the Government Entomologist in the Department of Agriculture Annual Reports, *Cape Parliamentary Papers* 1898, 1900, 1901, 1902, 1903, 1904. For corroboratory research in the Transvaal, Arnold Theiler: 'Rhodesian Tick Fever', *TAJ*, 2, 7 (1904), 421–39; 'Redwater', *TAJ*, 3, 11 (1905), 476–96; 'Diseases, Ticks and their Eradication', *TAJ*, 7, 28 (1909), 685–99.

ensuring that the animals entered in single file in a controlled way. Once in the tank, the cattle swam through several metres of arsenic solution and staggered out to dry. The dipwash left a residue that killed the ticks. Because brown ear ticks also burrowed into the ears, farmers were advised to hand-dress or manually clean out the ears.[8] Laboratory Dip remained the recommended formulae for killing all species of ticks until the 1940s.[9]

After Union (1910) when state veterinary services were unified under Arnold Theiler at Onderstepoort, the Diseases of Stock Act (1911) provided for universal compulsory measures. The state employed an army of stock inspectors who were responsible for the everyday administration of the dipping laws and livestock restrictions.[10] The government encouraged farmers to build plunge dipping tanks on their properties and the Land Bank provided low interest loans to assist them. In the African reserves, the Native Affairs Department and the Transkeian Territories General Council constructed the tanks and then recouped the costs through tax levies. It became compulsory for farmers to dip their cattle every five days in East Coast fever areas in an effort to stamp out the infection. East Coast fever nevertheless rampaged through the former Transvaal, Natal and the Eastern Cape killing up to 80 per cent of cattle in some districts. The Orange Free State with its cold winters, and the arid Karoo and northern areas of the Cape proved inhospitable to the brown ear tick, so cattle dipping was not obligatory in these areas.

Blue and bont ticks often occupied similar ecological zones to the brown ear tick and they also died in the dipping tanks. Better control of redwater, gallsickness and heartwater was achieved as a by-product. Controls over movement were simultaneously introduced to ensure that the disease did not reinvade 'cleaned' areas. This had a significant impact on animal husbandry. In the longer term dipping was successful against East Coast fever and in 1953 the Union Department of Agriculture declared South Africa free of this disease.[11] Over the previous 50 years, East Coast fever had claimed the lives of an estimated 5.5 million

[7] Mordechai Tamarkin, *Volk and Flock: Ecology, Identity and Politics among Cape Afrikaners in the Late Nineteenth Century* (Pretoria, University of South Africa Press, 2009).

[8] Arnold Theiler, 'Diseases, Ticks and their Eradication', *Agricultural Journal of the Union of South Africa (AJUSA)*, 1, 4 (1911), 491–508; 'Disease Ticks and their Eradication', *Journal of the Department of Agriculture (JDA)*, 2, 2 (1921), 141–59; Arnold Theiler and Charles Gray, 'Inquiries into Dips', *AJUSA*, 4, 6 (1913), 814–29.

[9] Editorial, 'Cattle Dipping at Nels Rust', *Natal Agricultural Journal and Mining Record (NAJMR)*, 5, 8 (1902), 253–7, and 5, 10 (1902), 517; Editorial ', The Conquest of the Tick', *AJUSA*, 5, 1 (1913), 1–12; Herbert Watkins-Pitchford, 'Dipping and Tick-Destroying Agents', *NAJMR*, 12, 1 (1909), 436–59; *An Illustrated Pamphlet of Tick Destruction and the Eradication of East Coast Fever* (Pietermaritzburg, P. Davis and Sons, n.d., c.1914).

[10] See e.g., 'Government Notice: East Coast Fever', *AJUSA*, 3, 5 (1912), 763–6.

[11] Karen Brown, 'Veterinary Entomology, Colonial Science and the Challenge of Tick-borne Diseases in South Africa during the late Nineteenth and early Twentieth Centuries', *Parassitologia*, 50, 3–4 (2008), 305–19.

cattle.[12] Although the campaign against East Coast fever had been difficult, and at times unpopular, the elimination of this disease showed that dipping could be an effective weapon. The eradication of East Coast fever did not, however, solve the problem of tick-borne diseases.[13] Redwater, gallsickness and heartwater proved intractable and remained endemic in many parts of the country. Veterinarians developed vaccines for these diseases during the twentieth century, but they were attenuated live vaccines that are hard to store and can bring about adverse reactions that need antibiotics to cure. Technical difficulties, as well as the cost of administering these drugs, have meant that most farmers continue to rely on externally administered poisons to tackle ticks and the diseases they carry.

In 1938 some white farmers from the East London district (Eastern Cape) reported dip failure against the blue tick and blamed the drug companies for providing sub-standard arsenic. The Department initially prosecuted farmers for failing to dip regularly. However, laboratory tests on livestock revealed that some blue ticks had developed a defence mechanism to withstand the arsenic. Mixing nicotine with arsenic proved effective against immature, but not adult, blue ticks so this was temporarily added to the dipping tank. (Tobacco dips had been used in the nineteenth century against scab in sheep). These developments marked the start of a war of attrition against the blue tick. Never again was a chemical to have 40 years of effective use before some ticks evolved a biological resistance.

In 1946 farmers in the Eastern Cape started to add BHC (Benzene hydrochloride) to their tanks.[14] This was the beginning of a period of experimentation and uncertainty in the world of dips. By the end of 1948, some farmers discovered that neither BHC nor arsenic killed all blue ticks. After 1948 a range of new acaricides entered the market. Tick resistance became an international phenomenon and many acaricides originated outside South Africa. DDT followed BHC into the dipping tank, but the blue ticks had successfully fought back by 1956. The state banned DDT in the 1960s because it could harm the environment. Organophosphates came onto the scene, although these began to fail by the late 1960s and their danger to animals and human health soon became apparent. L. H. Walsh, a Cape mohair producer, used Rachel Carson's *Silent Spring* to highlight the toxic effects of such dips.[15] During the

---

[12] Alexander Diesel, 'The Campaign against East Coast Fever in South Africa', *Onderstepoort Journal of Veterinary Science and Animal Industry (OJVSAI)*, 23, 1 (1948), 19–31; Raymond Alexander, 'Ooskuskoors', in the Annual Report of the Director of Veterinary Services, 1954–55, accessed 15 March 2003.

[13] Alexander, 'Ooskuskoors'.

[14] A. B. M. Whitnall, 'Historical Review of Insecticide Resistance in the Blue Tick 1939 to 1949', in proceedings of a symposium on The Biology and Control of Ticks in Southern Africa, Rhodes University, 1–3 July 1969 (Grahamstown, Institute of Social and Economic Research, 1969), 78–95.

[15] L. G. Walsh, 'South African Mohair Growers' Memorandum on the Problem of Tick Resistance to Insecticides', in proceedings of a symposium on The Biology and Control of Ticks in Southern Africa (Grahamstown, Institute of Social and Economic Research, 1969).

1970s new brands appeared such as Triatix (amitraz), still widely distributed by the state and retailers. These have remained in use because resistance to them is not universal. They are manufactured in powder form, easy to transport, and can be applied as sprays as well as dips.

The emergence of pyrethroids in the early 1980s led to the gradual replacement of the dipping tank in some areas with pour-on insecticides, like Deadline (flumethrin), applied along the animal's spine. It was not long before some ticks survived the pyrethroids too. The most recent acaricides include injectables such as ivermectin, which enter the blood stream and poison the ticks from within.[16] According to Arthur Spickett, who researches ticks at the Onderstepoort Veterinary Institute, by 2010 there were 90 acaricides on the market, containing 18 chemicals from five chemical groups. Some of the preparations have combinations of chemicals to enhance their killing-power.[17]

Resistance has developed unevenly and scientists have long speculated as to why and how the blue tick has been the most successful in defying the chemical onslaught. Unlike the other major species that transmit disease, the blue tick spends three weeks on a single host. Graeme Whitehead, writing in 1973, thought that this repeated exposure of one tick to weekly dipping facilitated survival through natural selection.[18] These 'superticks' passed on their resistance to their young. Blue ticks also propagated more rapidly throughout the year and this reproductive turnover helped to explain why they were first to produce a large resistant population.

Human agency was also an issue. From the start scientists complained that farmers did not keep the solutions at the right strength. Under-strength dips could have facilitated the survival of some ticks. Solutions in the tanks became diluted by rainwater or contaminated by mud, muck and faeces introduced by the cattle themselves. Watkins-Pitchford invented an isometer to measure the strength of the solution, but it was too complicated for many farmers or dipping foremen in the African areas to use.[19] The dips that followed arsenic were even more difficult to test on the farm. In 1969 Coopers and Nephews Ltd, a chemical company that manufactured dips, published surveys which highlighted the problem of weak dipwashes. For example, an investigation in Natal in 1965–66 revealed that 80 per cent of the tanks had an inadequate concentration of chemicals and 50 per cent contained solutions that were less than half the recommended strength. They concluded that this could be a major contributor to

[16]   Discussion with Arthur Spickett, Onderstepoort Veterinary Institute, 22 February 2010; G. B. Whitehead, 'The Problem of Acaricide Resistance in the Control of Ticks in South Africa', *Veterinary Entomology in South Africa, Entomology Memoir No. 44* (Pretoria, Department of Agriculture, 1975), 12–16.

[17]   Discussion with Arthur Spickett, Onderstepoort Veterinary Institute, 22 February 2010.

[18]   Graeme B. Whitehead, 'Resistance to Acaricides in Ticks in the Eastern Cape Province', *South African Medical Journal*, 47, 2 (1973), 342–4.

[19]   Watkins-Pitchford, *Illustrated Pamphlet of Tick Destruction*.

the development of resistance. By contrast, farmers believed that drug companies stipulated strong solutions in order to raise sales and profits, consequently some cattle owners deliberately under-dosed.[20] A survey in 2002 indicated that incorrect acaricide preparation remained a major obstacle to tick control in North West Province and the Eastern Cape.[21] Resistance seemed to speed up most noticeably after compulsory dipping for East Coast fever ceased in the 1950s. Although dipping did continue in the African reserves and homelands, it was not always systematic due to a lack of state commitment and local opposition. During the Mpondoland Revolt of 1960 rebels complained that the government was trying to kill their cattle with inadequate dips. The likelihood was that they too were suffering from tick resistance.[22] The chemistry and biology of tick resistance is complicated and not well understood. It is not clear whether resistance is more likely where ticks are exposed to a regular and orthodox regime of dipping or whether it results from irregularity and carelessness.

Questions about the long-term viability of dipping, as well as the costs, generated alternative ecologically-informed approaches to understanding and managing ticks. Between 1937 and 1948 Gertrud Theiler, daughter of Arnold Theiler, began to plot the national distribution of ticks as part of South Africa's zoological survey.[23] Theiler believed that by mapping out the distribution of ticks, noting their preferred habitats as well as seasonal changes in reproductive rates, she could provide an important tool for tick management. This would help entomologists and farmers demarcate the key areas of particular vector-borne diseases. Theiler discovered that although many of the dangerous species inhabited the summer rainfall areas, there were nonetheless some marked differences in their environmental preferences. Climate, soil types and local vegetation all interacted to shape distribution patterns. She found that the blue tick

[20]   G. E. Thompson, 'The Management of Cattle Dips and Sprayraces', in proceedings of a symposium on The Biology and Control of Ticks in Southern Africa (Grahamstown, Institute of Social and Economic Research, 1969), 51–60.
[21]   S. Mekonnen, N. R. Bryson, L. J. Fourie, R. J. Peter, A. M. Spickett, R. J. Taylor, T. Strydom and I. G. Horak, 'Acaricide Resistance Profiles of Single- and Multi-host Ticks from Communal and Commercial Farming Areas in the Eastern Cape and North-West Provinces of South Africa', *Onderstepoort Journal of Veterinary Research*, 69, 2 (2002), 99–105.
[22]   W. Beinart, 'The Mpondo Revolt through the Eyes of Leonard Mdingi and Anderson Ganyile', in Thembela Kepe and Lungisile Ntsebeza (eds), *Rural Resistance in South Africa: The Mpondo Revolts after Fifty Years* (Leiden, Brill, 2011), 91–114.
[23]   Gertrud Theiler: Introduction to the 'Zoological Survey of the Union of South Africa: Part 1', *OJVSAI*, 23, 1 (1948), 217–31; 'Zoological Survey of the Union of South Africa: Part 2: The Distribution of *Boophilus (palpoboophilus) decoloratus* – The Blue Tick', *OJVSAI*, 22, 2 (1949), 255–69; 'Zoological Survey of the Union of South Africa: Part 3: The Distribution of *Rhipicephalus appendiculatus* – The Brown Tick,' *OJVSAI*, 22, 2 (1949), 269–84; 'Ticks: Their Biology and their Distribution', *Journal of the South African Veterinary Medical Association,* 30, 3 (1959), 195–203; Brown, 'Veterinary Entomology'.

was the most widespread of the infective ticks and seemed to be the most environmentally adaptable. It could survive almost anywhere apart from the driest districts of the Cape. The bont tick was more localised to areas of favourable vegetation, notably bushes and thickets; in the Eastern Cape grasslands it was largely restricted to the thickly vegetated river valleys.

In the late 1970s Yigal Rechav, based at the Tick Research Unit at Rhodes University, Grahamstown, initiated a five-year survey of tick activity on four white-owned farms.[24] He confirmed that the bont tick had a patchy distribution and mobile frontiers in the Eastern Cape, probably related to ecological conditions. Abundant in some bushy coastal terrain, he thought they were spreading inland. Our informants also believed them to have spread in recent decades (see below). Heat and humidity encouraged their reproduction. In practical terms, Rechav hoped to find an alternative tick control strategy to dipping and suggested that tick numbers could be reduced by ecological interventions, such as the clearance of bush. His recommendation did not translate into policy but his findings tended to confirm the idea that transhumance was a strategy for reducing tick exposure at particular times of the year.

In recent research, Nkululeko Nyangiwe noted that the brown ear tick was becoming more numerous in the inland areas of the Eastern Cape, which he postulated might be due to the rise in game farming, as this species was particularly prevalent on buffalo.[25] Ironically, one of the reasons for the switch to game farming was that wildlife was assumed to be less vulnerable than cattle to tick-borne diseases. Higher rates of translocation of wildlife, and antelope in particular, can set up new nodes of infection in formerly disease free areas. Kudu numbers have increased rapidly in the Eastern Cape and they are known to be carriers of heartwater.[26] Kudu are not easily restricted by livestock fences. There has been no requirement to dip wild animals and it is difficult to treat them as they cannot easily be herded. Thus the expansion of protected areas for wildlife and the growth of game farming were likely to have been significant factors in the re-emergence of ticks.

All varieties of infective ticks abound in North West Province, especially in the eastern districts which are wetter and have denser vegetation.[27] Some species

[24] Yigal Rechav: 'Ecology of Ticks in the Eastern Cape', Annual Report of the Rhodes University Tick Research Unit 1978–79, 1–4; 'Tick Population Studies in Southern Africa', Annual Report of the Rhodes University Tick Research Unit 1979–80, 20–120; 'Dynamics of Tick Populations (Acari: Ixodidae) in the Eastern Cape Province of South Africa', Journal of Medical Entomology, 19, 6 (1982), 679–700.

[25] Nkululeko Nyangiwe, 'The Geographic Distribution of Ticks in the Eastern Region of the Eastern Cape Province', MSc Veterinary Science, University of Pretoria (2007), 104–6.

[26] J. S. van der Merwe, et al., 'Acaricide Efficiency of Amitraz/Cypermethrin and Abamectin Pour-on Preparations in Game', OJVR, 72, 4 (2005), 309–14.

[27] Malesela Sekokotla, 'Assessing Implementation of Veterinary Extension on Control of Cattle Parasites in Moretele District North West Province', unpublished MSc (Veterinary Sciences), University of Pretoria (2004), 19–22.

seem to be spreading. According to Arthur Spickett, the bont tick is particularly numerous and is moving into the Northern Cape where it had never been recorded before. He suggested that bush encroachment enabled the bont to move along the river valleys compounded by movements of tick-infested livestock and wildlife. Since 1994 controls regulating stock movements have been relaxed.[28]

Scientists are debating whether endemic stability may be the best way to manage tick-borne diseases. Animals could build up some resistance to diseases like redwater, gallsickness and heartwater by exposure to a small number of ticks. Livestock can produce antibodies to combat the pathogens. This strategy has long been discussed in relation to the control of malaria and trypanosomiasis. For example, in the early 1970s John Ford advocated low-level exposure of livestock to tsetse fly in order to instil partial immunities. Whether or not he was correct, his experience in the field in east and central Africa convinced him of the intractability of the problem and the impossibility of eradicating tsetse flies.[29] The idea of endemic stability in relation to ticks was developed in Australia where scientists also encountered growing resistance to dips.[30] Recent recommendations in South Africa suggest a range of 10 to 20 ticks per host, but there are disadvantages in this strategy as the bont tick in particular can cause damage to teats and hides.[31] Enhancing endemic stability is considered as part of an 'integrated approach' to tick management and control, using multiple strategies which include: exposing animals from a very young age to a small number of ticks; alternating dipping preparations to try to forestall resistance; rotating full dipping with strategic hand-dressing of infested parts; and the application of pour-ons. Effective endemic stability is also partly contingent upon access to adequate nutrition and water.[32]

A complementary strategy is to invest in breeds of livestock that are naturally less susceptible to tick-borne infections, such as indigenous Nguni cattle, because they have developed a resistance over centuries.[33] In the 1940s, J. C. Bonsma, who worked in the northern Transvaal, cross-bred exotic British cattle with indigenous Afrikaner varieties to create a high-yield Bonsmara animal

[28] Discussions with Arthur Spickett and Heloise Heyne, Onderstepoort Veterinary Institute, 22 February 2010 and 30 November 2012; Arthur M. Spickett, I. Heloise Heyne, Roy Williams, 'Survey of the Livestock Ticks of the North West Province, South Africa', *OJVR*, 78, 1 (2011), 12 pages (doi: 10.4102/ojvr.v78i1.305).

[29] John Ford, *The Role of the Trypanosomiases in African Ecology: A Study of the Tsetse Fly Problem* (London, Oxford University Press, 1971).

[30] G. A. Tice, N. R. Bryson, C. G. Stewart, B. du Plessis and D. T. De Waal, 'The Absence of Clinical Disease in Cattle in Communal Grazing Areas where Farmers are Changing from an Intensive Dipping Programme to One of Endemic Stability to Tick-borne Diseases', *OJVR*, 65 (3), 1998, 169–75.

[31] Malesela Sekokotla, 'Assessing Implementation', 21–28.

[32] Discussion with Arthur Spickett and Heloise Heyne, Onderstepoort Veterinary Institute, 30 November 2012.

[33] Arthur Spickett, factsheet 'Integrated Tick Management', received 22 February 2011.

that was also more resistant to tick infections.[34] Tick bites on Nguni cattle are also less damaging. Blue ticks cannot easily engorge on Nguni and so are unable to contract or transmit gallsickness and redwater.[35] Arthur Spickett argues that this results from a genetic immune reaction in these cattle. Ngunis are also less susceptible to malnutrition than many breeds as they can digest a wider range of grasses. This reduces the likelihood that their immunity will break down in times of environmental stress.[36] Fort Hare's Nguni Project, initiated in 1998, has bred herds of this strain and distributed both bulls and heifers in surrounding communities that meet basic conditions of controlled grazing.[37] The North West Province Provincial Government has also encouraged farmers to join in its Nguni Cattle Scheme set up in 2007.[38] The aim is to help black farmers establish themselves on a commercial basis. Advocates of pure-bred Nguni see their re-dispersal in the African areas as a major route forward in developing disease control. Some white farmers have adopted a similar strategy. African owners also see Brahmans as less susceptible to ticks. These originated from India, were bred in the United States in the first half of the twentieth century and imported to South Africa in the 1950s.

In an ironic twist, scientists now think irregular dipping may actually have led to the establishment of partial immunities where dipping has failed to control ticks. In 1998 an Onderstepoort team carried out investigations at a number of dipping tanks in North West Province and Mpumalanga. Procedures varied from site to site but the overall conclusion was that in communities where state-supervised dipping had been lax or non-existent for a long time, cattle had a stronger immunity to tick-borne infections. The implications of this important article have not been generalised and the nature of immunities to different tick-borne diseases, in specific areas, is by no means clear. Animals can survive in some numbers in areas with high tick infestation. If they did not, African herds in villages such as Mbotyi would be decimated. Since the great epizootics of the late-nineteenth and early-twentieth centuries there has not been a tick-borne epidemic in South Africa. But these endemic diseases still cause considerable losses. Systematic evidence on the cause of death is not collected in communal areas and some observers feel that, despite some possible partial immunity, tick-borne diseases remain the major reason for low productivity and relatively high rates of death.[39] Most of our respondents saw gallsickness as their major annual problem.

[34] Available at www.frontierbonsmaras.co.za/index.php/faq (accessed 8 May 2013).

[35] Discussion with Heloise Heyne, Onderstepoort Veterinary Institute, 22 February 2010, 30 November 2012.

[36] Arthur Spickett and Heloise Heyne, Onderstepoort, 30 November 2012.

[37] L. Musemwa et al., 'Nguni Cattle Marketing Constraints and Opportunities in the Communal Areas of South Africa: Review', *African Journal of Agricultural Research*, 3, 4 (2008), 239–45.

[38] Tefo Kepadisa, Department of Agriculture Mafikeng, 8 February 2010.

[39] Nkululeko Nyangiwe, 'The Geographic Distribution of Ticks'; Tice, et al., 'The Absence of Clinical Disease in Cattle'.

To sum up the section on science, the discovery around the turn of the twentieth century that ticks carried a number of infectious diseases led to compulsory dipping to destroy the vectors. Dipping coupled with control over livestock movements resulted in the disappearance of East Coast fever in livestock in the 1950s. However, its vector, the brown ear tick, was not eradicated. Although the arsenic dips initially knocked back other species, blue ticks developed a partial resistance. This is possibly the reason for the ubiquity of gallsickness. The most effective strategy for combatting ticks is now seen as a multi-pronged, integrated approach. Some animals clearly acquire immunity through exposure but this is generally achieved more by chance than a systematic strategy and it is not yet clear what constitutes a safe tick-load.

From state compulsion to individual responsibility – the changing role of the state in relation to dipping

East Coast fever shaped state responses to tick eradication until the 1950s. In subsequent years this commitment on the part of the state gradually eroded with complex and unpredictable outcomes. Over the long term, livestock owners have increasingly been left to take responsibility for their own herds and flocks. One primary reason for the gradual demise of systematic dipping in the African reserves and homelands was continued opposition to this practice. Resistance had been manifest from the very moment of implementation in the Transkei and East Griqualand from 1908 until 1914.[40] Dipping intersected with suspicion both of government and chiefs who acquiesced with such unpopular measures.[41] Driving the animals to the dipping tanks at least once a week was arduous, time consuming and labour intensive. Old-established patterns of transhumance and bridewealth transfer were curtailed by the regulations and the scientific justification for full immersion was not widely understood. The stock inspectors imposed deeply resented fines if farmers failed to dip regularly and concealed their cattle. In some districts stockowners accused the government of poisoning their cattle. Animals did occasionally die through imbibing over-strength dip, drowning in the tank or suffering scalding.[42]

Although dipping ceased to be compulsory on white-owned farms from the 1950s, it remained so in the African reserves. As Patrick Masika, now head of the Agricultural and Rural Development Research Institute at Fort Hare put it:

[40] Colin Bundy, '"We Don't Want Your Rain, We Won't Dip": Popular Opposition, Collaboration and Social Control in the Anti-Dipping Movement 1908–16', in W. Beinart and C. Bundy, *Hidden Struggles in Rural South Africa* (London, James Currey, 1987), 191–221.
[41] William Beinart, *The Political Economy of Pondoland 1860 to 1930* (Cambridge, Cambridge University Press, 1982), 104–12.
[42] Brown, 'Veterinary Entomology'.

In the homelands, however, compulsory dipping continued to be enforced, because animal health authorities viewed the cattle in communal areas as a source of infections, with tick-borne disease being of major concern. By continuing the dipping service, the department hoped to reduce and, if possible, eradicate the tick population in these areas.[43]

Dipping also provided a vehicle for government surveillance of African livestock.

While dipping was gradually accepted, opposition recurred sporadically. For example, during the Mpondoland Revolt of 1960, rebels attacked the dipping tanks as a symbol of state authority. They may have taken their cue from the stoning and infilling of tanks by rural dissidents in Natal during the previous year.[44] As mentioned above they believed that the state was intentionally distributing under-strength poisons in order to destroy their cattle. Inadequate dipping rather than dipping itself prompted their suspicions. Dipping was suspended in a few coastal districts for some months. Tim Gibbs found that antipathy to betterment (see chapter 1), which had been one precipitating factor in the Mpondoland Revolt, became tied up with hostility to dipping.[45] As noted, betterment was one of the most radical interventions into African rural life since dipping itself.[46] Stock culling, designed as another conservation measure, was a critical element in the early schemes. Because cattle were counted at the dipping tanks for culling, the two state projects became conflated in some people's minds.

Apartheid, specifically the homeland policy, also shaped the changing regulatory environment. In 1951, the National Party government passed the Bantu Authorities Act which aimed to install Tribal Authorities throughout the African-occupied rural areas of South Africa. They were to be the building blocks of new regional and territorial authorities, based largely on traditional leaders or chiefs. The aim of the National Party, defined more explicitly when Verwoerd became Prime Minister (1958–1966), was to create self-governing African territories where they intended that the majority of Africans would live. They planned to consolidate the scattered African reserves into more coherent spatial blocks called homelands. At the same time, stringent controls

[43] P. J. Masika, A. Sonandi and W. van Averbeke, 'Tick Control by Small-Scale Cattle Farmers in the Central Eastern Cape Province, South Africa', *Journal of the South African Veterinary Association* (*JSAVA*), 68, 2 (1997), 45–48.

[44] Tom Lodge, *Black Politics in South Africa since 1945* (London, Longman, 1983).

[45] Tim Gibbs: 'Cattle Dipping and Vet Services. Patterns of State Formation and Rural Unrest in the Transkei: From Popular Struggles to Populist Politics, 1950s to Today', unpublished research report, Oxford (2009); 'From Popular Resistance to Populist Politics in the Transkei', in William Beinart and Marcelle Dawson (eds), *Popular Politics and Resistance Movements in South Africa* (Johannesburg, Wits University Press, 2010).

[46] Chris de Wet, *Moving Together, Drifting Apart: Betterment Planning and Villagisation in a South African Homeland* (Johannesburg, Witwatersrand University Press, 1995).

attempted to constrain African urbanisation. In the long term, apartheid theorists hoped that white power could be conserved in the cities and rural farmlands of South Africa, while a measure of authority was devolved to the chief-led, and relatively conservative, homeland governments. In this way also, the apartheid state hoped to defuse the rising force of African nationalism.

Agricultural services, including dipping were amongst the first to be transferred to the homeland governments. In some areas it remained compulsory to dip; in others it did not. Each homeland financed the dipping operations differently. In QwaQwa and Ciskei for example, livestock owners paid a fee for each animal dipped, whereas in Transkei the homeland government collected an annual dipping levy.[47] Transkeian reports revealed a growing shortage of veterinary supplies and dipping foremen, as well as organisational problems and poor financial management. Even in the summer months when ticks were most prevalent, dipping became irregular. The political context also changed. Chiefs were reluctant to enforce unpopular weekly dipping and stringent collection of levies. As we will explain, few communities understood the need for such regular dipping. Once the central state abandoned responsibility for compulsory dipping, local authorities were no longer able or willing to act as coercively as before.[48] We can see the results in the surviving Mbotyi dipping registers. In 1954 they dipped about 37 times, in 1971 17 times, in 1976 25 times and in 1992 18 times.[49]

Throughout the apartheid period there was also some distrust of dipping in Bophuthatswana (now part of North West Province). Stockowners expressed reservations about stock censuses and vaccinations at the dipping tank.[50] Samuel Mothapo from Slagboom near Hammanskraal explained how the white veterinarians and (African) stock inspectors would order people to bring their animals to the communal dipping tank, not only to deal with ticks but to administer injections and to count them. He complained: 'the whites never explained what they were vaccinating for or why they took the censuses. Stock inspectors used to identify certain cattle, usually older animals, and brand them for collection the following day. They paid us a nominal fee for our animals, but it was far below market value.' Then, Mothapo claimed, 'they sold our marked animals for a higher price because they knew these cattle were not really "scrap"'.[51] In his

---

[47] Masika, et al., 'Tick Control by Small-Scale Farmers'; M. Hlatshwayo and P. A. Mbati, 'A Survey of Tick Control Methods used by Resource-poor Farmers in the Qwa-Qwa area of the Eastern Free State Province, South Africa', *OJVR*, 72, 3 (2005), 245–9.

[48] Masika, et al., 'Tick Control by Small-Scale Farmers'.

[49] Dipping registers, Mbotyi. The time span covered by the registers was not exactly the same as a calendar year.

[50] Albert Ilemobade, 'The Development of Strategies and Information for Technology Transfer on Ticks and Tick-borne Diseases to Developing Communities in some Parts of South Africa', unpublished report for the Director of the Onderstepoort Veterinary Institute, (n.d. – late 1990s). Thanks to Heloise Heyne and Arthur Spickett for a copy of this report.

[51] The translator gave this as 'scrap' but the term often used in South Africa for non-pedigreed mixed cattle is 'scrub'.

view, whites deliberately profited from African powerlessness. Confiscating live-stock angered many people in Slagboom and 'made us determined to hide our animals from the authorities'. Distrust of the Department of Agriculture has lingered, he suggested.[52]

Informants in some villages in North West Province claimed that the state, in cahoots with the pharmaceutical companies, deliberately spread ticks in the apartheid period so that they could make money from selling more drugs. Some also believed that the white government in Pretoria strove to destroy African cattle to enforce betterment or out of malevolence. In Fafung and in the Moretele areas, informants were adamant that some species of ticks were recent introductions and had arrived due to the machinations of drug companies. A group of farmers from Fafung recalled that that ticks have become a real menace since 1963 when they were dropped from planes; some claimed to have found ticks in envelopes in the veld.[53] A similar story came from farmers in Sutelong and Kgomo Kgomo although the timing was different.[54] Two informants from Kgomo Kgomo suggested that profiteering whites strewed packets of ticks across the veld in the 1980s. Sello Moiosane said he had found 'ticks in an enve-lope whilst out hunting'.[55] A neighbour, Simon Mokonyana, recounted that 'in the 1980s ticks came in planes and they appeared everywhere, in envelopes, plastic wrappers and bags'.[56] In the late 1970s and early 1980s scientists were collecting ticks in envelopes in order to investigate tick resistance in present day North West Province. It is possible that their activities triggered the rumours and perhaps farmers found dropped or discarded envelopes containing ticks in the veld.[57] Essentially these stories speak of suspicion of the state and the phar-maceutical companies.

Within South Africa the 1984 Animal Diseases Act declassified tick-borne diseases as notifiable infections, giving local authorities more power to decide whether or not to finance dipping in particular locations.[58] Since 1994 provin-cial governments have been responsible for dipping which, once again, has meant that there is no standardised policy even in areas with similar species of ticks and tick burdens. When QwaQwa re-joined the Free State in 1994, the Provincial Government deprioritised dipping and stopped procuring free dip for farmers.[59] The same happened in North West and Limpopo Provinces. In

---

[52] Samuel Mothapo, Slagboom, 10 November 2009.

[53] Fafung group meeting, 25 January 2010.

[54] Sutelong group meeting, 27 January 2010; Sello Moioisane, Kgomo Kgomo, 29 January 2010; Simon Mokonyana, Kgomo Kgomo, 29 January 2010.

[55] Sello Moioisane, Kgomo Kgomo, 29 January 2010.

[56] Simon Mokonyana, Kgomo Kgomo, 29 January 2010.

[57] Arthur Spickett and Heloise Heyne, 30 November 2012; K. R. Solomon, 'Acaricide resistance in ticks', *Advances in Veterinary Sciences and Comparative Medicine*, 27 (1983), 273–96.

[58] Ilemobade, 'The Development of strategies'. Only Corridor Disease, closely related to East Coast fever is still notifiable. It exists in buffalo but not in the domestic livestock population.

the Eastern Cape, by contrast, the Department of Agriculture initially supplied the chemicals. However, by 1996 budgetary constraints led to cuts and the state withdrew funding for dips and dipping foremen.[60] As explained by the state veterinary surgeon in Lusikisiki, who had been in government service since 1993, the Department wished not only to save money, but also to give livestock owners more responsibility and to encourage the formation of local dipping committees. In the 1990s, he recalled, the government was saying 'teach a person how to fish rather than give them fish'.[61] Government discourse, even under the populist Minister of Land Affairs and Agriculture Derek Hanekom, tended to view African-owned livestock as uneconomic because so small a proportion was sold on the market. Hence major expenditure could not be justified. Mbotyi stockowners were fully aware of this: 'Government said that these animals are just grazing on grass … so we must buy medicine ourselves'.[62]

In the early twenty-first century the Eastern Cape government, responding to popular pressure, placed greater priority on agriculture and reinstated some state subsidies. As the Lusikisiki veterinarian noted, livestock 'saved unhappiness' in the Eastern Cape.[63] The provincial Department of Agriculture's strategic plan for 2005–9 was a hugely ambitious document, which also recognised that 'the department is grappling with the challenge of expanding access to veterinary public health to reach the communal areas'.[64] Dipping was a major element in the government's 'six pegs policy' and a 2005 document projected 200 new tanks – though few if any had been built in Lusikisiki by 2008.[65] In principle, the state provided free Triatix 500 TR for weekly dipping in a plunge tank in summer and fortnightly in winter. Triatix was provided in powder form, which was relatively easy to distribute, and far cheaper than the pour-ons and injectables.[66] Some Animal Health Technicians (AHTs) also assisted in the organisation of dipping.

AHTs were employed in government district agricultural offices in the 1990s as intermediaries between state veterinarians and communities. To some degree they replaced the stock inspectors and dipping foremen who were largely dispensed with. They had higher qualifications, being in possession of two and

---

[59] Hlatshwayo & Mbati, 'Survey of Tick Control Methods'.

[60] Masika, et al., 'Tick Control by Small-Scale Farmers'.

[61] Kassim Kasule, Lusikisiki, 18 February 2009.

[62] Myalezwa Matwana, Mbotyi, 22 March 2008.

[63] Kassim Kasule, Lusikisiki, 18 February 2009.

[64] Province of the Eastern Cape, Department of Agriculture, Strategic Plan, 2005–2009, 41.

[65] Province of the Eastern Cape, Department of Agriculture, 'A People's Contract to Create Work and Fight Poverty: The Agrarian Way' (2005); Interview with Kassim Kasule, Lusikisiki, 31 March 2008.

[66] Discussion with Patrick Masika, Fort Hare, 23 February 2011; B. Moyo and P. J. Masika, 'Tick Control Methods used by Resource-limited Farmers and the Effect of Ticks on Cattle in Rural Areas of the Eastern Cape Province, South Africa', *Tropical Animal Health and Production*, 41, 4 (2009), 517–23.

three year degrees or diplomas from agricultural colleges or similar institutions. This enabled them to play a wider role in animal health provision, including inoculation, organising stock days as well as dipping. In North West Province and QwaQwa, farmers said that they discovered much about modern therapeutics from AHTs, even if they did not always accept the medical explanations for the usage of particular drugs. Although the AHTs represent an innovative response to the problems faced by the veterinary services in communicating with African stockowners, their engagement and coverage is uneven.

In practice, dipping remained patchy. Despite expansion of AHT positions, not all could be filled in Lusikisiki. Even with a full complement each official would have to manage three or four dipping tanks with an average of 7,000–8,000 cattle per officer. The onus was still on livestock owners to organise themselves into dipping committees. In some areas, the dipping committees functioned relatively efficiently. The AHT in charge of Hombe, a large settlement on the south-eastern side of Lusikisiki town, reported that the tank was in good condition, and they dipped about 300 animals twice a week. In Hertzog and Tamboekiesvlei, one of our research sites, Vimbai Jenjezwa found that dipping was maintained not least due to the commitment of an 81-year-old man, treasurer of the committee, who was the largest cattle owner with 53 head.[67] He owned a vehicle and was able to collect a pump from the nearby government training centre at Mpofu to empty and fill the tank with fresh water. They also benefited from the enthusiasm of a young AHT who helped procure dip. Even then, dipping was irregular and generally suspended when it rained, because stockowners thought that their animals caught cold in inclement weather and that the rain washed off the chemicals before they could take effect. Not all livestock owners dipped, some did not dip regularly, and even in this relatively well-organised committee, there were disputes about payments to the treasurer.

In Lusikisiki, many tanks were damaged and committees ineffective.[68] The state veterinarian's own view, certainly confirmed by the position in Mbotyi, was that dipping committees often reflected political and economic divisions within communities. Dipping in Mbotyi was sporadic and some people did not get access to the tank. Although dip was supposedly free, it was only available at the offices in Lusikisiki, about 30 km away on a difficult road, and had to be collected by an accredited member of the dipping committee. The committee had made a charge for the collection, for the organisation of dipping and for much-needed maintenance of the dipping tank. If dipping was to be effective, the tank had to be emptied or pumped out regularly, and each new

---

[67] Vimbai Jenjezwa, 'Stock Farmers and the State: A Case Study of Animal Healthcare Practices in Hertzog, Eastern Cape Province, South Africa'. Unpublished MA Thesis, University of Fort Hare (2010), 116–7.

[68] Edmore Fisani, Animal Health Technician, Lusikisiki, 20 February 2009; Nonyebo Kwinana, Animal Health Technician, Lusikisiki, 20 February 2009.

batch of dip mixed with clean water. Muddy and dirty water, as noted above, undermined the effectiveness of the chemicals. In 2008–9 there was a stand-off between those who claimed that they had paid towards dipping, but were not getting adequate access, and the committee, associated with the sub-headman, which refused to organise dipping unless the dues were paid.

Those not on the committee felt that the network around the headman either failed to use money collected or had effectively monopolised the dip.[69] The headman and his main adviser, in turn, were adamant that the Department did not provide dip in sufficient quantities, so they were not to blame; they suggested that officials in Lusikisiki were corrupt.[70] In February 2009, the local AHT confirmed that there was not, and had not been since December, free dip available in Lusikisiki for his area at least.[71] At the height of summer, the village stopped dipping. The headman also insisted that the government dip did not work: 'if you dip today then ticks are still there tomorrow, they are not falling from the animals and even if they fall, then within a week there will be many ticks back on the animal.'[72] Dipping was an issue about which feelings ran high and remained at the heart of village politics. It also highlighted weaknesses in the Eastern Cape administration's capacity to deliver services at both a financial and organisational level.

Tank dipping was organised very irregularly during the period of research in Mbotyi from 2008–12. In March 2012, during the final research trip, we contributed financially to facilitating a dipping day, the first for six months.[73] The hotel lent its petrol pump to extract the dirty water and replace it with clean water from the nearby stream. This is a tedious task if done by hand. Free dip was secured from the veterinary office in town and three bags of 1.5 kg powdered Triatix mixed into the 5,000 litre tank. Over 600 cattle were dipped. Those close to the dipping committee, mostly with larger herds, tended to go first. But it was not monopolised by them and over 30 herds were dipped in all.[74] A number of cattle were brought from more distant pastures. We were able to visit some of the herds in the days after dipping and those dipped earlier, roughly the first 400, were – contrary to the headman's assertions – distinctly clean of ticks. The Triatix solution seemed to be less effective on the subsequent herds – as noted by their owners. This was not surprising. It rained heavily during the night before dipping, and the animals brought mud into the tank. The cattle were held in a race of about 50 metres after dipping and generally

[69] Vuyisile Javu, Mbotyi, 30 March 2008.
[70] Sub-headman Malindi, David Ogle and others, Mbotyi, 1 April 2008.
[71] Mr Fusile, Animal Health Technician, Lusikisiki, 23 February 2009.
[72] Sub-headman Malindi, David Ogle and others, Mbotyi, 1 April 2008.
[73] Mbotyi, 15 March 2012.
[74] It was not easy to count herds, as some came together, and others became entangled together in the queue to dip, and our attention was occasionally distracted by conversations. It is likely that there were substantially more owners dipping.

Photo 2.1 Cattle in the dipping tank Mbotyi (WB, 2012)

they were only led out after a herd was completed. Many of the animals urinated and defecated on the sloping concrete floor and some of the run off flowed back into the tank. Dirty water greatly reduces the effectiveness of the acaricides.

Dipping is a village spectacle and both the older men on the dipping committee, as well as the youths driving animals to the tank, were absorbed and enthused by the event. They worked effectively to put most of the animals through in three hours. Echoes of the conflicts over costs surfaced during the day. Although the dipping committee's costs for collecting the dip, and filling the tank were covered, they requested that all of those who wanted to dip either assist in repairing the tank or pay R10. Some brought materials and others put in labour, although the repairs were not complete by dipping day and a broken section of the race enabled a few of the more boisterous animals to clamber over the side. Two old men, who had not been able to help, felt deeply aggrieved about having to pay the R10. The chair of the committee, Alfred Banjela, argued that they wished to save some money to finance a further dipping day. Some were sceptical that it would be used for this purpose.

As a result of these financial, political and administrative problems, livestock owners in Mbotyi had to fend for themselves. Most used 5-litre pump sprays. Some purchased dip from town at the chemist or from retail agricultural supply shops; a few went further afield to the agricultural cooperative in Kokstad where it was cheaper. Many also bought 'nips' of dip. The term 'nip' is derived from a 200ml bottle of brandy in which dip is sold. Informal retailers visited Mbotyi on pension day in bakkies (small pick-up trucks) or set up stalls in Lusikisiki town. They procured large quantities of chemicals at wholesale prices, divided it into unmarked bottles, and sold them off as nips. We do not know the mark-up charged, but at R60–70 a nip in 2008–9, this was still significantly cheaper than a similar quantity of liquid Triatix bought from the chemist or retailer. Of course there may have been significantly less of the active ingredient in the nips. Possibly some vendors were able to gain access to government Triatix powder illegally and sell it in liquid form. Leaving aside the question of corruption, a key problem was that the content and strength of the nips were not identified. They may have included chemicals that were adulterated or past their expiry date.

A number of those interviewed had a rough formula for dilution and felt that the nips were effective, at least in the short term. Mamcingelwa Mtwana, an old woman who had nurtured the few cattle left by her deceased husband, struggled to keep a family herd of about 10 head in 2011.[75] Although she had very limited education, and spoke only isiXhosa, she attempted to administer the most effective treatments, both plant and biomedical. She only bought medicines from hawkers, not branded medicines from retailers, even though she paid up to R200 for a bottle. She also used to buy an expensive pour-on from hawkers in Lusikisiki in an unmarked bottle – they subdivided bulk purchases so that impoverished livestock owners could afford small quantities. But she found the liquid nips, diluted and sprayed on the cattle, worked more quickly; they were 'strong' and 'finished the ticks'. In general, the treatment lasted a couple of weeks. But she, as in the case of others, felt that the informal retailers served a useful function.

A similar picture emerged from Peddie in the former Ciskei. Here interviewees commented on the apparent inefficacy of the dips as well as tensions between cattle owners and members of the dipping committee. Andrew Ainslie found that the local tank was in a state of disrepair and full of sludge.[76] Many did not bother to bring their cattle regularly to the communal tank. However, it was clear that a lack of knowledge as well as a lack of organisational structure undermined the whole procedure. Stockowners brought their animals to the tank erratically and there was no systematic way of recording which cattle had been dipped and which had not. In the absence of a dipping foreman, the strength of solution was not adequately tested. Farmers argued that the dip was too weak, yet on one occasion they filled the tank with three times the amount

[75] Mamcingelwa Mtwana, Mbotyi, 13 December 2011.

[76] Andrew Ainslie, 'Hybrid Veterinary Knowledges and Practices among Livestock Farmers in the Peddie Coastal Area of the Eastern Cape, South Africa', March 2009. Names are not used from Ainslie's research as anonymity has been requested.

of Triatix powder stipulated on the bag. As in Mbotyi tensions within the committee revolved around issues of money and patronage. Stockowners were not prepared to cooperate to empty the tank of water, clean it and maintain the crush pens. 'People here don't want to work together properly', one informant confirmed; so they acted individually, investing in their own sprays and pour-ons.[77] Some lamented that the situation had been better in the past when the whites and the state had organised the dipping. Government policy, centred on the dipping committees, had failed to enable cattle owners to take greater collective responsibility for disease control.

In North West Province some farmers were also nostalgic about the past and they too wanted a more top-down approach to tick management because they felt that they lacked the resources, as well as the educational and administrative capacity, to run things themselves. Despite suspicion of state veterinarians during the Bophuthatswana era, we came across no record of active opposition and violence directed against stock inspectors and dipping tanks, as in the Transkei in earlier decades. Tim Gibbs argues that the major impulses behind popular politics in the former homelands began to change around the 1960s. The rebellions against Bantu Authorities and betterment, he suggests, had largely been driven by animosity to the state and an attempt to retain a remnant of rural autonomy. But popular politics in the homelands era increasingly revolved around access to expanded resources available from the Bantustan governments.[78] Some informants from the former Bophuthatswana homeland looked back to Lucas Mangope's time in office as a golden age of free dips and vaccines.[79]

Yet, there were inconsistencies in dipping policy prior to 1994. A survey published in the *Onderstepoort Journal* in 1998 showed that even in the same veterinary area there were different procedures and highly uneven provision of tanks.[80] The reasons were not obviously explicable from an epidemiological perspective as this had once been an East Coast fever area. Since 1994 there has been little official involvement in dipping and livestock owners have had to find their own solutions. The North West Province government ceased to subsidise the provision of dips and did not encourage farmers to form dipping committees. The same applies to Limpopo Province, but not Mpumalanga. The network of dipping tanks in North West Province was far smaller than in the Eastern Cape and those that did exist were now discarded as relics of a former

---

[77] Ainslie, 'Hybrid Veterinary Knowledges', 25.

[78] Gibbs, 'From Popular Resistance to Populist Politics'; Tim Gibbs, 'Transkei's Notables, African Nationalism and the Transformation of the Bantustans, c.1954–c.1994', DPhil Thesis, University of Oxford (2010).

[79] Lott Motuang, Bollantlokwe, 10 November 2009; Group discussion, Mita Mogodu, Aletta Nare, Stoki Taedi, Mareetsane, 4 November 2009; Frederick Matlhatsi, Mabopane, 29 October 2009; Thomas Seemela, Madidi, 23 October 2009; Otsile Frederick Morwe, Dithakong, 3 November 2009. Shakung Group Meeting 19 January 2010.

[80] Tice, et al., 'The Absence of Clinical Disease in Cattle'.

age. In some villages, such as Madidi (north of Pretoria), where the tank had become defunct due to poor maintenance, some farmers were clamouring for help from the Department of Agriculture. Thomas Seemela, in his mid-80s, had tried a number of acaricides and was adamant that: 'the plunge tanks are far more effective than anything else. They ensure that the entire animal is covered in dip. Sprays and pour-ons are expensive and do not work so well. The animals remain covered in ticks.'[81] When asked about the possibilities of working together, cattle owners throughout North West Province frequently replied that the 'Tswana are too individualistic' and would only cooperate if the state, or white people, directed them.[82] This might seem perverse in view of South Africa's recent history and it did not imply that they wanted a restoration of white rule. They seemed to be expressing the necessity for a stable and well-informed administrative structure to assist them.

Many North West Province farmers knew about Deadline and Bayticol (flumethrin), poured in small quantities onto the backs of animals, but these were too expensive an alternative for most. In 2009 a one-litre bottle of Deadline retailed at R425 (c. £30 – May 2013) at a farmers' cooperative in southern KwaZulu-Natal and in 2011 the price had risen to about R450 a litre. This was nearly half the monthly pension in 2009.[83] For those in the Kat River valley, prices were even higher, at R156 for 200 ml in 2009.[84] Rural communities and poor people often have to pay more for commodities because they are isolated and purchase in small quantities. A litre covers about 20 cattle and it should be applied at least four times a year. In Mbotyi, some cattle owners said that they found using Deadline difficult because they had to tie animals up, and keep them still, to administer it effectively. Although there are often poles in or near the cattle kraals, or sturdy corner posts, this takes time and effort. Stockowners also perceived that Deadline washed off more quickly in rainy weather. Moreover, instructions for acaricide application appeared in English or Afrikaans and were inaccessible to many stockowners who had limited literacy or familiarity with these languages. Administration of these expensive pour-ons was therefore very occasional, haphazard, dependent on trial and error, or on the advice of neighbours. Stockowners often used cheap disinfectants such as Jeyes Fluid which they applied either as a spray or as a swab for cleaning ears and abscesses. Supona spray (chlorfenvinphos), highly valued in Mbotyi for clearing ears, was far too expensive for most to purchase.

Another strategy pursued by a small number of individual African farmers has been to take advantage of the Nguni Cattle Schemes. In North West

[81] Thomas Seemela, Madidi, 23 October 2009.

[82] Shakung group meeting 19 January 2010; Fafung group meeting 25 January 2010; Isaac Modiselle, Madidi, 27 January 2010; Charles Matome, Sutelong, 28 January 2010; Molekwa Mabe, Mabeskraal, 1 February 2010; Hendrick Metwamere, Mantsa, 12 February 2010.

[83] Personal experience (2009) and www.vetproductsonline.co.za (2011).

[84] Jenjezwa, 'Stock Farmers and the State', 115.

Province this was specifically designed for 'emergent' farmers with 300 to 400 hectares of grazing which was well watered, fenced and properly rotated. Grantees also had to have some biomedical knowledge of animal diseases and the capital to invest in vaccines and dips. Theoretically, applicants who met the exacting criteria received 12 Nguni cows and one bull to build up a herd, which was to be kept separate from other cattle to prevent inter-breeding. After five years the farmer repaid the state with 10 Nguni heifers, but could keep the rest of the offspring.[85] According to Tefo Kepadisa, a livestock specialist based at the Mafikeng Department of Agriculture, the scheme also empowered women.[86] Dineo Morule, a particularly successful women farmer from Ramatlabama on the Botswanan border, won a prize for her cattle rearing in 2009.[87] Morule had great faith in the Nguni Cattle Scheme and praised its aims and its outcomes. She had managed to acquire 24 Nguni cows from the government and built a herd of 72 cattle. She rented a 530 hectare farm for R7,000 a year, and was in the process of negotiating its purchase with the Department of Land Affairs. However, the scheme helped only a small number of people. Morule was the wife of Gideon Morule, president of the North West Farmers' Association. They were both well educated, well connected and had worked in the public sector. The Nguni Cattle Scheme was too small an initiative to become a strategy for building up tick-resistant herds for the masses. At a grassroots level, the clamour was for a return to state-run dipping rather than efforts to transform their herds. It would be very difficult to control reproduction on communal lands. Insofar as they had heard about these government schemes in Mbotyi, stockowners thought they were far too restrictive and they preferred the larger Brahmans.[88]

In sum, the legacy of dipping has been mixed. Policies varied from one part of the country to another depending on the local epidemiological picture and the politics of the region. After the eradication of East Coast fever in the 1950s, followed by the imposed devolution of agricultural services to the homelands, dipping became patchier in terms of its geographical distribution and its actual enforcement. In the 1990s provincial governments tended to withdraw such services on the grounds of costs or in an attempt to encourage farmers to take responsibility for the health of their animals. Dipping committees in the Eastern Cape have not been an overall success and North West Province did not even attempt such a strategy. Lack of knowledge about the scientific rationale for dipping and about prescribed chemical concentrations undermined effective control. In theory, pour-ons and injectables have made it easier to manage ticks; however their costs are prohibitive for most smallholders. Stockowners said that they find it difficult to submerge individual interests in communal provision,

---

[85] Tefo Kepadisa, Mafikeng, 8 February 2010.
[86] Tefo Kepadisa, Mafikeng, 8 February 2010.
[87] Dineo Morule, Ramatlabama, 10 February 2010.
[88] Report-back meeting to stockowners in Mbotyi, 5 December 2012.

and an institutional framework for new community-led veterinary controls remains elusive.

Local knowledge about ticks and tick control

In 1997 Patrick Masika and his co-workers published the results of an investigation into African farmers' ideas about the cause of gallsickness and redwater. The field sites were in the Eastern Cape, where they found that tick-borne diseases were the most common cause of death after malnutrition. The analysis suggested that many stockowners could not differentiate between the two diseases and largely identified them from post-mortems, in particular the presence of an enlarged gallbladder. Most farmers associated these infections with seasonal changes in the grass cover or over-consumption of certain vegetation. Masika concluded that only 2 per cent of the 138 interviewees linked gallsickness to ticks, and 12 per cent regarded redwater to be tick-borne. [89] Our research over a decade later in the Eastern Cape and elsewhere suggested that little had changed. Most African livestock owners, whether from the Eastern Cape, North West Province or the former QwaQwa homeland did not regard ticks to be vectors of diseases. Ticks were a scourge and a major preoccupation for stockowners. Ticks figured large and early in many of the interviews, especially in Mbotyi. But informants rarely conceived them as spreading potentially fatal infections. This was surprising given the decades of state-organised dipping. Biomedical knowledge was not transferred with the dip.

During Brown's first field trip to North West Province in 2009 she interviewed more than 60 farmers of whom only two believed that gallsickness was tick-borne and neither named the blue tick as a vector.[90] As in Masika's study, most stockowners associated gallsickness and the symptoms of a swollen gallbladder with some type of digestive complaint, caused by changes to the veld (See chapter 3). In Mbotyi, those interviewed were of a similar view. They identified these diseases, and their symptoms, and they used African names for them, but they did not associate them directly with ticks. We discuss understandings of this and other diseases in chapter 3.

Farmers in North West Province associated heartwater more clearly with ticks even if they did not identify the bont tick as the carrier. A few respondents recognised that heartwater could infect sheep and goats as well as cattle.[91] Testimonies suggested that heartwater was a more recent introduction to North West

[89] P. J. Masika, A. Sonandi and W. van Averbeke, 'Perceived Causes, Diagnosis and Treatment of Babesiosis and Anaplasmosis in Cattle by Livestock Farmers in Communal Areas of the Central Eastern Cape Province, South Africa', *JSAVA*, 68, 2 (1997), 40–44.

[90] Thomas Seemela, Madidi, 23 October 2009.

[91] Mosweu Labius, Disaneng, 6 November 2009.

Province and this may explain why people had a more scientifically informed understanding of the disease. Our impression was that they had gleaned information about it from veterinarians, stock inspectors and more recently the AHTs. Whereas Setswana speakers had their own name for gallsickness (*gala*), they had adopted the Anglo-Afrikaans word heartwater/*hartwater*, suggesting it was not an infection that had been familiar to them for a long time. Kgomoto Goapele, a 40-year-old farmer from Mantsa near Mafikeng, related: 'heartwater is now the main problem disease here. Cattle shiver and tremble and lots of snot comes out of their noses. The disease came with the new ticks, the *kgofa e tala* (the bont tick).'[92] His views correlated with the findings of tick researchers at Onderstepoort, who found that bont ticks were moving westwards into the more arid zones of North West Province and the Northern Cape. Stockowners may have been reluctant to shift their explanations of old-established diseases, but were more open to veterinary opinion about the aetiology of infections that were less familiar to them – possibly recent introductions.

A few informants in North West Province who had heard that ticks propagated diseases translated this to mean that they were responsible for all ailments, not just heartwater and gallsickness, but also anthrax, blackquarter and lumpy skin.[93] They were also aware that ticks thrived in particular environments. Places like Fafung appeared to be especially pathogenic, partly because of the plethora of ticks. Fanti Pitso explained that the Mafikeng area was a much healthier place to rear animals as there were far fewer ticks than there were to the east. The zone from Rustenburg eastwards was seen as more dangerous due to higher levels of rainfall which enabled ticks to flourish.[94] A number of farmers complained that ticks had increased in recent years. They attributed this to denser populations of both people and cattle, especially in the Odi/Moretele districts north of Pretoria. Some directly ascribed the rise in livestock populations to the apartheid policy of forced removals. Interviewees in Moiletswane, for instance, explained that during the 1960s and 1970s there was a big influx of people from the Free State and Mpumalanga due to the Group Areas acts, the clearing of African 'black spots' and white landowners forcing 'surplus' tenants off their land.[95] This view showed an awareness that diseases could travel to hitherto unaffected parts of the country.

Informants who associated ticks with disease were often unsure how they actually killed animals. The concept of germ, bacterium, virus or parasite was

[92] Kgomotso Goapele, Mantsa, 12 February 2010.
[93] Molelekwa Mabe, Mabeskraal, 1 February 2010; Jacob Tau Diole, Mabeskraal, 4 February 2010; Modisaotsine Taikobong, Magogwe, 11 February 2010; Reuben Ramutloa, Disaneng, 11 February 2010; Bassie Monngae, Mabeskraal, 1 February 2010.
[94] Fanti Pitso, Mafikeng Municipality, 2 November 2009; Lott Motuang, Bollantlokwe, 10 November 2009.
[95] Group Meetings Moiletswane, 20 and 21 January 2010; L. Platzky and C. Walker, *The Surplus People: Forced Removals in South Africa* (Johannesburg, Ravan Press, 1985).

not part of their vocabulary. A minority of our Setswana-speaking informants explained that ticks injected a poison (*bolthole*) into livestock that ultimately killed them.[96] Simon Mathibedi from Mmakau (near Brits) was categorical that 'Africans have always known about ticks and that ticks impart a poison into animals that makes them sick'.[97] However, this does not seem to have been the case from the accumulated testimonies from North West Province. The idea of an injectable poison was the nearest respondents came to explaining a concept of corporal invasion by external agents.

Regardless of whether they conceived of a link between ticks and certain diseases, every respondent wanted clean animals and did their utmost to destroy acarids for a variety of reasons. Ticks could penetrate and damage hides, ears, tails, anuses, genitalia and teats. They identified the bont tick as an especially vicious variety because its deep mouth parts could gorge out the skin and they also tended to gather around the teats, making it difficult for calves to suckle. Bonts created deep lesions that attracted parasitising flies, leading to sepsis and myiasis. Heavy tick infestations could cause intense irritation which veterinarians call 'tick worry'. This resulted in a reduced appetite and with it failing strength and failing health. Informants portrayed ticks as sucking animals dry, depriving them of their life force. Ticks could also strip cows and bulls of their libido threatening the survival of the herd.[98]

Stoki Taedi identified various species of ticks from pictures and observed that different species attached themselves to different parts of the body. Blue ticks (*tampane*), she explained, 'grow so big and they are the ones that often cause blindness and death. The bont ticks (*kgofa*) cling to the anus and suck the blood, poisoning it. They are very dangerous as they kill cows.'[99] Nick Lebethe explained that ticks caused a vitamin A deficiency as they suck the blood. He believed that vitamin A kept animals alive and gave them energy. 'Good health means accumulated energy.'[100]

Given that every interviewee in North West Province talked about ticks, the vocabulary for them was surprisingly limited. By and large they were just known as *kgofa* which simply means tick and informants did not differentiate between species. If they used descriptive words they drew on colours, but even then there was no consistency. For example stockowners described the brightly

[96] Simon Mathibedi, Switch (Mmakau), 9 November 2009; Stoki Taedi, Mareetsane, 4 November 2009; Otsile Frederick Morwe, Dithakong, 3 November 2009; Andrew Mabaso, Mmabatho, 5 November 2009; Kgomotso Goapele, Mantsa, 12 February 2010; Laurence Bodibe, Ramatlabama, 10 February 2010.

[97] Simon Mathibedi, Switch (Mmakau), 9 November 2009.

[98] E.g., group discussion, Mareetsane, 4 November 2009; Nick Lebethe, Bethanie, 26 October 2009; Itumeleng Sedupane, Mareetsane, 4 November 2009; Tollo Josiah Mahlangu, Mabopane, 29 October 2009; Thomas Seemela, Madidi, 23 October 2009; Molekwa Mabe, Mabeskraal, 1 February 2010.

[99] Stoki Taedi, Mareetsane, 4 November 2009.

[100] Nick Lebethe, Bethanie, 26 October 2009.

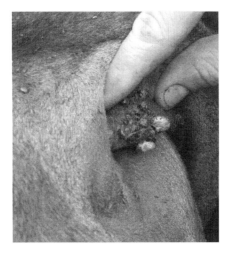 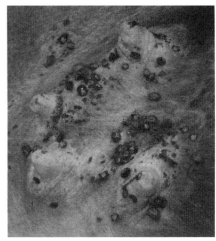

Photo 2.2 Ticks on the anus, Mbotyi (WB, 2012)

Photo 2.3 Ticks destroying the udders, Mbotyi (WB, 2012)

coloured bont ticks variously as *kgofa e thamaga* (literally meaning multi-coloured tick) or *kgofa e tala* (green tick, referring to the bright yellowy green marking on its outer casing). Stoki Taedi said all big ticks in Mareetsane were called *tampane*, otherwise they were *kgofa*.[101] Similarly in QwaQwa people used *mokhasa* as their one word for ticks. Ticks were a problem in QwaQwa but not to the same extent as they were in North West Province or the Eastern Cape. The cold winters limited tick reproduction and the main species was the blue tick. Sesotho-speaking farmers, like their counterparts elsewhere, regarded gall-sickness and redwater as grassland infections.

In the Eastern Cape there appeared to be a richer vocabulary for ticks. In Peddie, Ainslie interviewed a farmer with over 100 head of cattle, who offered four words for five different types of ticks: *indlanga* for the bont tick; *inkezane*, the tiny pepper tick; *ikhalane*, the engorged blue tick that was abundant in the kraals; and *umkasi* which referred to both the little red ticks with a hard thorax and the greenish ticks with red dots.[102] Ainslie also found that some farmers did link ticks with disease but they were unable to specify which infections they caused. As in North West Province most informants associated gallsickness – named locally *inyongo* – with grasses. However, one farmer attributed redwater to the blue tick and he had observed that cattle succumbed when they were covered with too many of these parasites. He had also noticed that there were far more ticks in the forest than on the hillside grazing lands (*etafeni*), but stockowners allowed their animals to enter the woodlands due to better pasture. There was

---

[101] Stoki Taedi, Mareetsane, 4 November 2009.
[102] Ainslie, 'Hybrid Veterinary Knowledges', 21.

thus a balance to be struck between optimising nutrition and protecting livestock from diseases. An elderly woman knew that ticks made cattle sick, but she attributed this to pus and boil-like infections on the animals caused by blowfly abscesses (*impethu*). Another informant associated ticks with redwater and believed that they had to be controlled because they sucked blood so the cattle could not develop. He also confirmed that they seriously damaged the udders so that the calves would die. If the ticks parked on the ears they could invade the brain, generating strange behaviour and causing 'confusion in the cow'.[103]

In Mbotyi informants distinguished *umkhasi*, a small tick; *ikhalane* – a generalised term applied especially to the engorged brown ear tick; and the bont tick, which caused most anxiety. As in parts of North West Province farmers identified the bont as a recent pest, from around the mid-1990s in 'Mandela's time'. The Mpondo called it by the vivid name *qologqibe* (or *qwelagqibe*) which can roughly be translated as 'finisher' of livestock.[104] Informants were clear that the *qologqibe* tick was different to the other local varieties. It bit deeper and went through the hide of the cow. It could not be removed. When the blood sack fell off the head remained embedded. It caused open sores which would not heal and in which flies laid their maggots. These, one man thought, crawled into the orifices and killed the livestock. When they bit into the udders, they caused scabs which closed the teats so that calves could not feed, nor could the cattle be milked. In one case, a cow was slaughtered because its urinary tract had been closed by ticks. Calves were especially susceptible to the bont tick and they were also disadvantaged because of this difficulty in feeding. *Qologqibe* bit in between the hooves of goats, causing lameness and they were very difficult to extract by dipping or spraying.[105]

A variety of explanations emerged, not mutually exclusive, for the bont tick's recent prevalence. Some people, such as Richard Msezwa, associated the advent of the tick with the demise of compulsory and frequent dipping. Msezwa recalled: 'when I was a boy herding you seldom saw ticks in the veld. If you saw a tick you would get excited and call the other herdboys over. Now if you walk through grass your trousers become black with ticks.'[106] He was in his 60s in 2008 and was probably referring to his boyhood in the 1950s when dipping was still universally enforced. Sidwell Caine, of a similar age, agreed that when he was a boy, there were few ticks about. In fact, he thought that as the cattle carried the dip with

---

[103] Ainslie, 'Hybrid Veterinary Knowledges', 23.

[104] *Umqolo* means, amongst other things, the backbone or spine of an animal; *ukugqiba* is the verb to finish or end. A few people said *qwelagqibe* and that the words were used interchangeably. *Ukuqwela* can have the meaning to empty or to finish or strike down.

[105] According to Arthur Spickett, Onderstepoort, 30 November 2012, the tick found in goat's hooves is more likely to be of the genus *Hyalomma*, or the bont-legged tick because bont ticks prefer softer, smoother skin. However, Nkululeko Nyangiwe, 'The Geographic Distribution of Ticks', found many more bont ticks on goats than bont-legged ticks.

[106] Interview with Richard Msezwa, Mbotyi, 25 February 2008.

them through the grass, both ticks and snakes would stay away because of the poison.[107] This idea was echoed in Peddie: 'the cattle were chased into the veld still dripping with dip, to kill more ticks in the grass.'[108] Caine thought that the decline of dipping, and of snake-eating birds like the ground hornbill, not only contributed to the explosion in tick numbers but had also opened the way for a return of snakes, especially black mambas, which bit the cattle.

Myalezwa Matwana remembered that the *amakhalane* (probably referring to the brown ear tick) came back first after the demise of dipping and were easily distinguishable by their large black blood sacks.[109] *Qologqibe*, he thought, came later from the farms (*eplasini*) and were imported from outside the area due to the greater traffic in live animals. People brought cattle in from the farms around Kokstad or further afield in KwaZulu-Natal, where they were working, because livestock were less expensive there. Others noted that migrant workers in Gauteng clubbed together, purchased cheap livestock from suppliers who specialised in that trade, and then hired a van and trailer to drive them down to Mpondoland. They thought that the *qologqibe* was brought from outside in that way. It is also possible that the bont tick had spread incrementally through both cattle movements and inconsistencies in dipping. As we noted above, its boundaries had been mobile in the past, shaped by ecological factors and new patterns of transmission. A veterinary surgeon working in Port Edward, immediately north of Mpondoland, believed that its range had increased in recent years.[110] A study in the coastal Centane district of the Eastern Cape suggested a similar recent spread.[111] This process may have been compounded by tick resistance to acaricides.

Mbotyi certainly provides an environment suitable for the bont tick, with adjacent grassland, forest and more open patches of bush and trees. It is likely that bont ticks and heartwater had been present at some stage in earlier years. Bonts can be harboured by wildlife and Simphiwe Yaphi's explanation suggested that they re-emerged from this source when conditions became more favourable.[112] Now in his late 30s, he remembered seeing them as a boy on a *leguaan*, a type of monitor lizard, which was killed for consumption.[113] *Leguaans* are common in the area and move around water sources across the boundaries

---

[107] Sidwell Caine, Mbotyi, 12 February 2009. Caine is descended from traders but lives in the communal areas.

[108] Ainslie, Hybrid Veterinary Knowledges', 24.

[109] Myalezwa Matwana, Mbotyi, 1 March 2009.

[110] Interview Port Edward, 17 February 2009.

[111] Busani Moyo, 'Determination and Validation of Ethno-Veterinary Practices used as Alternatives in Controlling Cattle Ticks by Resource-limited Farmers in the Eastern Province, South Africa', MSc dissertation in Animal Science, Fort Hare University (2008), iii.

[112] Simphiwe Yaphi, Mbotyi, December 2011.

[113] It is possible that this was a different species of bont tick, *Amblyomma exornata*, which favours monitor lizards, rather than *Ablyomma hebraeum*. I. G. Horak, I. J. McKay, B. T. Henen, Heloise Heyne, Margaretha D. Hofmeyr and A. L. de Villiers, 'Parasites of Domestic and Wild Animals in South Africa. XLVII. Ticks of Tortoises and Other Reptiles', *OJVR*, 73, 3 (2006), 215–27.

of forests and settlement. Bont ticks could have remained endemic in the forest fauna and crossed from wild animals when dipping diminished and ecological opportunities arose. Many think that bush pigs (*ingulube*) have multiplied in recent years and they too are (or were until recently) killed for food. They emerge from the forests to forage on maize fields, where cattle also graze. Tortoises, guinea fowl and even egrets – usually associated with tick control – have also been noted as possible vectors.

Like those in North West Province, informants in Mpondoland were aware that ticks clustered round areas of softer skin where blood vessels were closer to the surface. They knew and had words for gallsickness, redwater, and heart-water, but they associated these diseases neither with ticks in general, nor with particular species (see chapter 3). Nongede Mkhanywa, a *sangoma* (traditional healer) who was very knowledgeable about plants and treatments, explained: 'the ticks suck the blood from animals and they die; there is no particular disease that causes death but they suck the blood …The ticks also go to the ears, and bite on the testes, and flies come and lay their eggs in the wound which goes rotten and causes death.'[114] Myalezwa Matwana, an elderly man recognised as a specialist on livestock diseases, affirmed that in large numbers, ticks were a major problem because they 'kill the cattle by sucking the blood' (*sela igazi*) – in large numbers they 'finish the blood' (*phela igazi*).[115] He noted that the flesh of cows that seemed to die from tick infestation could be grayish (*ingwevu* – meaning the colour and also an old grey-haired person) because the ticks had finished the blood. In some cases the meat was so unpleasant (*imbi*) that they fed it to dogs. He knew the symptoms of different tick-borne diseases, but he insisted that the causes of the diseases that were generally recognised by scientists as tick-borne were seasonal and environmental in their origins. While he and others were clear about the impact of ticks on udders, they did not seem to have a word for the disease mastitis, which is widespread in the coastal Eastern Cape.

In Mbotyi such understandings also affected attitudes to dipping. The inter-views suggested strongly that livestock owners had not internalised the logic of regular dipping or spraying, despite the long history of this practice. Generally people sprayed or dipped to get rid of ticks, and they would wait until the ticks had begun to accumulate again before repeating the operation. They judged the effectiveness of sprays by their capacity to kill the ticks. They did not artic-ulate an understanding of the importance of regular dipping in order to break the life-cycle of ticks or the transmission of tick-borne diseases. Richard Msezwa felt that government dip was not working well, but occasionally they came across a solution which did kill ticks on the spot. 'You know a good dip', he joked, 'because cattle will smell it on dipped animals and run away. That's a good dip. The cattle are scared of going into the tank because it is so strong that

---

[114] Sidwell Caine, 12 February 2009.
[115] Myalezwa Matwana, Mbotyi, 1 March 2009.

it is itchy on their skin.'[116] He said he had sprayed with such a dip but he had bought it in an unmarked container and did not know what it was.

The impact of ticks was highly visible on cattle and horses. These gathered around the ears, the anus, the udder of cows and the soft flesh in between, but they could be seen elsewhere. The ears were perhaps worst affected by brown ear ticks although bont ticks also embedded themselves there. They caused bleeding and in a number of cases, the ears of cattle were partly or wholly eaten away. Walking behind cattle, you could see a cluster of ticks around their anus, and hindquarters, many swollen with blood so that they looked like a coating of drooping, red-black pendants. The ears of a few horses were badly infested and drooped because of tick damage. Ticks crept into the skin below the manes, and some manes were cropped close to the body, so that they could be better treated. Many stockowners were not rigorous in cleansing their animal of ticks, or could not afford to be.

In the drier North West Province, with cooler nights, livestock did not seem to be so dreadfully shrouded with ticks as the coastal area of the Eastern Cape. Climatic factors played some part, but our impression is that more efforts were made to manage ticks and stockowners in Setswana-speaking areas were more actively involved in prevention. Every single informant we met stated that they wanted clean animals and always did their best to keep them tick-free. We could not record dipping practices over time in any detail, but there was a verbal commitment to tackling ticks. Many of those who lacked the financial resources to buy biomedicines made their own preparations with local plants. By far the most common in North West Province were aloe trees and shrubs (*Aloe spp*). According to Simon Mathibedi: 'Africans had to deal with ticks long before the Europeans came. The medical power of aloes is very old knowledge.'[117] Some stockowners grew aloes in their gardens for medicinal purposes. In Garankuwa, not far from the Magaliesberg, people went into the mountains to procure aloes which they transplanted in their own yards. Some stockowners simply cut the leaves into pieces and threw them into the water troughs to make a weak infusion. Others made stronger decoctions which they fed directly to the animals, or else they ground dried aloe leaves into a powder and dosed them with a spoon. Farmers gave aloes to their goats as well as to their cattle. Protagonists of aloes claimed that it made the blood bitter which had the dual effect of repelling new ticks and encouraging ticks that were already lodged in the skin to drop off. Sometimes they would add salt to the water trough or aloe mixture because they believed it increased potency. The sap from aloes was also a soothing unguent which could be made into poultices to treat wounds caused by bont ticks and blowfly-strike.[118]

[116] Richard Msezwa, 19 February 2009.

[117] Simon Mathibedi, Switch (Mmakau), 9 November 2009.

[118] Joseph Gumede, Garankuwa, 21 October 2009; Norah Mabena, Garankuwa, 21 October 2009; Zachariah Maubane, Mafikeng, 2 November 2009; Chris More, Kgabalatsane, 13 November 2009;

In North West Province *nata* (botanical name unknown) was also mentioned as a tick repellent. Described as a large tree like the *marula* (*Sclerocarya birrea*), it grew in the Mabeskraal area and its leaves purportedly made the blood bitter like the aloes.[119] Stockowners supplemented plants with other measures. To dislodge ticks they cut them off with scissors or a razor blade.[120] Alternatively, they dabbed the ticks with cigarettes or smeared them with ash from burnt plants.[121] Petrol or Jeyes Fluid was good for cleansing the skin and disinfecting bites and some sprayed their animals with Jeyes Fluid as if it were a dip.[122]

Scientists have tested a number of popular products for possible repelling qualities and hold little store by the powers of aloes and garlic as acaricides or preventatives. In 2007, for example, extensive testing of aloes produced negative results as far as the scientists were concerned.[123] Nevertheless, many people remained convinced about the therapeutic powers of aloes and continued to use them on their own or alongside modern treatments. Elias Pooe from Shakung (north of Pretoria) argued: 'aloes are far better than dips because they work almost immediately and stock are clear of ticks in less than a day… Dips take longer, much longer, to work and they never get rid of all the ticks.'[124] In Fafung, informants said they only ever used aloes because they could not normally afford modern dips and they found that the aloes worked just as well.[125] Many Tswana stockowners who did purchase dips said they would spray their cattle whenever they saw ticks, but they also dosed them with aloes for added protection against tick strike. In fact aloes served multiple purposes. Farmers not only used aloes as a repellent, but also gave them to their animals as a laxative in the belief that free-flowing bowels were the key to good health. The purgative

(cont) Elias Pooe, Shakung, 19 January 2010; Anna Pooe, Shakung, 22 January 2010; Martha Rakgomo, Mabeskraal, 1 February 2010; George Malatsi, Mabeskraal, 5 February 2010; Reuben Ramutloa, Disaneng, 11 February 2010; Martha Rakgomo, Mabeskraal, 1 February 2010.

119 Martha Rakgomo, Mabeskraal, 1 February 2010.

120 Mosweu Labius, Disaneng, 6 November 2009; Lott Motaung, Bollantlokwe, 10 November 2009; Anna Pooe, Shakung, 22 January 2010.

121 Patrick Molefe, Mmabatho, 5 November 2009; Johannes Nkosi, Garankuwa, 21 October 2009; Joseph Gumede, Garankuwa, 21 October 2009; Lott Motaung, Bollantlokwe, 10 November 2009.

122 Soloman Mokebe, Kgomo Kgomo, 29 January 2010; Anna Pooe, Shakung, 22 January 2010; Soloman Mokebe, Kgomo Kgomo 29 January 2010; Bosie Morapadi, Mabeskraal, 2 February 2010; Henry Ramafoko, Mabeskraal, 2 February 2010; Philemon Ramafoko, Mabeskraal, 3 February 2010; Pagiel Sehunoe, 4 February 2010; John Smit, farmer and pharmacist at Mafikeng Provincial Hospital, 10 February 2010.

123 A. M. Spickett, D. van der Merwe and O. Matthee, 'The Effect of Orally Administered *Aloe Marlothii* Leaves on *Boophilus Decoloratus* Tick Burdens on Cattle', *Experimental and Applied Acarology*, 41, 1–2 (2007), 139–46. Currently researchers are looking at the possibility that a fungus, which lives on aloes, may have acaricidal properties.

124 Elias Pooe, Shakung, 19 January 2010.

125 Fafung group meeting, 25 January 2010.

properties of aloes were held to be particularly beneficial for curing gallsick-ness.[126] (See chapter 5.)

In QwaQwa there was also a strong reliance on local plants to deal with ticks. Sesotho-speaking farmers placed great emphasis on plants that they thought made the blood bitter, and they dosed accordingly. Aloes were less common than in North West Province as the climate is much colder. Popular alternatives were leaves from the *lebejana* bush (*Asclepius decipiens*) or the *mofifi* tree (*Rhamnus prinoides*, dogwood or *blinkblaar*), and a mixture made from the ground bulbs of *hloenya* (*Dicoma anomala*) and *poho tshehla* (*Phytolacca heptandria* or *Xysmalosium undulatum*).[127] They normally administered these plants through the mouth of animals as a drench, although they sometimes sprayed their cattle with *mofifi* liquid as a dip. According to Daniel Khonkhe, *mofifi* had been the staple tick repellent before the arrival of the dipping tank.[128] Like aloes, *mofifi* had a purgative effect and it was used to treat constipation and gallsickness.[129] Some would dip their animals first either in the tank or else using pour-ons, and then dose them with *mofifi* because they thought this strengthened the effect of the chemicals.[130]

Although there was no subsidised dipping in QwaQwa, there were still a number of tanks in operation, which the AHTs oversaw. Some farmers managed the tank themselves and were not using recommended products. According to a group of men from the Ntsoeu Livestock Company in QwaQwa's main city Phuthaditjhaba, farmers filled their local tank with a mixture of diluted Jeyes Fluid, potassium permanganate and *lebejana* leaves.[131] Jeyes fluid was very popular in QwaQwa as both a repellent and an acaricide.[132] So too was motor oil.[133] Informants believed that these products worked. There may be analogies with recent scientific ideas about multiple, integrated strategies that reduce the dangers of resistance. Suggestions of tick resistance did not emerge from inter-views in QwaQwa.

[126] Zachariah Maubane, Mafikeng, 2 November 2009; Gabriel Letswalo, Moiletswane Commu-nity Centre, 20 January 2010; Philemon Ramafoko, Mabeskraal, 3 February 2010; Molelekwa Mabe, Mabeskraal, 1 February 2010; Henry Ramafoko and Elsie Ramafoko, Mabeskraal, 2 February 2010; Emily Diole, Mabeskraal, 2 February 2010; Majeng Motsisi, Mabeskraal, 3 February 2010.

[127] Ntsoeu Livestock Company, Phuthaditjhaba, 11 October 2010; Thomas Mahlaba, Tseki, 15 October 2010.

[128] Daniel Khonkhe, Ha-Sethunya, 13 October 2010.

[129] Paulus Mosia, Ha-Sethunya, 13 October 2010; Sethunya Mathu, Ha-Sethunya, 13 October 2010.

[130] Ntsoeu Livestock Company, Phuthaditjhaba, 11 October 2010.

[131] Ntsoeu Livestock Company, Phuthaditjhaba, 11 October 2010.

[132] Ephraim Mofokeng, Phuthaditjhaba, 15 October 2010; Abel Lebitsa, Mountain View, 14 October 2010.

[133] Ephraim Mofokeng, Phuthaditjhaba, 15 October 2010; Seketsa Mokoena, Lusaka 1 March 2011.

In a recent thesis, Busani Moyo from the School of Agriculture at Fort Hare looked at some of the products and plants in use at his field sites in Centane District in the Eastern Cape. There, as in Mbotyi, there were overwhelming complaints that Triatix did not work effectively, with 95 per cent of his 59 informants arguing to that effect. As a result farmers also complemented the dipping tank with Jeyes Fluid and motor oil. Some stockowners allowed chickens into the kraal as a deliberate tick control measure. Chickens fed off recumbent cattle and also ingested ticks that had fallen onto the ground. We saw this also in cattle kraals in Mbotyi and in North West Province. A small minority used *Aloe ferox* (bitter aloe/red aloe), known locally as *ikhala*. They soaked the leaves in water and applied the solution with brushes or by flicking with branches from trees. Alternatively, cattle owners took powdered bark from the *uBhaqa* or sneezewood tree (*Ptaeroxylon obliquum*), and soaked it overnight to make a dip. Other locally available repellents were *khakibos* (*Targetes minuta*), which has a strong smell, and the colourful shrub weed *Lantana camara,* both of which are exotics from Latin America. Moyo found that knowledge about plants was very patchy and only about 7 per cent of those interviewed revealed awareness of floral preparations to tackle ticks.[134] In Hertzog, a mixture of aloe and *ubuhlungu* (*slangbossie, Melianthus?*), sometimes with other plants, was widely used for gallsickness but not specifically as a tick repellent.[135]

Those interviewed in Mbotyi did not recall any plant remedies for ticks. Although local people have herbal medicines for a number of different animal diseases, this did not include tick repellents. We questioned key informants on this point, and they suggested that they had never had a local treatment for ticks and particularly not for the bont tick. It may be that compulsory and universal dipping was in place for so long, since East Coast fever in the 1910s, that the memory of remedies used before has faded. Myalezwa Matwana, in his 70s, and generally regarded as the most knowledgeable about treatments, hinted at this: 'I never learnt medicines for ticks. Maybe it was because ticks were cured (*aphela* – finished) by dipping. We were dipping often at that time.'[136]

Zipoyile Mangqukela, the oldest man interviewed, about 90 in 2008, said that you needed poison for ticks, and plant-based medicine had to be edible, not toxic.[137] Like others, he was quite clear that there were no local remedies for ticks: 'we never had a traditional *muthi* for ticks; I never heard about it.' It is possible however, that even he had lost knowledge about plants remedies. He mentioned that many years ago he had included the *uqangazana* tree, probably *Clerodendrum glabrum*, in a remedy for redwater (chapter 6).[138] This plant was

[134] Moyo, 'Determination and Validation'; Moyo & Masika, 'Tick Control Methods used by Resource-limited Farmers'; discussion with Busani Moyo, Fort Hare, 23 February 2011.

[135] Jenjezwa, 'Stock Farmers and the State', 121ff.

[136] Myalezwa Matwana, Mbotyi, 1 March 2009.

[137] Zipoyile Mangqukela, Mbotyi, 29 March 2008.

[138] Zipoyile Mangqukela, Mbotyi, 29 March 2008.

widely used in other parts of South Africa as a tick and insect repellent. Another informant mentioned a solution made from peach leaves (*impitchi*), an introduced plant, and the leaves of the *uqangazana* tree to treat tick wounds. These were mixed with salt and paraffin to increase their effectiveness. Perhaps this plant had been incorporated as a tick repellent in coastal Mpondoland in earlier generations. To clean and treat tick wounds respondents also used disinfectants like Mathubula and Jeyes Fluid mixed with paraffin, which together penetrated the skin. Aloe is scarce around this coastal village and we saw it added only as a supplementary chicken feed to loosen their bowels. When we talked about the possible value of aloes in a report-back meeting, livestock owners were open to this idea, and thought they might procure leaves from areas further inland.[139]

Conclusion

Innovative veterinary research and prescriptions resulted in a rigorous dipping programme in South Africa for half a century, accompanied by tight controls over the movement of infective animals. Many African stockowners grew to accept dipping as part of everyday life. Despite earlier opposition to state intervention, informants in most areas looked back to a time when diseases appeared to be less virulent, ticks less abundant and the state more proactive in supporting poorer farmers. Organised dipping ceased in many areas because of gradual withdrawal by the state and the inability of farmers to work together. African livestock owners differed in their approaches to prophylaxis and in recent years have acted more individually. Their techniques could include dipping, spraying, use of local plants, disinfectants and newer pour-on insecticides.

Although many farmers did not routinely dip and ticks were widespread, it is likely that herds have developed some immunity to tick-borne diseases through on-going exposure to infection. In Mpondoland, for example, tick-borne diseases, particularly gallsickness, do cause deaths annually but many local livestock also survive. Informants were clear that animals brought in from outside the area were more susceptible. The favoured type of introduced livestock were Brahman cattle, both on account of their size and because stockowners perceived them to be sufficiently resilient to local hazards. The route forward for the control of ticks and tick-borne diseases is likely to be multifaceted, especially in the light of tick resistance to acaricides in some areas.

Most of those interviewed did not link specific diseases with ticks, but they were aware of the infections that scientists ascribe to acarids such as gallsickness, redwater and heartwater. They also had names for these diseases and could distinguish between them. However, they looked for alternative explanations that were probably deeply rooted in older African medical ideas. The patterns

---

[139] Group meeting, Mbotyi, 5 December 2012.

of local knowledge about disease transmission inhibited regular dipping because most livestock owners dipped or sprayed to kill ticks and waited till these had accumulated again before the next treatment. Treatment was insufficiently frequent to break the life-cycle of ticks or prevent diseases. In the following chapter on environmental understandings of ill health, we will give a fuller account of how farmers conceptualise these and other key diseases. Stock-owners also saw transhumance, illustrated in chapter 4, as in part a strategy to avoid ticks and more general dangers to animal health. In chapters 5 and 6 we discuss in greater detail the medicinal plants and other remedies used for prevention and treatment.

3

'The Grave of the Cow is in the Stomach'[1]

Environment & Nutrition in the Explanation & Prevention of Livestock Diseases

Introduction

Richard Molebalwe from Bethanie (near Brits, North West Province) told us that 'the grave of the cow is in the stomach'. He was not alone in regarding the stomach and the digestive organs as the key to good health. Stockowners often described the primary symptoms of sickness as lack of appetite, diarrhoea or constipation. Many farmers carried out post-mortems to determine the cause of death before preparing the meat for consumption. They examined the state of the gallbladder – was it inflamed or were the fluids an unusual colour? Were the intestines and stomach congested or blocked? Were the inner linings of the gut smooth or rough? Was the digestive tract full of worms, stealing the food and sapping the animal of its life force? Good health, by contrast, was manifest in a rapacious appetite and weight gain resulting in a shiny coat and gleaming eyes. An aesthetically attractive cow fed well and defecated regularly. Healthy cattle bred freely and calved with abandon. A stockowner's wealth was tied to the health of his animals' innards.

Consequently, diet was at the heart of understandings about wellness and sickness. Stockowners believed many diseases, but in particular frequently encountered infections such as gallsickness and blackquarter, emanated from the veld. Poor grasses, watered by too much or too little rainfall portended disaster for a flock or herd. In North West Province, worms and toxic flora parasitised the grasslands and brought doom to the unwary animal that ingested them. So too did rotting carcasses, polluted streams and the indigestible plastic bags and wrappers that littered the grazing lands. Ticks were seen as an envi-

---

[1]   Richard Molebalwe, Bethanie, 26 October 2009.

ronmental problem also. As we discussed in chapter 2, few farmers associated ticks with diseases, but they sucked the blood and deprived animals of their strength. They invaded the hides and the bodily orifices generating such irritation that they distracted livestock from grazing and maintaining optimal health.

The state of the natural environment was at the heart of these descriptions of disease. There is a growing literature that explores African relationships with the environment. Much of it falls into two main strands: one explores the adaptability and resilience of local knowledge and the ability of Africans to manage their environment in productive and sustainable ways; the second, which is interlinked, is a critique of colonial interventions in framing and managing the African environment, which may have destroyed or undermined sustainable agricultural systems in some areas.[2] However, historians and social scientists have largely ignored the environmental aspects of African thinking about disease. In the case of veterinary ideas, at least, we argue that popular conceptions of the natural world framed African stockowners' understandings of sickness and good health. We did not lead our informants in this direction – or at least not to start with. It emerged out of the interviews themselves, when they volunteered explanations for diseases and transhumance. Once such ideas were opened, however, we did pursue them. In response to our questions, informants tended to express rather instrumentalist approaches towards natural resources that had to be managed and used to the advantage of livestock and their owners. They did not explicitly express a conservationist ethos that emphasised romantic or spiritual ideals about the landscape. Some informants in North West Province mentioned problems of environmental degradation, such as loss of grazing, reduced availability of medicinal plants or pollution by mines. In QwaQwa too there were concerns about overgrazing. In Mbotyi, where younger men especially are involved in ecotourism, the discourse of environmental protection was stronger. But most stockowners were worried about access to natural resources on an everyday basis, rather than conservation at a regional level.

Our observations and interviews across the research sites suggested that nutrition and the condition of the veld have long been at the heart of African explanations and conceptualisations of animal disease. They remain a central element in the concerns of stockowners. Farmers' knowledge about the aetiology of infections is largely based on what they have encountered in the visible world, rather than on an understanding of the microscopic realm of germs, invisible organisms and minuscule parasites. In many texts on African ideas about disease and misfortune, supernatural causes come to the fore. We found a few specific diseases and conditions that informants ascribed to witchcraft,

[2]   For overviews see, William Beinart, 'African History and Environmental History', *African Affairs*, 99, 395 (2000), 269–302; William Beinart and JoAnn McGregor, *Social History and African Environments* (Oxford, James Currey, 2003); William Beinart, Karen Brown and Daniel Gilfoyle, 'Experts and Expertise in Colonial Africa Reconsidered', *African Affairs*, 108, 432 (2009), 413–33.

the supernatural, or offending the ancestors (dealt with in chapters 7 and 8). But in relation to livestock, at least, it is the natural and seasonal changes in the physical world of grasslands and water supplies that most strongly influenced ideas about health and sickness. Unpredictable environmental disturbances were also invoked to explain specific diseases.

Until recent times these explanations were shared more widely. David Arnold notes that in Europe two hundred years ago many people regarded such features as rivers, forests, undergrowth and marshlands as dangerous places, harbouring perilous 'miasmas', polluting agents that filled the air and brought disease to the vulnerable. Beyond Europe, certain environments, especially those that western Europeans designated as 'tropical' were especially hazardous. Heat speeded up putrefaction and contamination of the soils and the air. Unhealthy vapours or 'malaria' could bring about epidemics that affected humans and live-stock.[3] Mick Worboys has demonstrated how ideas about miasmas became more complex during the nineteenth century with the increase in public health inter-ventions in Western Europe. It was these lively debates surrounding the origins of poisons and microbes which paved the way for the development of germ theories towards the end of the nineteenth century.[4] In her book on the history of anthrax, Susan Jones drew examples from France and North America to show how farmers constructed particular landscapes as pathogenic. Until scien-tists unravelled the aetiology of anthrax in the late-nineteenth century, stock-owners regarded it as a mysterious telluric disease that emanated from the soils. Animal that grazed on what the French called *les champs maudits,* the accursed fields, could drop dead suddenly and without warning.[5] All these authors have demonstrated how ideas about infection were not monolithic. Scientists and the general public have continually contested explanations of disease and the mechanisms of transmission; popular conceptions about environmental causes of disease survived into the era of germ theory. Those dealing with tropical diseases such as malaria and trypanosomiasis needed to understand complex ecological interactions.

Our testimonies signalled some similarities between African interpretations and the European conceptions of 'miasmatic' or environmentally-related disease causation. We are not arguing that African ideas are static or stuck in the nine-teenth century. Rather we are trying to emphasise the universality of environ-mental contexts in the historical understanding and treatment of diseases. We

[3] David Arnold, *The Problem of Nature: Environment, Culture and European Expansion* (Oxford, Blackwell, 1996), 9–38; David Arnold, 'The Place of 'the Tropics' in Western Medical Ideas since 1750', *Tropical Medicine and International Health*, 2, 4 (1997), 303–13.

[4] Michael Worboys, *Spreading Germs: Disease Theories and Medical Practice in Britain 1865–1900,* Cambridge Studies in the History of Medicine (Cambridge, Cambridge University Press, 2000), especially 20–42.

[5] Susan Jones, *Death in a Small Package: A Short History of Anthrax* (Baltimore, John Hopkins University Press, 2010), 14.

found that our informants held ideas, expressed in different forms and language, which were analogous to elements in pre-germ theory conceptions of infection. Few livestock owners demonstrated an awareness that some diseases could be contagious and could spread to other animals within the close confines of the kraal, or amongst a herd grazing on the veld. Seketsa Mokoena from QwaQwa, for example, explained that farmers 'normally allow sick and healthy animals to mix together. You just medicate for the sick. It is the same for people. We do not separate the sick.'[6] At times farmers believed that individual dispositions or characteristics could have a bearing on the propensity to contract certain diseases. Some Sesotho and Setswana-speaking stockowners noted that blackquarter often attacks the youngest and the fattest calves in a herd.[7] The belief in personal susceptibility also shared similarities with humoural explanations about the aetiology of sickness. This idea, dating back to ancient Greece, suggested that the human body was made up of four humours – blood, phlegm, yellow bile and black bile. Disease was a manifestation of an imbalance in the humours. The state of the body was a factor in creating the conditions for disease.[8]

The notion of pathogenic landscapes resonated with the ideas of some of South Africa's early veterinary scientists. The Cape's first state veterinary surgeon, William Branford, who arrived from England in 1876, believed that the Colony was seething with disease. Working in the eastern Cape, Branford's postmortems revealed that internal parasites were decimating flocks. Worms and liver flukes had made their home in the internal organs of the unfortunate sheep and goats. Branford associated worms with degraded or overgrazed pastures and with kraaling livestock at night to protect them from predators such as jackals. The kraals were full of pernicious, miasmatic dung and slime that could breed all manner of infection.[9] His successor, Duncan Hutcheon, was more open to germ theory. But he was likewise convinced that the environment was important in explaining the distribution of diseases. He defined horsesickness and bluetongue, now known to be spread by tiny *Culicoides* midges, as 'malarial' and 'miasmatic' infections. At that time, farmers associated both diseases with a mixture of climatic and ecological circumstances. Heavy rains, hot humid summers, exposure to morning dew or damp night air, could all

---

[6] Seketsa Mokoena, Lusaka, 1 March 2011.

[7] David Mphuthi, Phuthaditjhaba, 14 October 2010; Lott Motuang, Bollantlokwe, 10 November 2009; Laurence Bodibe, Ramatlabama, 10 February 2010; Johannes Mabena, Moiletswane, 20 January 2010; Simon Mokonyana, Kgomo Kgomo, 29 January 2010; David Modiga, Garankuwa, 24 October 2009; Tollo Josiah Mahlangu, Mabopane, 29 October 2009.

[8] Robert M. Stelmack and Anastasios Stalikas, 'Galen and the Humour Theory of Temperament', *Personality and Individual Differences*, 12, 3 (1991), 255–63; Gerasimos P. Syklotis, George D. Kalliolias and Athanasios G. Papavassillou, 'Pharmacogenetic Principles in the Hippocratic Writings', *Journal of Clinical Pharmacology*, 45, 11 (2005), 1218–20.

[9] William Beinart, *The Rise of Conservation in South Africa: Settlers, Livestock and the Environment, 1770–1950* (Oxford, Oxford University Press, 2003), 128–57.

Photo 3.1 Large cattle kraal at Bollantlokwe, NWP (KB, November 2009)

Photo 3.2 Small cattle kraal in a back yard in Madidi, NWP (KB, October 2009)

bring danger. *Lamsiekte* (botulism) rendered cattle crippled and prostrate. It struck in mysterious ways then associated with certain types of veld. Hutcheon thought it was a deficiency disease but for some it seemed to be a disease, like anthrax, directly emanating from the soil or grassland. It was only in the 1920s that Arnold Theiler proved that the disease was botulism, spread by animals ingesting toxic bacteria (*Clostridium botulinum*) that inhabited rotting carcasses and bones in the veld. Scientists also demonstrated that livestock ate such infectious material to compensate for mineral deficiencies in the grasses.[10] Even then, however, scientists recognised environmental factors in explaining the aetiology of the disease.

Our interviews revealed that many African stock owners continued to invoke seasonal and ecological reasons for infection. In North West Province, they also conceived of some areas as especially pathogenic. Stock owners, for instance, regarded the village of Fafung, as a particularly inhospitable place. Local people talked about the prevalence of ticks, contaminated water, poor soils and in particular the presence of an exceptionally toxic weed called *mohau* (*gifblaar*, *Dichapetalum cymosum*). Josiah Letseka, a farmer from Mmakau recalled that his family left Fafung in the 1970s because they had lost so much stock to poisonous plants.[11] Simon Mathibedi, who had lived all his life in Mmakau, explained that 'Fafung is known as the place where animals die'.[12] Fafung had entered the local rural consciousness as a dangerous place to farm livestock, mainly because of *mohau*. To make matters worse, the area was quite arid and it was difficult to grow crops.[13]

Old and new diseases

Farmers were also aware that the epidemiological situation was not static and new diseases could appear without warning. In North West Province, informants clearly identified some diseases as being endemic before European colonisation. *Lamsiekte* was one example, as well as the ubiquitous gallsickness, blackquarter and anthrax. We can confirm this from early European travellers who recorded infections that appeared unusual to their eyes. For example, in the 1780s the French naturalist, François Le Vaillant, described *spongsiekte* (or blackquarter) and he discussed how Afrikaner and Khoikhoi stock owners tackled this 'terrible scourge among horned cattle'.

---

[10] Daniel Gilfoyle, 'A Swiss Veterinary Scientist in South Africa: Arnold Theiler and the Explication of Lamsiekte in Cattle', unpublished paper presented at the conference on 'Imperial Culture in Countries without Colonies', University of Basel, 23–25 October 2003.

[11] Josiah Letseka, Mmakau, 13 November 2009.

[12] Simon Mathibedi, Mmakau, 9 November 2009.

[13] Peggy Setshedi, Fafung, 27 October 2009.

Photo 3.3 Cattle in landscape around Fafung, NWP, known for its poisonous plants and ticks (KB, October 2009)

> This plague which spares nothing, causes speedy destruction; and happy is he who does not lose by it the half of his herd. It is a kind of leprosy, that may be communicated in an instant; and the flesh of such animals as are attacked by it, swells in an extraordinary manner, and grows spongy and lurid. One would say that it was bruised in a state of decomposition; it becomes filled with reddish viscous humour; and is so disgusting, that even the dogs will not approach it.[14]

Lungsickness, redwater, rinderpest and East Coast fever were all new diseases in the nineteenth and early-twentieth centuries. But our informants had not known rinderpest and East Coast fever as they had been eradicated from South Africa. Redwater had become endemic and appeared to be an old disease called *mbendeni* or *ihlwili* in Mbotyi.[15] Nonjulumbha Javu had a clear view about old scourges and new diseases, which he believed had increased in number.[16] Gall-sickness, redwater and anthrax (which was rare) were old diseases. *Amathumba* (tumours or swellings), cattle-blindness, cancer (*umhlaza* or *intsumpa*) and a disease where the lungs of cattle or goats stuck to the thorax were new. The latter caused acute coughing before death. Tapeworms were getting worse. He said that he did not have traditional remedies for any of these new diseases.

There is a parallel in our interviews in North West Province. African stock owners often regarded heartwater as a new disease although it was almost certainly old in South Africa as a whole. Accounts of heartwater date back to the 1870s –

---

[14] Francois le Vaillant, *Travels into the Interior Parts of Africa by Way of the Cape of Good Hope in the Years 1780, 1781, 1782, 1783, 1784 and 1785* (London, G. G. & J. Robinson, 1790), 79–80.

[15] Sidwell Caine, Mbotyi, 23 February 2008 and 12 February 2009; Vuyisile Javu, Mbotyi, 3 March 2008; Zipoyile Mangqukela, Mbotyi, 29 March 2008.

[16] Nonjulumbha Javu, 19 December 2011.

possibly to the 1830s – but it was only in recent decades that most Tswana informants learnt to recognise it (see chapter 2).[17] As we noted, this is evident in their use of the Afrikaans or English word for the disease. Like the Mpondo they struggled to find local medicines to prevent and treat new infections. All our informants agreed that lumpy skin disease was a recent arrival. It gradually spread southwards through central Africa during the 1920s and 1930s and scientists first recorded it in South Africa in the 1940s. However, outbreaks have been sporadic and many stockowners only recalled encountering it in the first decade of the twenty-first century.[18] In recent years lumpy skin disease has emerged as a major problem in parts of all our research sites. Interviews revealed how the naming of maladies could present challenges – especially where the common medical term had not penetrated. Mbotyi informants did not refer to lumpy skin specifically but included it in generic terms for skin diseases or tumours, such as *amathumba* or *umhlaza*.[19] Daniel Khonkhe, from Ha-Sethunya in QwaQwa described it as an 'epidemic of sores' that broke out in cattle and 'came from out of the blue'.[20]

To a significant degree, new diseases were interpreted through frameworks already familiar to livestock owners. Farmers presented environmental suggestions for the appearance of lumpy skin and associated the disease with exceptionally heavy rains.[21] In North West Province some respondents knew that midges and mosquitoes propagated lumpy skin as Animal Health Technicians (AHTs) had advised them of this and encouraged them to protect their cattle by vaccination.[22] Others thought that tick bites triggered the initial lesions and flies laid their maggots in the wounds, creating the tell-tale lumps.[23] None of the respondents conceived that a type of virus was the cause, and that climatic factors mattered when it came to explaining the proliferation of insect vectors.[24]

[17] A. Provost and J. D. Bezuidenhout, 'The Historical Background and Global Importance of Heartwater', *Onderstepoort Journal of Veterinary Research*, 54, 3 (1987), 165–9.

[18] J. A. W. Coetzer, 'Lumpy Skin Disease', in J. A. W. Coetzer and R. C. Tustin (eds), *Infectious Diseases of Livestock with Special Reference to Southern Africa* (Oxford, Oxford University Press, 2004), vol 2, 1269–76.

[19] Sidwell Caine, Mbotyi, 22 and 23 February 2008; Vuyisile Javu, Mbotyi, 3 March 2008; Nonjulumbha Javu, Mbotyi, 19 December 2011; Nowelile Satsha, Mbotyi, 14 March 2012. *Umhlaza* was sometimes translated as a cancer of the skin. In two of these interviews, the translator, but not the informants, used the term lumpy skin disease.

[20] Daniel Khonkhe, Ha-Sethunya, 13 October 2010.

[21] Otsile Morwe, Dithakong, 3 November 2009; Matthews Moloko, Magogwe, 11 February 2011; Lekunutu Ramafoko, Mabeskraal, 4 February 2010; George Malatsi, Mabeskraal, 5 February 2010.

[22] Thomas Seemela, Madidi, 23 October 2009; Lott Motaung, Bollantlokwe, 10 November 2009; Moiletswane group interview, 20 January 2010. Mike Kenyon, 'Approaches to Livestock Health and Sickness in Masakhane' (unpublished report, July 2009).

[23] Noah Motaung and Philip Mokwena, Fafung, 27 October 2009; Piet Hlongwane, Sutelong, 27 January 2010; Simon Mokonyana, Kgomo Kgomo, 29 January 2010.

[24] Otsile Morwe, Dithakong, 3 November 2009; Lekunutu Ramafoko, Mabeskraal, 4 February 2010; Matthews Motshwrateu Moloko, Magogwe, 11 February 2010.

Environmental contexts shaped much thinking about new diseases as well as old.

Identifying infection: (a) the problem of naming

Before we discuss explanations of disease, it is important to explore some of the difficulties in naming and recognition. Vernacular terms for animal ailments were in everyday use in the villages we visited. Immersed in discussions of animal diseases, these words become familiar to us and in this section we will introduce the most common diseases we encountered. Our research assistants were generally able to translate these local words into English terms. Some names related to quite specific diseases like blackquarter and in this case it was probably justifiable to transpose directly from the vernacular to the English/scientific name. However, there were a significant number of names that informants used to describe internal or external symptoms which were quite general, and probably covered more than one disease or scientifically categorised infection. For example *inyongo* in isiXhosa was translated as gallsickness and the term certainly includes the symptoms of tick-borne gallsickness (anaplasmosis). However, the concept as used by informants in Mbotyi also included seasonally occurring symptoms that were less specific and might have been related to other diseases or conditions. The same applied to the term *gala* in North West Province and *nyoko* in QwaQwa.

A similar reservation should be made about a term like *amathumba*. In Mbotyi this was used to describe a variety of tumours and swellings which could be caused by a range of different factors from boils that were evident around calves' eyes to ruptures in the skin of mature animals. In this case, the local term is not easily translatable into a scientific term for the disease. The same applies to *umhlaza* which was sometimes referred to as cancer and could apply to warts and poxes, as well as lumpy skin disease – for which *amathumba* was also used.[25] In Mbotyi the term *kohleha* denoted coughing but so far as we could record, it was not associated with a particular disease such as pneumonia or bovine tuberculosis. In North West Province informants used *pholotso* as a general term for miscarriage or abortion. We heard no evidence of separate terms for abortions from different causes or at different stages of pregnancy. Overall, interviewees did not differentiate between abortions from brucellosis and other infectious diseases, and those that may have resulted from malnutrition or violence in the kraal. *Pholotso* also referred to abortions attributed to the polluting effects of women. In QwaQwa informants talked of *boketa* which

[25] Sidwell Caine, Mbotyi, 22 and 23 February 2008; Vuyisile Javu, Mbotyi, 3 March 2008; Nonjulumbha Javu, Mbotyi, 19 December 2011; Nowelile Satsha, Mbotyi, 14 March 2012; observation and discussion, dipping day, 15 March 2012; Elsam Mfecale, 15 March 2012.

literally translates as emaciation. Informants tended to see it as a disease in itself rather than a condition that arose due to inadequate grazing. In general then, local names for diseases which we were able to record tended to have a rather wide reference point. African stockowners had a different classificatory system that was much less specific than the biomedical nomenclature, although some people had adopted common names from English/Afrikaans as well.

As mentioned, we are not trained in veterinary medicine so it was sometimes difficult to identify diseases that informants described or showed to us. Another limitation was the lack of uniformity in naming diseases. Throughout our interviews, it was apparent that names for specific infections could diverge. While there were socially shared general notions of disease names and causation, especially in particular localities, there was no standardised African nomenclature for ailments, descriptions of symptoms or interpretations of post-mortems. Diversity not only arose between different research sites, but could also occur in a single village as neighbours sometimes used different names to describe the same collection of symptoms. Moreover, while we came across occasional experts, especially in Mbotyi, our interviews reflected a wide range of opinion and non-specialist knowledge within African communities.

This diversity became highly apparent not only in interviews but also in observing farmers workshops organised by the Department of Agriculture. At a farmers day in Kgabalatsane (near Garankuwa, North West Province) in November 2009, for example, the lack of commonality in naming revealed some of the problems in holding discussions between farmers and biomedical specialists such as veterinarians, AHTs and employees of the pharmaceutical companies. At the meeting, the Pfizer pharmaceutical representative declared that the 'official' Setswana word for blackquarter was *serotswana*.[26] Many farmers found it easy to identify this disease because the legs turned black and, as Le Vaillant described, putrefaction after death was notably rapid and the meat foul smelling. However, it was only in Mabeskraal that all respondents used the term *serotswana* for blackquarter. Farmers often mentioned *serotswana* in the Mafikeng area, but its usage was not ubiquitous. In Ramatlabama, about 25 km away from Mafikeng, the term *sephatlho* was interchangeable with *serotswana*.[27] Most informants from the eastern part of North West Province tended to call blackquarter *leotwana*.[28] At the Kgabalatsane farmers' day, mostly attended by people from this area, *serotswana* was probably meaningless. According to informants in Shakung, not far from Kgabalatsane, *leotwana* referred to a cattle disease that revealed spongy, smelly meat at post-mortems: a disease akin to blackquarter. *Serotswana,* by contrast, meant anorexia caused by poor, thin grasses which

---

[26] Pfizer representative, Kgabalatsane Information Day, 12 November 2009. Pfizer Setswana Poster also noted *serotswana*.

[27] Laurence Bodibe, Ramatlabama, 10 February 2010.

[28] Mapule Bokaba, Brits, 9 November 2009; Elizabeth Serema, Thetele (Mmakau), 9 November 2009; Frederick Matlhatsi, Mabopane, 29 October 2009.

meant cattle fed on soil to survive. It was a form of nutritional deficiency rather than a disease that caused sudden death and notably pungent decomposition.[29]

It was not only blackquarter that exposed discrepancies and difficulties in naming. A Pfizer representative used the word *semmee* as a direct Setswana translation for heartwater.[30] Yet every respondent from North West Province who knew about heartwater called it heartwater. The only informant who was familiar with the term *semmee* was Anna Lebelwane from Mmakau. She explained that it related to sores on the mouths of goats acquired by browsing on thorny vegetation. She pronounced *semmee* in a long drawn out fashion to show that it echoed the sound of a crying goat, tormented by its cuts and swellings. This condition was possibly orf, caused by a poxvirus, not heartwater spread by ticks.[31] Clearly, the process of naming presented all sorts of problems in identifying, preventing and treating diseases. For Lebelwane there was no need to use antibiotics like Terramycin (oxytetracycline), a common treatment for heartwater, for *semmee*. She treated the goats' sores with a poultice made of warmed aloe leaves.[32]

We came across similar discrepancies in the Eastern Cape. A poster at the veterinary office in Lusikisiki clearly named heartwater as *umkhondo* in isiXhosa. In parts of the central Eastern Cape including Peddie, located 300 km away, the term was indeed used for heartwater. But for people in Mbotyi *umkhondo* had a completely different meaning associated with other symptoms and supernatural causations (chapter 7). They called heartwater *amanzana*. *Umkhondo* also referred to the agapanthus plant which is used medicinally, especially as a measure against supernatural agents.[33] In the Eastern Cape, *inyongo* was widely translated as gallsickness, yet Tony Dold and Michelle Cocks, interviewing near Alice, also found it applied to heartwater.[34]

Provincial veterinary departments and commercial companies have made some attempt to standardise names for diseases in the different African languages; we saw these in pamphlets and posters and heard them at stock days.[35] This is particularly important for treatments where, for example, drug companies recommend specific medicines for different aliments. However, our evidence from the village level is that standard terms have not penetrated or been taken up everywhere. With these reservations in mind, in Table 3.1 we have tried to summarise the most common names for the different diseases in the research sites.

[29] Elias Pooe, Shakung, 19 January 2010.
[30] Pfizer representative and Pfizer Poster advertising different types of Terramycin (oxytetracycline) for different conditions, Kgabalatsane Information Day, 12 November 2009.
[31] Anna Lebelwane, Thethele (Mmakau), 13 November 2009.
[32] Anna Lebelwane, Thethele (Mmakau), 13 November 2009.
[33] Discussion, Tony Dold, Michelle Cocks, Luvuyo Wotshela, William Beinart, Grahamstown, 26 March 2012.
[34] A. P. Dold and M. L. Cocks, 'Traditional Veterinary Medicine in the Alice District of the Eastern Cape Province, South Africa', *South African Journal of Science*, 97, 9–10 (2001), 375–9.
[35] Kgabalatsane Information Day, 12 November 2009; Hertzog Stock Day, 19 November 2008.

Table 3.1 Common disease names in research sites

English	Sesotho	Setswana	isiXhosa (Mbotyi)
Abortions/ Miscarriage	*Folotsa*	*Pholotso*	*Ukupile* (from *ukupha*, to come out)
Anthrax		*Lebete* (spleen) *Kwatsi*	*Ubende* (spleen) or *inyama makhwenkwe* or *isifo somkhwenkwe*
Blackquarter	*Serotswana*	*Serotswana* *Leotwana* *Ramokutwane* *Ramukutwane* *Letsogwane* *Sephatlho*	*isiDiya (isiDiya – rare in Mbotyi)*
Botulism		*Mokokomalo* *Magetla* (shoulder)	
Footrot	*Mokaka*	*Tlhakwana*	
Gallsickness	*Nyoko*	*Gala*	*Inyongo (Inyongo)*
Heartwater		*Heartwater*	*Umkhondo* (*amanzana*)
Lumpy Skin Disease	*Skin en vel*	*Boletswe ba Letlano* *Sekgwakgwa*	*Ungqakaqha* (*umhlaza*)
Orf		*Semmee* *Dikakana* *Selongwane*	
Redwater		*Motlapologo khibidu* *Omo khibidu*	*Amanzabomvu* (*mbendeni* or *ihlwili* – the latter referring to blood in urine)
Scab	*Lekgwakgwa*	*Lepalo*	(*ibuwa*)

Identifying infection: (b) contested clinical symptoms and post-mortem findings

When identifying diseases farmers firstly noted the outward clinical symptoms. There was no evidence that stockowners took temperatures and they did not refer to fevers. In North West Province, descriptions were often very general, focusing on defecatory activity. Constipation was a common indicator of ill health and a free-flowing gut pointed to a well-functioning system.[36] Symptoms could be vague and informants mentioned similar descriptions for a number of diseases. They often described healthy animals as strong, whereas the sick were weak. In Mbotyi informants used the English word 'strong' to denote disease resistant and healthy animals. In part this was simply one of many borrowings from English. However, when this usage was discussed in detail, they suggested that isiXhosa words did not adequately capture the notion of strong and weak in relation to disease. For example, the word for weak, *inqinile*, referred to an animal that was weak and thin, and *ityebile* to one which was well-fed and fat. Strong and weak had a slightly different connotation relating to health and disease or lack of health.[37] Myalezwa Matwana noted that they used the term *ukubhuxaka* to describe a beast that was 'unhappy, not grazing well and its ears point downwards'.[38] They had no particular illness, but this could be a sign of *umkhondo* (chapter 7). He also used the term *ingukela* to refer to a 'strong' animal that was particularly fertile and resistant to disease – one that you would keep in your herd and never sell or exchange.[39]

In North West Province, stockowners often described sick beasts as lethargic, reluctant to graze and lagging behind the rest of the herd. Nick Lebethe, for instance, explained that a cow with *magetla* (botulism) 'sleeps standing up and falls to the ground if it walks'.[40] Richard Molebalwe believed that blackquarter was caused by 'a worm that spits stuff into the stomach of the cow, depriving it of its power (*maatla*) and slowing it down'.[41] Aaron Aobeng identified *gala* (gallsickness) in animals that 'refuse to go out to graze and just hang around in the shade'.[42] If healthy animals ate *mohau*, Peggy Setshedi found they 'became weak and staggered around before they died'.[43] In Mabeskraal respondents also

---

[36] Laurence Bodibe, Ramatlabama, 10 February 2010; Jeremiah Ramakopeloa, Mabeskraal, 5 February 2010.

[37] Discussion with Sonwabile Mkhanywa and others, Mbotyi, 24 March 2012.

[38] Myalezwa Matwana, Mbotyi, 7 December 2011.

[39] Myalezwa Matwana, Mbotyi, 1 March 2009.

[40] Nick Lebethe, Bethanie, 26 October 2009.

[41] Richard Molebalwe, Bethanie, 26 October 2009.

[42] Abraham Meno, Mareetsane, 4 November 2009.

[43] Peggy Setshedi, Fafung, 27 October 2009.

spoke of dizziness associated with anthrax and gallsickness. John Modisane, for example, said that an animal suffering from *lebete* (anthrax) 'becomes very dizzy, bangs into objects like trees and fence posts and then collapses under the heat of the sun'.[44] Kgomotso Goapele from Mantsa, by contrast, stated that a cow with anthrax would 'shiver and shake and drop down dead'.[45] Informants thought that livestock that were especially thin due to malnutrition were more vulnerable to infection; but in certain cases, such as blackquarter, fat animals that grazed well could be especially susceptible.[46]

Sometimes stockowners could only determine the disease after death by carrying out post-mortems – usually as part of the butchering process, but also when animals were not eaten. In this respect, especially, local medical knowledge about animals differed from that about people because no similar procedure was pursued on human bodies. It may be in part, that explanations for animals diseases are, as a consequence, more strongly based on observation and natural causes than for human diseases. In Mbotyi there were specialist slaughterers whom neighbours called upon to supervise post-mortems and sacrifices. Two of the older men we interviewed, Myalezwa Matwana and Sidwell Caine, were well known for their skill in this area and had been dissecting and observing carcasses for years. We watched Matwana officiate at the slaughter of a cow for a large ceremony attended by over one hundred villagers to mark the initiation of a female *sangoma* (diviner/healer) and also at a funeral.[47] His weekends were frequently spent at such events. Matwana and Caine had a very thorough knowledge of the internal organs of animals and were attuned to detecting abnormalities. The gallbladder was often the most important organ in the identification of disease and *inyongo* was immediately recognisable by gall spewed out over the other organs. 'We look at the gall when the animals die', Nongede Mkhanywa, a *sangoma*, affirmed, 'we first check the gall.'[48]

Gall played a role in health as well as disease. Used in ceremonies it was linked to the slaughter of animals and the ancestors. In Mpondoland, for example, people anointed the bodies of young children with gall after a slaughter of goats to connect them with their ancestors and to promote good

[44] John Modisane, Mabeskraal, 2 February 2010; Emily Diole, Mabeskraal, 2 February 2010; Henry Ramafoko and Elsie Ramafoko, Mabeskraal, 2 February 2010.

[45] Kgomotso Goapele, Mantsa, 12 February 2010.

[46] Samuel Mothapo, Slagboom, 10 November 2009; John Modisane, Mabeskraal, 2 February 2010; Soloman Matome, Mabeskraal, 3 February 2010; Tollo Josiah Mahlangu, Mabopane, 29 October 2009; David Modiga, Moiletswane, 20 January 2010; Johannes Mabena, Moiletswane, 20 January 2010; Phillimon Mathaba, Kgomo Kgomo, 29 January 2010; Konese Mofokeng , Phuthaditjhaba , 11 October 2010; Johannes Motaung, Makwane Tebang, 1 March 2011; Tollo Josiah Mahlangu, Mabopane, 29 October 2009; Zachariah Maubane, Mafikeng, 2 November 2009; Chris More, Kgabalatsane, 13 November 2009; Paulus Mosia, Ha-Sethunya, 13 October 2010.

[47] Mbotyi, 1 March 2009; Mbotyi, 24 March 2012.

[48] Nongede Mkhanywa, Mbotyi, 27 March 2008.

health.[49] *Sangoma*s who had finished their training were similarly smeared with gall, and dried gallbladders formed part of their regalia. Perhaps the prevalence of gallsickness and other diseases which appeared to affect the gallbladder reinforced the significance of gall in African culture and influenced their interpretation of disease. Based on interviews in Zululand in the 1960s, Axel-Ivar Berglund explained that the gallbladder was especially important to Zulu informants who believed that it was the place where the shades of the dead lived in the beast. Ancestors fed off the gall which they enjoyed because it was bitter. The presence of the ancestors was vital for good health.[50]

In North West Province and QwaQwa, respondents did not explain a link between the gallbladder and the ancestors, but they nonetheless believed that the state of this organ was a major indicator of wellness and sickness. They ascribed diseases like *gala* to too little or too much gall which affected the functioning of the digestive tract and impacted on general health. Abraham Meno, who grazed a few cattle at Mareetsane, south of Mafikeng, for instance, was always struck by the 'size of the swollen gallbladder and the dryness of the stomachs' in cows that had succumbed to gallsickness.[51] Ernest Phage from Garankuwa believed that the gallbladder 'stops sickness' and if this organ failed to work the animal would die.[52] A number of interviewees gave more detailed explanations as to how they thought the gallbladder functioned. For some a blocked gallbladder meant that bile could not enter the intestines to digest the food, so the animal would be weakened and expire. Alternatively the gallbladder might secrete too much gall, or burst, turning the internal organs yellow. This too would prevent absorption of the nutrients and lead to death.[53] Some associated an enlarged gallbladder with heartwater too.[54] Our informants did not recognise that malnutrition could also affect the gallbladder.

Slaughterers were on the look-out for other key physical symptoms. Animals that died of heartwater revealed a large amount of foamy fluid around the heart and lungs.[55] A swollen liver was evidence of redwater.[56] When livestock died from mamba bites – common around Mbotyi – the blood in the heart was solid.[57] In

---

[49] Nongede Mkhanywa, 21 March 2009 and observation at slaughter of goats for children; observation, slaughter for *sangoma*, Mbotyi, 1 April 2009.

[50] Axel-Ivar Berglund, *Zulu Thought-Patterns and Symbolism* (Bloomington, Indiana University Press, 1976), 111.

[51] Abraham Meno, Mareetsane, 4 November 2009.

[52] Ernest Phage, Garankuwa, 21 October 2009.

[53] Francina Gumade, Mmakau, 30 October 2009; Mita Mogodu, Mareetsane, 4 November 2009; Abraham Meno, Mareetsane, 4 November 2009; Kgomotso Goapele, Mantsa, 12 February 2010; Ernest Phage, Garankuwa, 21 October 2009.

[54] Richard Molebalwe, Bethanie, 26 October 2009.

[55] Mosweu Labius, Disaneng, 6 November 2009; Patrick Sebeelo, Lokaleng, 3 November 2009; Andrew Mabaso, Mmabatho, 5 November 2009.

[56] Sidwell Caine, Mbotyi, 22 and 23 February 2008.

[57] Nongede Mkhanywa, Mbotyi, 27 March 2008.

North West Province, farmers detected fatalities ascribed to toxicoses either by uncovering undigested plant material in the stomachs, or through observing damage to the intestines. *Gifblaar* and tulp (*Homeria pallida* – see below), for instance, were said to destroy the lining of the gut, making it smooth and divesting it of its folds and rough edges.[58] Many farmers complained about finding the modern scourge of the veld – plastics – when opening cows' stomachs. Plastics referred not only to bags but also to discarded food wrappers which attracted animals due to their smell. If an animal swallowed a wrapper it could block the intestines, causing death.[59] Litter is ubiquitous in the more densely settled rural areas, with bags trapped in bushes and fences. Rubbish removal services are rare.

Informants in Peddie also mentioned that cattle ate plastic bags and other detritus from the veld which caused blocked bladders and urinary tracts (*valeka isinyi*).[60] At one post-mortem, slaughterers found a rope in the cow's gut, mixed with plastic. In another animal they unveiled a hard ball of hair and plastic in the stomach, which one man called *uboyawenkomo* (the hair of a cow). Stockowners treated cattle that ingested such noxious materials with a litre of Coca-Cola, but it was only possible to diagnose the problem with certainty at post-mortem. Stockowners praised Coca-Cola because it caused burping and extreme flatulence which they believed helped to flush out deleterious substances from the gut.

Some farmers in Peddie clarified their diagnosis by tasting the meat. Animals that died of gallsickness were edible. However, they said that meat from victims of heartwater or blackquarter had a tainted, sour taste and became stringy like mincemeat when cooked. As noted, some in Mbotyi said meat from animals attacked by the bont tick (and hence possibly with heartwater) was considered grey and bloodless. Although anthrax is a zoonotic disease and can kill humans, people from Peddie claimed that they would eat the meat of an animal that had died of anthrax, so long as they dosed themselves with a herbal brew called *intswele lomlombo,* that purportedly made the body strong so it could fight this disease.[61] In the late-nineteenth century, Andrew Smith, who taught at Lovedale Mission Station (near Alice) from 1867 to 1887, recorded plants that were used to counteract anthrax poisoning in meat (chapter 6).[62]

[58] Noah Motaung and Philip Mokwena, Fafung, 27 October 2009; Andrew Mabaso, Mmabatho, 5 November 2009.

[59] Don Modiane, Garankuwa, 11 November 2009; Simon Mathibedi, Mmakau, 9 November 2009; Caroline Serename, Kgabalatsane, 13 November 2009; Lott Motuang, Bollantlokwe, 10 November 2009 ; David Modiga, Garankuwa, 24 October 2009 ; Michael Mlangeni, Mabopane, 24 October 2009; Anna Lebelwane, Mmakau, 13 November 2009.

[60] Andrew Ainslie, 'Hybrid Veterinary Knowledges and Practices among Livestock Farmers in the Peddie Coastal Area of the Eastern Cape, South Africa' (unpublished report, 2009), 20.

[61] Ainslie, 'Hybrid Veterinary Knowledges'.

[62] Andrew Smith, *A Contribution to South African Materia Medica* third edition (Cory Library Rhodes University, 2011, first published Lovedale Press, 1895), 55–59.

For anthrax and blackquarter, the symptoms were generally clear before post-mortem. According to Tswana respondents, the common feature at death was the visible effect of unknown causal agents on the blood. With anthrax the blood appeared dark and spurted out of the orifices in a dramatic fashion. Inside there might also be foamy blood oozing from the spleen and liver.[63] One of the Setswana words for anthrax is *lebete* which means spleen. The Afrikaans word, *miltsiekte*, also refers to this organ, as does the isiXhosa word *ubende*. In Mbotyi, informants said the symptoms seemed to suggest that the animal had been struck on the back of the head with a club (see chapter 7).[64] With blackquarter the leg muscles become spongy and, according to Aaron Aobeng from Mareet-sane, this gave them a burnt appearance 'as if struck by lightning'.[65] During post-mortem, the colour and the smell of the carcass was a clear indication that a cow had died of blackquarter and there was no need for a thorough necropsy to determine the cause of death.[66]

Ideas about contagion, isolation and immunity

Not all post-mortems led to an identification of the disease and many animals died of unknown infections and unknown causes. If a stock owner deduced that his livestock had contracted a particular infection, he probably would not isolate the ailing animal as a disease-control measure. Sometimes stock owners kept sick livestock in the kraal during the daytime when the rest of the herd were grazing in the veld. However, this was generally done only if the animal could not walk, for purposes of observation, or to provide treatment. Inform-ants did not claim that they were separating potentially infectious stock from the rest, and at night the sick and healthy animals would share the same kraal. Even if a number of animals in a flock or herd died of the same disease, stock-owners would not necessarily assume that the infection had spread from one animal to another either by direct contact, or else indirectly by sharing the same water or veld.

Richard Msezwa in Mbotyi confirmed: 'I don't isolate the sick animal. I let it go free with the other animals but get medicine quickly for treating....There

[63] Samuel Mothapo, Slagboom, 10 November 2009; Kgomotso Goapele, Mantsa, 12 February 2010; Pagiel Sehunoe, Mabeskraal, 4 February 2010; Tollo Josiah Mahlangu, Mabopane, 29 October 2009.

[64] Zipoyile Mangqukela, Mbotyi, 29 March 2008; Nongede Mkhanywa, Mbotyi, 27 March 2008 and 7 December 2011.

[65] Aaron Aobeng, Mareetsane, 4 November 2009.

[66] Otsile Frederick Morwe, Dithakong, 3 November 2009; Thomas Seemela, Madidi, 23 October 2009; Samuel Mothapo, Slagboom, 10 November 2009; Kgomotso Goapele, Mantsa, 12 February 2010; Martha Rakgomo, Mabeskraal, 1 February 2010; Majeng Motsisi, Mabeskraal, 3 February 2010; Pagiel Sehunoe, Mabeskraal, 4 February 2010.

are none of the diseases that we isolate for.'[67] After further discussion between him and the translator, Sonwabile Mkhanywa, they did make an exception for rabies (*amarabi*), which they believed could be transmitted between dogs, but not to livestock. This knowledge is likely to have come from anti-rabies campaigns which have included inoculation for dogs. Msezwa added that they slaughtered pigs that were coughing because this might spread the disease, but they would not slaughter cattle or goats that coughed – 'you organise *muthi* for that'.

Sidwell Caine thought that *amathumba* – the general word for some swellings – might be able to pass between animals but he confirmed, 'if an animal is well enough to walk, we don't separate it'.[68] In areas with higher levels of literacy and education, ideas of infection were more evident. Mike Kenyon found that some stockowners at Masakhane, near Alice, separated animals that were suffering from lumpy skin and blackquarter (*isiDiya*). One respondent believed that lumpy skin disease could spread amongst cattle through contact with pus from the sores or saliva of a sick cow. This is a relatively new disease to the area and, as in North West Province, they may have received this information from AHTs or similar sources. Another thought blackquarter (a soil-borne infection) was a communicable air-borne disease and it was necessary to isolate cattle that displayed the symptoms.[69] In QwaQwa Moses Mahlamba claimed that diseases had only become contagious in recent times, for reasons he could not explain. He said that he would like to isolate sick animals but was unable to do so due to lack of space in the kraal.[70] Similarly Johannes Mofokeng, also from QwaQwa, had become convinced that *serotswana* (blackquarter) was infectious and it spread because infected animals contaminated the veld. He too admitted that he lacked the space to separate sick from healthy animals.[71] QwaQwa is a particularly densely populated area and it could be that greater concentrations of livestock in recent decades have facilitated the spread of diseases between and within herds. Observations of this phenomenon, or information from AHTs, could have alerted some farmers to the possibilities of contagion.

Although communal grazing makes it easier for diseases to spread between livestock, this too did not appear to be a major concern regarding the grazing arrangements. Informants in North West Province often complained that they did not like communal landholding because it was overgrazed or they were unable to control breeding patterns, or prevent theft, but not because it enhanced the likelihood of sickness.[72] Similarly, in Mbotyi, concerns about the

---

[67] Richard Msezwa, Mbotyi, 10 February 2009; Sikhumba, Mbotyi, 18 February 2009.

[68] Sidwell Caine, Mbotyi, 12 February 2009.

[69] Kenyon, 'Approaches to Livestock Health'.

[70] Moses Mahlamba, Mountain View, QwaQwa, 2 March 2011.

[71] Johannes Mofokeng, Lusaka, 3 March 2011.

[72] Peggy Setshedi, Fafung, 27 October 2009; Abraham Meno, Mareetsane, 4 November 2009; Group Interview, Moiletswane Community Centre, 20 January 2010; Modisaotsine Taikobong, Magogwe, 11 February 2010.

common grazing lands centred around theft and access to the best seasonal grazing. In QwaQwa informants emphasised the problem of competition for grazing resulting from denser settlement.[73] Across interview sites as a whole, informants did not strongly associate disease with the mixing of herds from different homesteads. Rather they saw the cause of the disease resulting from the seasons and veld itself.

While all animals could be susceptible to such diseases, informants recognised that not all suffered from them equally. This suggests the idea (as noted above) that an individual animal's disposition was important in determining which fell sick and which did not. When farmers in QwaQwa spoke of particular livestock, such as the plumpest calves, being most susceptible to diseases like *serotswana*, they highlighted the importance of individual characteristics in explaining the pathology of certain diseases. Johannes Motaung felt that isolating the sick was pointless as the 'fat animals always get sick'. When it came to blackquarter, the likelihood of contracting it or not depended on whether 'an animal can adapt to the superfluity of green grazing during the summer months'.[74]

It is not clear why many stockowners had little understanding of the infectious nature of diseases. Chris Andreas has shown that Xhosa chiefs whose territories were still independent of the Cape government took action against the arrival of the cattle disease lungsickness in the Cape in 1853. They imposed quarantines and forbade the introduction and movement of cattle from outside, suggesting that they understood this new disease spread directly between animals.[75] Mpondo chiefs, over 200 km further to the north east, also regulated the passage of wagon traffic into their territory during the lungsickness epizootic of the 1860s.[76] IsiXhosa-speakers had a name for introduced animals, *imofu*, probably derived from the Dutch/Afrikaans word *mof* which means foreign or boorish. In Afrikaans it was applied to Germans and introduced animals such as merino sheep and Friesland cattle. In the 1850s the Xhosa who lived near the Cape Colony associated *imofu* cattle with the arrival of lungsickess. This term is still used in Mbotyi to indicate exotic breeds, such as

[73] Charlie Tshabalala, Phutadijhaba, 11 October 2010; Ephraim Mofokeng, Phuthaditjhaba, 15 October 2010; April Nhlapo, Lusaka, 28 February 2011, Michael Molibeli, Ha-Sethunya, 2 March 2011, Johannes Mofokeng, Lusaka, 3 March 2011.

[74] Johannes Motaung, Makwane Tebang, 1 March 2011; Konese Mofokeng, Phuthaditjhaba, 11 October 2010; David Mphuthi, Phuthaditjhaba, 14 October 2010.

[75] Christian B. Andreas, 'The Spread and Impact of the Lungsickness Epizootic of 1853–57 in the Cape Colony and the Xhosa Chiefdoms', *South African Historical Journal*, 53, 1 (2005), 50–72; Christian Andreas, 'The Lungsickness Epizootic of 1853–57: an Analysis of its Socio-economic Impact and the Ensuing Reactions in the Cape Colony and in "Xhosaland"', MA Thesis, University of Hanover (2003), 127–38 and unpublished chapters from his draft doctoral thesis. See also Jeff B. Peires, 'The Central Beliefs of the Xhosa Cattle-Killing', *Journal of African History*, 28, 1 (1987), 43.

[76] William Beinart, *The Political Economy of Pondoland, 1860 to 1930* (Cambridge, Cambridge University Press, 1982), 24.

Jerseys or Frieslands, although it no longer seems to be associated directly with disease.[77] How the Xhosa conceptualised contagion through introduced cattle is more difficult to unravel. Andreas postulated that they might have viewed lungsickness and foreign cattle as a form of pollution (see chapter 7). Nor are the reasons for the regulation of stock movements entirely clear. They may have been advised by missionaries or traders. More likely, they demonstrated an appreciation that cattle and diseases from outside were particularly perilous.

In Mbotyi, livestock owners still preferred locally raised livestock. They expressed a sense that animals became attuned to ecological conditions, especially if exposed to the local environment from birth. Even though many farmers did not regard sickness within their herds as contagious, experience taught them that animals brought in from elsewhere could be particularly susceptible to disease. Sidwell Caine, for example, mentioned that cattle such as Jerseys or Shorthorns were 'a big problem – they die quickly'. The exception was Brahmans: 'they don't die here, they survive'.[78] Informants believed that stock from white-owned farms were notably vulnerable. Except when they were purchasing directly for consumption, local people tended to buy livestock from nearby sources. This has led to an interesting anomaly in prices. Cattle prices were actually higher in Mbotyi than they were, for similar quality bovines, in the heartlands of the Eastern Cape. Local buyers were prepared to pay a premium for animals that they believed could survive in Mbotyi.

Their preference for local livestock was also tied to understandings of immunity. This term or idea was not used specifically, nor conceived in biomedical terms. A few interviewees, who had been exposed to HIV/AIDS education, had some sense of the medical understanding of immunity in humans which they expressed as the operations of 'soldiers of the blood'.[79] However, some respondents in Mbotyi thought that animals which survived disease might have some protection against it. They were divided as to whether animals that had contracted gallsickness or redwater, for example, could succumb again. In fact, observation may not have been able to resolve this question easily. Generally, veterinary research suggests that calves are less affected by gallsickness (anaplasmosis) than mature cattle and that early infection does provide some protection against further symptoms. However, infected animals remain carriers and the symptoms can reappear if livestock experience acute stress such as malnutrition or other diseases. Given the conditions in communal grazing lands, and the possibility of *inyongo* being used to describe symptoms that were broader than anaplasmosis, it would be very difficult to identify patterns of immunity and recurrence without laboratory tests.

The significance attached to locally bred animals illustrated how concepts of exposure to the local environment, and the inherent strength of some animals,

[77] Nongede Mkhanywa, Mbotyi, 25 February 2009.
[78] Sidwell Caine, Mbotyi, 12 February 2009.
[79] Nongede Mkhanywa and Sonwabile Mkhanywa, Mbotyi, 25 February 2009.

underpinned the purchasing choices of local farmers. Notable exceptions included the importing of Brahmans and also of horses. The death rate of horses from African horse sickness and other ailments was high in Mbotyi, but so was the demand. Men used them for personal transport and herding and local owners provided horses for visitors to the Mbotyi River Lodge. As important, horse-racing events were popular in coastal Lusikisiki. Cecil Malindi, brother of the sub-headman, regularly purchased horses from Newcastle and Vryheid in KwaZulu-Natal to train them for the races at Lambasi.[80] This involved altering their gait to a type of rapid trot, rather than a canter or gallop, which was favoured in the area and used in competitive races. Called *uhambe* in isiXhosa this style of movement is translated locally as trippering or trippeling (from Afrikaans) and seems to be particularly effective for covering rough ground at pace. Malindi was able to buy horses far more cheaply in northern KwaZulu-Natal but he had to dose and inject them frequently if they were to survive the move to Mbotyi. As he noted: 'Mbotyi is too hot, and full of disease for horses.' If horses are not treated, especially from September to December, when horse sickness, colic, boils (*amathumba*) and worms strike, 'it is goodbye'. Even so, he lost two horses during the research period, including the swift and popular Shaka.

Environment and the seasonality of diseases

'Disease does not spread between animals but it has its time or its season. It will come again in that season. Each disease has its own time'.[81]

In discussing environmental causes of disease, we use the term broadly to refer to climate, seasons, pastures, plants, terrain, and changing conditions resulting from industrialisation and pollution. Seasonal changes were at the forefront of environmental explanations for disease causation. In every site we visited the disease calendar revolved, to some degree, around the seasons. Stockowners tended to associate specific conditions with particular times of the year. Changes in temperature and rainfall patterns alter the grasslands and make animals susceptible to infections brought on by cold and heat. In South Africa the climate outside the Western Cape is characterised by dry, chilly, winters and hot, wetter summers. The winter frosts and winds can leach the vegetation of its moisture creating a brown and dusty landscape, broken around September or October by the first spring rains. Winter is potentially a time of malnutrition and on the highveld, especially in the mountains of QwaQwa, it is also a season of frosts and glacial winds. In summer the rains bring more verdant vege-

[80] Cecil Malindi, Mbotyi, 13 December 2011.
[81] Nongede Mkhanywa, Mbotyi, 25 February 2009.

tation which consists of both nutritious grasses and – especially in North West Province – noxious plants. Heat and humidity spawn ticks and worms. Water gathering in the kraal causes footrot and *umkhondo*. Summer can be a particularly difficult time for livestock not only because of a greater abundance of ticks, but also due to the threat of diseases like gallsickness and blackquarter.

Gallsickness

Gallsickness was possibly the most universal disease we encountered. Informants associated it with a variety of external symptoms: constipation, a poor appetite, reluctance to graze and lethargy.[82] Descriptions of the symptoms were not homogenous and although many stockowners commented on the dry nose, some believed a wet runny nose was an indicator of gallsickness.[83] Typically, infected animals strayed from the rest of the herd: 'the animal does not move with the other cattle and wants to go to the bushes alone'.[84] Occasionally, foam collected at the mouth. Some informants in Mbotyi mentioned that the animals shivered.[85]

Stockowners ascribed these various symptoms to an enlarged gallbladder. In North West Province informants thought that this stopped the digestive organs functioning. In severe cases, Majeng Motsisi from Mabeskraal said, 'gall drips from the nose and the gallbladder swells so much that it bursts inside the cow'.[86] Many informants confirmed their diagnoses through necropsies which revealed that the yellow gall had spread over other organs such as the liver or intestines. As noted in our chapter on ticks, very few farmers understood gallsickness as a tick-borne disease. Some Mpondo stockowners commented on noticing animals suffering from gallsickness or *inyongo,* with no ticks on their bodies. (This is quite possible as the incubation period from infection to symptoms is at least a couple of weeks and can be much longer).

---

[82] P. J. Masika, A. Sonandi and W. van Averbeke, 'Perceived Causes, Diagnosis and Treatment of Babesiosis and Anaplasmosis in Cattle by Livestock Farmers in Communal Areas of the Central Eastern Cape Province, South Africa', *Journal of the South African Veterinary Association*, 68, 2 (1997), 40–44; Ronette Gehring, 'Veterinary Drug Supply to Subsistence and Emerging Farming Communities in the Madikwe District, North West Province, South Africa', MMed Vet (Pharmacology), University of Pretoria (2001), 66; Deon van der Merwe, 'Use of Ethnoveterinary Medicinal Plants in Cattle by Setswana-speaking People in the Madikwe Area of the North West Province', MSc Thesis, University of Pretoria (2000), 41.

[83] Patrick Sebeelo, Lokaleng, 3 November 2009.

[84] Francina Gumade, Mmakau, 30 October 2009; Thomas Seemela, Madidi, 23 October 2009; Kgomotso Goapele, Mantsa, 12 February 2010; Bosie Morapadi, Mabeskraal, 2 February 2010; Ntsoeu Livestock Company, Phuthaditjhaba, 11 October 2010; Maria Khoboko, Lejwaneng, 12 October 2010; Daniel Khonkhe, Ha-Sethunya, 13 October 2010; Paulus Mosia, Ha-Sethunya, 13 October 2010; Abel Lebitsa, Mountain View, 14 October 2010; Thomas Mahlaba, Tseki, 15 October 2010.

[85] Mamcingelwa Mtwana, Mbotyi, 13 December 2011; Mayakalisa Jikijela, Mbotyi, 12 December 2011.

[86] Majeng Motsisi, Mabeskraal, 3 February 2010.

The most common reason stockowners gave for gallsickness was changes in the texture of the grasses. In Mpondoland, they associated *inyongo* with fresh rich grass in the spring, when the rains came. Some believed the disease originated in the morning dew, especially if it infected goats. It was also linked to spider webs dispersed around the veld at ground level in the early morning. Goat owners felt that one way of preventing this disease was to keep animals in the kraal until the heat of the sun burnt off the dew and spider webs.[87] In Peddie and Masakhane farmers also linked it to the first flush of growth that came with the spring rains or after the veld had been burnt.[88] In the Kat River area Vimbai Jenjezwa found that *inyongo* was the most common disease in the region and some stock owners differentiated between two types of gallsickness. One was described as 'normal' gallsickness manifest at post-mortem in a hard and dry stomach. The second type was called 'black' gallsickness, which presented as a shortage of breath, followed by dizziness and red eyes. Cattle also foamed at the mouth and their head and ears hung down. This distinction goes back to the nineteenth century, but it was not made in Mbotyi.[89] Most respondents associated both kinds of gallsickness with fresh grasses which resulted in increased secretions of gall. However, *inyongo* could also occur in the winter when the veld began to dry out. This sudden change in the consistency of the grasses was seen as detrimental to the functioning of the gallbladder. In the Kat River a similar interpretation applied to heartwater: 'They say that disease related to *inyongo* that's called heartwater is caused by ticks. It's not that. It's the change from dry to green grass. The stools are stiff and the cattle get heartwater.'[90]

Setswana-speaking informants were not united about which types of grasses triggered *gala* (gallsickness). Some attributed it to ingesting dried grasses because constipation was a major symptom and desiccated shards were assumed to block up the digestive tract.[91] Other stockowners felt it was wet grasses that caused both constipation and an over-production of bile, preventing the stomachs from working properly.[92] Alternatively it was the seasonal alterations in vegetation, the

---

[87] Zipoyile Mangqukela, Mbotyi, 29 March 2008; Myalezwa Matwana, Mbotyi, 1 March 2009; Richard Msezwa, Mbotyi, 10 February 2009; Nongede Mkhanywa, Mbotyi, 27 March 2008 and 25 February 2009.

[88] Ainslie, 'Hybrid Veterinary Knowledges'; Kenyon, 'Approaches to Livestock Health'.

[89] Smith, *Contribution to South African Materia Medica*, 154–7.

[90] Vimbai Jenjezwa, 'Stock Farmers and the State: A Case Study of Animal Healthcare Practices in Hertzog, Eastern Cape Province, South Africa', unpublished MA Thesis, University of Fort Hare (2010)'; V. Mzele, 10 June 2009.

[91] Itumeleng Sedupane, Mareetsane, 4 November 2009; Abraham Meno, Mareetsane, 4 November 2009; Laurence Bodibe, Ramatlabama, 10 February 2010; Matthews Motshwrateu Moloko, Magogwe (near Mafikeng), 11 February 2010; Kgomotso Goapele, Mantsa, 12 February 2010; Jeremiah Ramakopeloa, Mabeskraal, 5 February 2010.

[92] Andrew Mabaso, Mmabatho, 5 November 2009; Stoki Taedi, Mareetsane, 4 November 2009; Patrick Sebeelo, Lokaleng, 3 November 2009.

unpredictable and at times sudden shift from green to arid veld.[93] Cattle that moved from a dry to a wet area or vice versa were especially susceptible to gall-sickness.[94] Thus *gala* mirrored the changing seasons: green grasses caused it in spring, dry ones in winter.[95] In QwaQwa the majority of informants said that *nyoko* arose from the mixture of dry and wet grasses, brought about by the revolving seasons and it was especially virulent in the spring from August to October.[96] However, a few farmers ascribed the disease to particular kinds of vegetation. Kikuyu grass (an introduction from East Africa) and a type of sweet-veld called *mohlwa* upset the digestive system and caused *nyoko*.[97]

Blackquarter

Stockowners explained blackquarter, a common disease in North West Province and Free State, in an analogous way. Many in QwaQwa regarded it as the most serious disease in that area and one informant had lost 28 cattle to blackquarter in 2007 alone.[98] Sesotho-speaking informants called the disease *serotswana* and they had observed that it struck in the summer months and, like gallsickness, they associated it with transformations in the veld from dry to moist conditions.[99] Zulu Mokoena, for example, 'observed a direct link between lush vegetation and sickness'. Apart from the blackened legs and spongy muscles, he believed that the pathogenesis of blackquarter was similar to gallsickness. If a cow contracted *serotswana* 'the gallbladder struggles to cope with green grasses, so they remain undigested in the stomach. Constipation and death follow'.[100] His account showed the importance of the gallbladder even in interpreting diseases that stockowners had not identified as *nyoko* and which bore unique and distinctive characteristics, like blackened forelegs. Alternatively, *serotswana* could arise due to the desiccation of the grasses in early winter as the winds that blew across the Maluti mountains from Lesotho became cooler. The cold air also chilled the animals which made them more susceptible to digestive problems.[101] In winter,

[93] Fafung group meeting 25 January 2010; Charles Matome, Sutelong, 28 January 2010; Bosie Morapadi, Mabeskraal, 2 February 2010; Emily Diole, Mabeskraal, 2 February 2010; Henry Ramafoko and Elsie Ramafoko, Mabeskraal, 2 February 2010.

[94] Moses Tobosi, Mafikeng, 6 November 2009.

[95] Kgomotso Goapele, Mantsa, 12 February 2010.

[96] Ntsoeu Livestock Company, Phuthaditjhaba, 11 October 2010; Chadiwick Mbongo, Lejwaneng, 12 October 2010; Daniel Khonkhe, Ha-Sethunya, 13 October 2010.

[97] Maria Khoboko, Lejwaneng, 12 October 2010; Sethunya Mathu, Hasethunyu, 13 October 2010; April Nhlapo, Lusaka, 28 February 2011; Paulus Motsoeneng, Lusaka, 3 March 2011; Johannes Mofokeng, Lusaka, 3 March 2011.

[98] Johannes Motaung, Makwane Tebang, 1 March 2011.

[99] Daniel Gambo, Lusaka, 28 February 2011; Zulu Mokoena, Lusaka, 2 March 2011; Johannes Motaung, Makwane Tebang, 1 March 2011; Ntsoeu Livestock Company, Phuthaditjhaba, 11 October 2010.

[100] Zulu Mokoena, Lusaka, 2 March 2011.

[101] Michael Molibeli, Ha-Sethunya, 2 March 2011; Johannes Mofokeng, Lusaka, 3 March 2011.

there was more dew on the grass in the mornings, and it took much longer to evaporate. In this case, the cold wet grasses had a detrimental effect on the multiple stomachs and the gallbladder.[102] Thomas Mahlaba argued that particular types of grass such as *seboku* generated digestive problems that led to bloat, rigidity and the development of spongy muscles. *Seboku* grew tall around September, when the rains started, and it appeared appetising to stock. He described *serotswana* here as a form of toxicosis.[103] This is particularly surprising because *seboku* usually refers to *rooigras*/redgrass, or *Themeda triandra*, which is generally seen as the most desirable pasture grass in South Africa. It may be that this seasonal association, rather than the grass itself, underpinned his understanding of the disease.

In North West Province, explanations of blackquarter varied, but there was an overwhelming telluric theme. The timing of blackquarter outbreaks was linked to the arrival of cold winds, blazing sunshine and the presence or absence of substantial rainfall. Spring and autumn were the most dangerous times because of the noxious mix of fresh and old grasses.[104] As in QwaQwa, some ascribed blackquarter to toxic plants. In Mabopane, for instance, farmers believed that a shrub called *ramukutwana* made the cattle sick because it produced a lethal foamy substance inside the stomach.[105] The association of foam with blackquarter emerged in a number of testimonies in which informants assumed that a particular type of worm or insect left a spumous substance on the veld which poisoned the livestock. Bosie Morapadi from Mabeskraal believed: 'the foam putrefies the animal from within. It makes the blood rotten and turns the muscles black and spongy. The meat is horrible and black.'[106] This was the most common explanation for *serotswana* in the village of Mabeskraal.[107]

[102] Johannes Mofokeng, Lusaka, 3 March 2011.

[103] Thomas Mahlaba, Tseki, 15 October 2010.

[104] Francina Gumade, Mmakau, 30 October 2009; Chris More, Kgabalatsane, 13 November 2009; David Modiga, Garankuwa, 24 October 2009; Frederick Matlhatsi, Mabopane, 29 October 2009; Itumeleng Sedupane, Mareetsane, 4 November 2009; Samuel Mothapo, Slagboom, 10 November 2009; Lott Motuang, Bollantlokwe, 10 November 2009; Mapule Bokaba, Brits, 9 November 2009; Johannes Mabena, Moiletswane, 20 January 2010; Kgomotso Goapele, Mantsa, 12 February 2010.

[105] Frederick Matlhatsi, Mabopane, 29 October 2009.

[106] Bosie Morapadi, Mabeskraal, 2 February 2010.

[107] Richard Molebalwe, Bethanie, 26 October 2009; Abraham Meno, Mareetsane, 4 November 2009; Fafung group meeting, 25 January 2010; Piet Hlongwane, Sutelong, 27 and 28 January 2010; Simon Mokonyana, Kgomo Kgomo, 29 January 2010; Jeremiah Ramakopeloa, Mabeskraal, 5 February 2010; Martha Rakgomo, Mabeskraal, 1 February 2010; Henry Ramafoko and Elsie Ramafoko, Mabeskraal, 2 February 2010; Lekunutu Ramafoko, Mabeskraal, 4 February 2010; Kemoreng Monebi, Mabeskraal, 1 February 2010; Bassie Monngae, Mabeskraal, 1 February 2010; Bosie Morapadi, Mabeskraal, 2 February 2010; George Malatsi, Mabeskraal, 5 February 2010; Jacob Tau Diole, Mabeskraal, 4 February 2010; Lucas Moatlhodi, Mabeskraal, 3 February 2010; Molelekwa Mabe, Mabeskraal, 1 February 2010; Pagiel Sehunoe, Mabeskraal, 4 February 2010.

The fact that certain communities, like Mabopane and Mabeskraal, had distinct explanations suggested that the observation of local environmental factors, as well as an exchange of information between stockowners, had led to some regional variations in understanding the aetiology of this disease, although environmental factors played a central role in all.

A number of informants in North West Province linked blackquarter to carcasses.[108] This bore some parallels with scientific evidence about the source of blackquarter. The aetiological agents of blackquarter are the bacteria *Clostridium chauvoei* and *Clostridium novyi*. They multiply in the absence of oxygen and release toxins similar to the *Clostridium botulinum* which causes *lamsiekte*. These toxins are found in carcasses and the bacteria can survive several years in the ground so it is possible for animals to contract the disease through skin lesions as well as through grazing. [109] Some farmers had multi-causal explanations that not only included carcasses but also other agents. Samuel Mothapo, who lived in Slagboom in North West Province, had some interesting ideas about animal health in general. He had worked as a railway mechanic in Johannesburg for many years and in his retirement he had invested in livestock because he had loved herding as a child. Mothapo had relatively large holdings – 35 cattle, 20 goats and a number of donkeys he kept for transport. He had observed that blackquarter could arise 'from ingesting bones from the veld, feeding on a mixture of grasses and suffering from a lack of salt in the diet which makes the legs swell and turn black'.[110] People would normally eat the meat of an animal that died of blackquarter and in Mothapo's village of Slagboom, there was the belief that if a man urinated in or near the cattle kraal after eating contaminated meat, he would pass on the infection to cattle. Although humans cannot contract blackquarter, Mothapo believed that some diseases could be zoonotic and also cross species barriers. Hence he mentioned there were local taboos against urinating and defecating near to the cattle kraal.

Blackquarter did not appear to be a problem in Mpondoland and local AHTs interviewed at the Lusikisiki veterinary office said they had not come across it.[111] The isiXhosa word for this disease, *isiDiya*, was barely mentioned in interviews in Mbotyi. Perhaps the annual inoculation has made a long term impact in keeping it at bay; alternatively the mix of grasses around the village may diminish the likelihood of livestock feeding on carcasses. Mbotyi informants did speak of an autumnal disease *nonkhwanyana* which they believed caused stiffness in the front limbs and affected the gallbladder. It did not occur every

---

[108] Itumeleng Sedupane, Mareetsane, 4 November 2009; Samuel Mothapo, Slagboom, 10 November 2009; Lott Motuang, Bollantlokwe, 10 November 2009.

[109] N. P. J. Kriek and M. W. Odendaal, '*Clostridium chauveoei* infections' in Coetzer & Tustin (eds), *Infectious Diseases of Livestock*, vol. 3 (1857).

[110] Samuel Mothapo, Slagboom, 10 November 2009.

[111] Discussion at Lusikisiki Veterinary Office, 19 March 2012: AHTs Ndlela, Fusile, and Kgoa with Roger Davies and William Beinart.

year and outbreaks were sporadic. There were cases in 2006 when it attacked younger cows, affecting their forelegs so they were unable to walk. Post-mortems revealed blood and swelling in the upper leg.

We were unclear as to exact character of this disease. It is possible that they had contracted three-day stiff sickness, a viral disease. There was a national outbreak in the autumn of 2006 which accorded with the timing mentioned by respondents.[112] Local AHTs were uncertain about the identity of *nonkhwanyana* but thought it may be this disease. Sidwell Caine explained that it generally occurred from March to May, when the grass was seeding and drying, and when the cooler winds began to blow. Outbreaks occurred if cattle grazed at particular places, notably Ngquka, a grassland area to which he took his animals in summer. He thought this disease could also occur if cattle fed on *ngongoni* grass at Lambasi (chapter 4), the dominant sourveld in that area which is associated with *Aristida junciformis*. However, despite these observations, Caine mused that 'sometimes it just seems to attack' without a known cause.[113] Three-day stiff sickness is not generally lethal and animals can recover by themselves.

'Blood diseases'

Some Tswana farmers classified blackquarter and anthrax as blood diseases (*bolowetse ra madi*).[114] Whatever caused the animal to fall sick infected the blood, contaminating it and rendering it 'dirty'. Diagnosis of blackquarter could be made by cutting the ears and examining the colour of the blood and the speed at which it spurted out. Dark blood was particularly noxious and they deliber-ately bled their cattle to prevent infections and cure blackquarter.[115] In North West Province, blood disease could be both specific, as in the case of black-quarter, or it could be a more vague term for describing poor health. Anna Pooe gave a particularly detailed explanation about the role of blood in the generation of ill health. She was an elderly traditional healer from Shakung and kept goats which she could slaughter to protect people from witchcraft or to connect them with the ancestors during the healing process. She was an expe-rienced goat farmer and preferred to breed white goats for her practice. Pooe explained: 'dirty blood is not only a cause and symbol of disease, but it also reduces reproductive capacity. Blood purification is needed for humans and animals to remove the dirt that ruins health and leads to disease.'[116] Redwater

[112] Kassim Kasule, Lusikisiki, 18 February 2009.

[113] Sidwell Caine, Mbotyi, 23 February 2008 and 12 February 2009.

[114] Tollo Josiah Mahlangu, Mabopane, 29 October 2009; Samuel Mothapo, Slagboom, 10 November 2009; David Modiga, Moiletswane, 20 January 2010; Kemoreng Monebi, Mabeskraal, 1 February 2010; Soloman Matome, Mabeskraal, 3 February 2010; Elphars Dinne, Mabeskraal, 4 February 2010.

[115] Charles Matome, Sutelong, 28 January 2010; Philemon Ramafoko, Mabeskraal, 3 February 2010; George Malatsi, Mabeskraal, 5 February 2010; Reuben Ramutloa, Disaneng, 11 February 2010.

[116] Anna Pooe, Shakung, 22 January 2010.

was not familiar to many Tswana stock owners, but Pagiel Sehunoe from Mabeskraal was unusual in recognising the disease. He believed that the red urine (*motlhapologo khibidu*) showed that the cow was 'dirty' and in need of purification with traditional medicines.[117] There were similarities in Mpondoland, where stock owners spoke of weak blood (*igazi liweak*) and they too used herbal preparations to strengthen it. This weakness did not seem to be linked to any particular disease and was one generic way of explaining why livestock looked sickly or had a predisposition to catch infections.[118]

These testimonies linking clean blood with good health and dirty or weak blood with sickness, malnutrition or infertility resonate with some of the anthropological literature and suggest a surviving belief in the role of blood in explanations of disease. H. Stayt, in his ethnography of the Venda (1931), wrote about farmers slitting the ears and sending the blood to the 'medicine man' for diagnosis.[119] Monica Hunter described how, in the 1930s, Mpondo farmers burned herbs in the kraal every two to three months 'to make well the blood of cattle'.[120] This tradition of burning medicine continues and in Mbotyi was associated with reintegrating livestock into the kraal after a period at the summer grazing grounds.[121] In North West Province farmers linked such practices with preventing specific diseases rather than dealing with generic conditions of the blood (chapter 6). The significance of the blood might also relate to ideas about human health. The Kriges, who worked in the north-eastern Transvaal in the 1930s, argued that blood lay at the heart of Lovedu descriptions of diseases. Bloodletting was an important part of Lovedu therapeutics: too much blood made people sick, causing pains and swellings.[122] Bloodletting had been a popular form of therapeutics in western medicine up until the nineteenth century. Practitioners believed that phlebotomies flushed out disease, prevented inflammation and promoted vigour and mental agility.[123]

Worms and grubs

Worms, identified as a scourge in Cape livestock by the first government veterinarian in the 1870s, remain a major problem in Mbotyi in the early twenty-first century. Aside from the lack of regular treatment, this probably resulted from the method of keeping animals. When livestock were in the village, their owners kraaled them at night. They did not remove the dung which accumulated in the

---

[117] Pagiel Sehunoe, Mabeskraal, 4 February 2010.

[118] Nongede Mkhanywa, Mbotyi, 25 February 2009.

[119] H. Stayt, *The Bavenda* (London, Oxford University Press, 1931), 40.

[120] Monica Hunter, *Reaction to Conquest: Effects of Contact with Europeans on the Pondo of South Africa* (London, International Institute of African Languages and Cultures, 1936), 66.

[121] Myalezwa Matwana, Mbotyi, 7 December 2011.

[122] E. J. Krige and J. D. Krige, *The Realm of a Rain-Queen: A Study of the Pattern of Lovedu Society* (London, Oxford University Press, 1943).

[123] Shigehisa Kuriyama, 'Interpreting the History of Bloodletting' *Journal of the History of Medicine and Allied Sciences*, 50, 1 (1995), 11–46.

kraal — again an echo of the nineteenth-century practices amongst settler farmers that drove the early veterinarians to distraction. While dung is occasionally collected to smear floors, or to put on gardens, this is generally seen as too labour intensive. Respondents claimed that fresh dung on gardens spread weeds. When the dung is thick and dry, or fully rotted as manure, it may not be too congenial for worms; but when fresh and wet it provides an ideal reproductive environment.

Moreover, some livestock owners whom we visited took their animals out of the kraal, but let them graze in their fenced homestead sites for a short time before sending them out to the pastures. During this period they could check over their animals and enjoy them while the animals could 'relax'.[124] The result was that fresh dung collected on the grass around the homesteads and provided an ideal medium for worms' eggs which were then imbibed by other animals. Horses, kept to a greater extent than cattle and goats around the homesteads, may have been especially susceptible to worms for this reason. Worms were called *izilo* — a word connected with thinness from fasting or abstaining — or *iskelem* (from the Afrikaans *skelm* — a rogue or rascal) in Mbotyi. Stockowners recognised that worms reduced energy and power. While there were plant remedies, Ivomec (ivermectin) and Dectomax (doramectin) were increasingly favoured. Tapeworms have been a problem in cattle for some decades in the Transkeian area and they are one constraint on the marketing of meat.

We found, more generally, that informants were not very specific about the identity of worms and insects, nor about their impact. Sonwabile Mkhanywa from Mbotyi provided an alternative to a grassland explanation when he discussed the cause of *nonkhwanyana*. He had noticed larvae attached themselves to tall old grasses and attributed the disease to small insects, or possibly spiders, that were nesting in the vegetation at that time of the year.[125] He made a strong seasonal correlation between specific kinds of grass and the arthropods that thrived in them. It is interesting to note that stockowners linked diseases more with a rather non-specific idea of unidentified insects or spiders than with ticks.

A number of informants in North West Province and QwaQwa also referred to a worm that lived in the grass and poisoned livestock. When asked about toxic flora, many farmers throughout North West Province immediately mentioned a dangerous worm that sheltered in the grasses. The name of the worm varied from place to place. Patrick Sebeelo from Lokaleng near Mafikeng reported that if a cow ate the *mohamba ka ndlwana* it 'foamed at the mouth, became sleepy and died within three hours'. The worm was hard to spot in the grass, but it left a slimy track across the veld in the morning. It was unsafe to release stock onto the grazing lands until the sun had evaporated the trail. *Mohamba ka ndlwana* was most common during the summer from January to

---

[124] Visit and discussion with Cecil Malindi, 13 March 2012.
[125] Sonwabile Mkhanywa, Mbotyi, February 2009.

April and mortality rates could be high at that time of the year.[126] This concern about tracks in the veld was evident also in Mbotyi as a cause of *umkhondo* (see chapter 7), but in North West Province it was offered entirely as a natural rather than a supernatural phenomenon. Itumeleng Sedupane, from nearby Mareetsane, explained that *sebokwana sa tlhaga* lived in the grass and multiplied in the stomachs of cattle. They corroded the intestines, causing rapid death.[127] In the Winterveld, 300 km away, Michael Mlangeni found a worm – *ndlozi* – in the intestines of all species of livestock: 'the worm poisons the animals; they drop dead suddenly in the kraal and post-mortems reveal a blackened liver.'[128] In QwaQwa some stock owners spoke of similar worms that covered themselves in foam and camouflaged themselves to prevent predation. One informant recounted how the worm protected itself with *moseme* – a particularly hardy type of grass that the Sotho used to make thatched roofs. If a cow ingested this worm it would surely die.[129]

Stockowners also associated worms with horse sickness. Very few of our informants in North West Province and QwaQwa had horses. They were far more likely to keep donkeys for draught. Horses were expensive and susceptible to diseases like African horse sickness. In the Mafikeng area however there were a few farmers who did keep horses. They called the disease *sterf* – the Afrikaans word for death. They did not attribute it to midges but to worms. Zachariah Tlatsane from Lokaleng, for example, explained that *sterf* was caused by a worm that 'is found everywhere around the grazing lands'.[130] Kgomotso Goapele had observed that *sterf* usually broke out in March. It was a highly fatal disease and the only way to prevent it was to add *manyana* (medicinal *dagga*; marijuana) to the feed in February as it protected equids from the 'dangerous worms that appear in the morning dew'.[131] The idea that horsesickness was caused by a dew or substance on the grass correlates with veterinary reports from the 1850s which suggested that the disease struck when farmers allowed their horses to graze before the sun had burnt off the morning dew. The dew was said to contain spiders' webs that were toxic to horses.[132] It is possible that avoiding the morning dew minimised bites from the midges that spread the disease. In Mbotyi, where a number of men kept horses, African horse sickness (*isimoliya*) is a major annual scourge, but it is associated with the seasons rather than with insects or substances in the grass. It frequently made an appearance

---

[126] Patrick Sebeelo, Lokaleng, 3 November 2009.

[127] Itumeleng Sedupane, Mareetsane, 4 November 2009.

[128] Michael Mlangeni, Mabopane, 24 October 2009.

[129] Johannes Mokotla, Ha-Sethunya, 2 March 2011.

[130] Zachariah Tlatsane, Lokaleng, 9 February 2010.

[131] Kgomotso Goapele, Mantsa, 12 February 2010.

[132] Karen Brown, 'Frontiers of Disease: Human Desire and Environmental Realities in the Rearing of Horses in Nineteenth and Twentieth-Century South Africa', *African Historical Review*, 40, 1 (2008), 30–57.

in mid-summer, around December.[133] One informant thought that it was carried by flies, and that its incidence and intensity was in part shaped by the wind.

The species of these worms was unclear. Michael Raito, an AHT at North West University in Mafikeng, commented that many people believe that the *mohamba ka ndlwana* killed animals and they were particularly numerous in hot weather. However, there have been no proper laboratory tests to identify the worm and check what might have been killing the livestock.[134] It was possible that army worms (*Spodetera exempta*) could be responsible especially on veld which has a lot of kikuyu grass. Army worms are caterpillars that develop into moths. They survive in the higher rainfall areas of North West Province and the Free State and appear in the summer and early autumn. Scientists have discovered that when the worms feed on kikuyu grass (*Pennisetum clandestinum*) specifically, they damage its re-growth making it toxic to cattle. However the mechanism by which the grass kills the grazers remains a mystery.[135] If this was a cause of poisoning, it represented a relatively new disease because this grass was introduced in the first decade of the twentieth century and may have taken some time to reach these rural areas.

In QwaQwa, sheep scab – a condition brought about by the gnawing of mites – was a common problem and many stock owners dipped their animals regularly to prevent it. Sotho farmers were aware that some sort of grub caused *lekgwakgwa* because they had seen the sheep either vigorously trying to scratch at the skin with their hooves or rubbing up against rough objects. There were evidently signs of intense itching and irritation that observers could attribute to a kind of biting bug.[136] Dipping for scab was enforced for many years in South Africa, and knowledge about its cause may have spread; it was also a condition rather than a disease so it was easier to spot and understand from an epidemiological perspective than tick-borne ailments.

Poisonous plants

In his thesis that dealt with local knowledge in the Madikwe District of North West Province, Deon van der Merwe talked about a familiar toxic plant, *mogau* (the Afrikaans spelling of *mohau*) or *gifblaar*. He mentioned that some farmers believed *mogau* was a poisonous weed, but others associated it with a pupa that surrounded itself with twigs made from *Panicum maximum* grass. The symptoms

---

[133] Cecil Malindi, Mbotyi, 13 December, 2011; Sidwell Caine, Mbotyi, 8 December 2011.

[134] Michael Raito, NWU Mafikeng, 5 November 2009.

[135] T. S. Kellerman, J. A. W. Coetzer, T. W. Naude, C. J. Botha, *Plant Poisonings and Microtoxicoses of Livestock in Southern Africa* (Oxford, Oxford University Press, 2005), 191–4; P. G. Alcock, *Rainbows in the Mist: Indigenous Weather Knowledge, Beliefs and Folklore in South Africa* (Pietermaritzburg, Intrepid Printers, 2010), 408–9.

[136] Maria Khoboko, Lejwaneng, 12 October 2010; Abel Lebitsa, Mountain View, 14 October 2010; Chadiwick Mbongo, Lejwaneng, 12 October 2010.

stockowners described resembled *gifblaar* poisoning.[137] *Gifblaar* kills by incapacitating the heart. Symptoms are varied, but include salivation, breathing difficulties, bellowing and staggering. Cattle often drop dead after exercise or drinking water as the liquid speeds up the absorption of toxins.[138] In our interviews, stockowners in North West Province had also observed that animals that fed on *gifblaar* wandered around in circles with foam dripping from their lips. Death was notably sudden, pointing to *mohau* as its cause at certain times of the year.[139] Some of our informants believed that *gifblaar* was itself toxic, whilst others claimed that a worm or larva nesting in the plant was the actual cause of death.[140]

Stockowners from Fafung, as already noted, were especially familiar with *gifblaar*. They had two words for the plant: they used *mohau* when the leaves were green and fresh and *legolo* when they were brown and desiccated. Both could be toxic.[141] Noah Motaung believed that the 'poison lies in the tiny hairs that cover the outer layer of the leaves. Once the heavy rains come they wash away these hairs and the plant stops being dangerous to livestock.' In a bad year, people could lose five cows at a time to *mohau* poisoning, with most cases occurring between August and October.[142] *Mohau* was a curse of spring. The plant sprouted with the early primaveral rains, but did not seem to be toxic all the time.

Farmers in Fafung have found it very difficult to deal with *mohau* because it grows like an underground tree, with thick roots that extend many metres into the ground. It was almost impossible to uproot effectively as fragments broke off and it has proved to be largely impervious to herbicides. They have tried to burn it but it only grew back with renewed vigour. It was impossible to fence off patches of *mohau* because people stole the fence posts.[143] The only solution was to drive animals away from the plant until a deluge came. After the rains, informants found that the livestock could return to infested veld, without being poisoned. This also relieved the pressure on the grazing lands that were free of this poisonous plant.[144] Grazing management was shaped by the incidence of poisonous plants, as transhumance had been in former years (see chapter 4). Zachariah Maubane, who had a farm near Bela Bela in Limpopo Province, reported similar problems with *gifblaar* there. In the past white farmers had tried to kill the plant with chemicals but they had not been very successful. Tracts of land had been abandoned. As in Fafung, fatalities occurred in the spring and the

---

[137] van der Merwe, 'Use of Ethnoveterinary Medicinal Plants in Cattle', 88.

[138] Kellerman, et al., *Plant Poisonings and Microtoxicoses*, 150.

[139] Josiah Letseka, Mmakau, 13 November 2000.

[140] Eva Sesoko, Madidi, 23 October 2009.

[141] Report-back meeting, Fafung, 29 November 2012.

[142] Noah Motaung, Fafung, 27 October 2009.

[143] Report-back meeting, Fafung, 29 November 2012.

[144] Simon Mathibedi, Mmakau, 9 November 2009; Noah Motaung, Fafung 27 October 2009; Philip Mokwena, Fafung, 27 Octrober 2009; Fafung group meeting, 25 January 2010; Paulus Mmotsa, Fafung, 25 January 2010.

plant lost its toxicity after three days of heavy rains. Maubane suggested that the water must dilute the toxic principles sufficiently to make *mohau* harmless to livestock.[145]

Overall, interviewees were not particularly familiar with toxic plants. The only other variety that came up frequently in interviews was tulp – known as *teledimo* in North West Province and *tele* in QwaQwa. The variety of tulp that farmers mentioned had a thin stem, a long, narrow wispy leaf and yellow flowers (*Moraea pallida*). The stem itself looks very much like grass and because it tended to sprout with the first spring rains, it often stood out as the greenest plant in the veld when much of the pastureland continued to bear the dull brown hue of a dry winter. This made the plant attractive to cattle, especially younger animals that had not experienced tulp before. According to Andrew Mabaso, from Mafikeng:'sick cows become dizzy, and then they get bloated and start to limp. Suddenly they collapse and drop down dead.'[146]

Cases of toxicoses were particularly problematic in areas where there was a shortage of grazing such as Koppies. The white commercial farmers we interviewed had no real problem with poisonous plants, but it was a concern for African stock owners who pastured their livestock on the communal lands outside the township of Kwakwatsi. Isaac Mzima found that poisonous plants were the biggest problem farmers had to contend with, the most dangerous varieties being tulp and *dubbeltjie* (devils thorn – *Tribulus terrestris*). Stock owners lost many animals each year to both weeds.[147]

Innocent Setshogoe, who used to be a stock inspector in the Brits area, explained that scientists often ascribed the encroachment of poisonous plants to overgrazing and a lack of proper veld rotation. Animals are selective grazers and the unpalatable plants they reject can quickly colonise the veld. Setshogoe argued that exposure to poisonous plants increased in the communal areas where it was difficult to implement a consistent programme of veld rotation in fenced camps. Furthermore, the number of livestock in many villages made it difficult to manage the veld and allow grasslands a chance to recover.[148]

One particular type of poisoning that our African informants did not discuss was diplodiosis, caused by the fungus *Diplodia maydis*. In the Koppies region, Afrikaner informants complained that diplodiosis was a major threat to their herds. Cattle contract the disease when left to graze on the harvested maize fields, picking it up from discarded maize cobs which have become black and rotten with mould. In the northern Free State it was common for farmers to

[145] Zachariah Maubane, interviewed in Mafikeng, 2 November 2009.

[146] Andrew Mabaso, Mafikeng, 5 November 2009; Thomas Mahlaba, Tseski, QwaQwa, 15 October 2010.

[147] Isaac Mzima, Kwakwatse, 8 October 2010; Karen Brown, 'Poisonous Plants, Pastoral Knowledge and Perceptions of Environmental Change in South Africa, c. 1880–1940', *Environment and History*, 13, 3 (2007), 307–32.

[148] Innocent Setshogoe, Bethanie, 26 October 2009.

depasture their livestock on harvested fields to supplement grazing during the winter months when the veld was dry. *Diplodia* attacks the nervous system and animals will die if they are not quickly removed from the contaminated fields, or dosed with treated feed that binds the toxins together so they can be safely excreted. Infected cows can produce brain-damaged calves and sheep will abort, so the economic consequences go beyond the current generation of livestock. The years 2009 to 2010 were a particularly bad time for diplodiosis in the Koppies area.[149] However, none of our African respondents from Koppies' contiguous township Kwakwatsi referred to it. Perhaps this reflects the different grazing systems of commercial farmers and African stockowners in this area. The latter only have access to communal grazing lands, not arable fields; any available land was used for livestock. On the one hand this protected animals from this particular form of toxicosis; on the other hand the lack of maize production limited the strategies that African livestock owners could adopt to supplement grazing during the winter months. The disease was not mentioned in the Eastern Cape or North West Province, where African farmers have long used the maize stubble as a winter feed. Perhaps this is a result of more thorough hand harvesting and recycling of maize cobs.

Malnutrition and depraved appetites

In North West Province many farmers conceived of winter as a time of scarcity and danger for their animals. Malnutrition could damage their herds. Fodder as well as vitamin and mineral supplements were too expensive for stockowners to purchase regularly – or, as in Mbotyi, they saw no necessity for such expenditure. Underfeeding constituted the antithesis of the ideal situation in which an animal's good health was reflected through a hearty appetite, a shining coat and free-flowing bowels. In North West Province, stockowners blamed malnutrition for a range of health problems. Don Modiane who farmed in Kraaipan, near Mareetsane, explained that a lack of water and grazing caused sickness especially in the dry season: 'dry grasses prick the inside of the cow's mouth making it so sore that the animal refuses to graze. The cow becomes weak, it falls over and dies.'[150] Some farmers speculated on the impacts of a lack of water and forage on the internal organs of the body. Samuel Mothapo described how a poor diet brought about a kind of 'blood disease':

> cattle become constipated and stagger around as if they are blind. They bump into things and appear drunk as if they have too much blood. Dry grasses and inadequate water stop the stomach working properly; the composition of the blood changes and drives animals mad.[151]

[149] Piet van Zyl, Private Vet, Heilbron, 7 October 2010; our Afrikaner informants chose to remain anonymous so their names are not cited here.
[150] Don Modiane, Kraaipan, 11 November 2009.
[151] Samuel Mothapo, Slagboom, 10 November 2009.

Worms and other internal parasites were a particularly irksome problem in the winter as they further undermined the strength of livestock already weakened by insufficient grazing.[152]

In Mbotyi we were struck by the lack of any provision of water near the kraals.[153] Water supply was plentiful in the three streams that flowed through the village and in the grazing lands. People said that the animals found their own water, or they would be led to drink when they went out in the mornings. There were cases however, especially when animals were ill and kept in the kraal, that water supply seemed inadequate. In the drier North West Province, some stockowners made deliberate provision of water for their livestock. In the wealthier households, there were stone water troughs in the kraals. Elsewhere, people improvised, using cut-out jerry cans, bath tubs and other vessels that could hold water. Herders also drove the animals to rivers to drink and water at the homestead was the supplementary rather than main source of supply. People worried about dry winters and drought years as they knew insufficient water contributed to poor health in livestock.

Stock owners in North West Province also associated malnutrition with dystocia as the cows became too weak to expel the calves from the uterus.[154] There did not appear to be any local knowledge of brucellosis in cows and other venereal diseases such as trichomoniasis and vibrosis which infect bulls. There had been inoculation campaigns and testing for brucellosis in many parts of South Africa, such as North West Province, in the past, but there was no historical memory of or connection between birthing problems and diseases. Myalezwa Matwana in Mbotyi was particularly concerned with abortion (*ikupile*) although he did not consider it a common problem.[155] He associated it not with disease, but drought, intense heat or a lack of clean running water. He had noticed that when a cow died from what local people perceived to be drought and cut it open, 'the stomach is dirty, full of mud and dirt'. Matwana made medicines to facilitate healthy growth of the calf foetus. These *muthi* would specifically help the calf to 'grow hairs in the stomach' – as a lack of hair was associated with abortion. AHTs in coastal Mpondoland had not come across brucellosis.[156]

In North West Province, respondents were also aware of the concept of a 'depraved appetite' which drove livestock to eat unsuitable foodstuffs, such as plastics, and bones from animal carcasses. *Lamsiekte* was familiar to a number of

[152] Andrew Mabaso, Mmabatho, 5 November 2009; Thomas Seemela, Madidi, 23 October 2009; Chris More, Kgabalatsane, 13 November 2009.

[153] Roger Davies, March 2012.

[154] Chris More, Kgabalatsane, 13 November 2009; Zachariah Maubane, Mafikeng, 2 November 2009.

[155] Myalezwa Matwana, Mbotyi, 7 December 2011.

[156] Discussion at Lusikisiki Veterinary Office, 19 March 2012, AHTs Ndlela, Fusile, and Kgoa with Roger Davies and William Beinart.

farmers in the region and they called it either *magetla* (literally shoulder disease) or *mokokomalo* (*mokokotlo* refers to the spine and spinal cord). Sick cattle limped and were unable to walk properly. The shoulders had a hunched up appearance, giving the disease the name *magetla*. The ears drooped and the face bore a generally depressed look. In the advanced stages the cow just squatted on the ground as if paralysed. Inappetence and constipation often accompanied these symptoms.[157] Few farmers understood that botulism was a bacterial disease ingested by gnawing on bones, but many had deduced that the cause lay in some form of veld deficiency, such as poor grasses. Otsile Morwe from Dithakong near Mafikeng had an explanation that echoed scientific findings: '*magetla* is a deficiency disease that comes about if cattle lack salt and calcium in their diets.' He believed that the soils in the Mafikeng area were deficient in minerals which meant that the grasses were poor, forcing livestock to search for missing nutrients elsewhere.[158] Nick Lebethe, from Bethanie, believed that poor grasses deprived animals of energy and vitamins which 'made the joints loose', dismembering the body and causing paralysis.[159]

In QwaQwa, informants regarded malnutrition, along with blackquarter, as the most significant pastoral problem they faced. Some described emaciation or *boketa* as a disease; stockowners simply watched their animal starve to death, lacking the means to do anything about it. June was often a particularly bad month. Alan Malinga explained how 'the cattle's skin sags; they stagger and lose their balance and then they die'.[160] The Sotho ascribed *boketa* to the dry grasses of winter that deprived livestock of their strength and their health. Farmers recounted how denser settlements were eating into the rangelands, ensuring that more and more animals were dying of malnourishment every year.[161] Some complained about arson attacks in the mountains which further depleted the amount of available grazing. Arsonists were allegedly people who did not possess any livestock and were jealous of those who did. This activity was indicative of tensions within communities which informants often expressed in terms of inter-communal envy, or in the language of insiders versus outsiders. People were continually moving into the QwaQwa region, often from other parts of the Free State, creating additional pressure in terms of access to local resources, housing and jobs.[162] Many Sotho farmers regarded arson, along with stock theft

---

157 Peter Kembo, Mafikeng, 2 November 2009; Patrick Sebeelo, Lokaleng, 3 November 2009; Abraham Meno, Mareetsane, 4 November 2009; Nick Lebethe, Bethanie, 26 October 2009; Zachariah Tlatsane, Lokaleng, 9 February 2010; Ernest Medupe, Mafikeng, 10 February 2010; Piet Hlongwane, Sutelong, 27 and 28 January 2010.
158 Otsile Morwe, Dithakong, 3 November 2009.
159 Nick Lebethe, Bethanie, 26 October 2009.
160 Alan Malinga, Phuthaditjhaba, 13 October 2010.
161 Charlie Tshabalala, Phutadijhaba, 11 October 2010; April Nhlapo, Lusaka, 28 February 2011; Michael Molibeli, Ha-Sethunya, 2 March 2011.
162 Maria Khoboko, Lejwaneng, 12 October 2010; Paulus Mosia, Ha-Sethunya, 13 October 2010; April Nhlapo, Lusaka, 28 February 2011; Michael Molibeli, Ha-Sethunya, 2 March 2011.

and witchcraft, as deliberate acts that victimised both individuals and communities and jeopardised a peaceful co-existence in the rural areas.

QwaQwa's winters could be bitter and the cold mountain winds, frosts and snows alone could reduce an animal's resistance to infection. Farmers spoke of *mohatsela* or cold. The term referred to the stiffness and shivering that affected animals when exposed to very chilly weather. Informants showed an awareness of how changing climatic patterns, or at least perceptions of temperature variations, affected the overall health of livestock. Charlie Tshabalala, for example, believed *mohatsela* was far less of a problem nowadays than it had been in the past because 'the winters are much milder than they used to be'.[163] Michael Ncobuga spoke of a disease called *mankitila* which affected sheep. Their necks swelled up and they salivated. Ncobuga found his sheep contracted this every June and he was not convinced that the winters were becoming warmer.[164] Both respondents emphasised that climate and seasonal changes in temperature could give rise to conditions that were environmentally specific.

Industrial pollution and animal disease

Not all the conditions and diseases informants spoke about could be understood in terms of the natural environment. In North West Province there were accusations that the government and mining companies were also responsible for poisoning stock. Francina Gumada, a retired school teacher, described some of the horrors of living near a big mining complex near Brits, where chrome and vanadium were extracted from the rolling koppies. Gumada lived in the impoverished village of Mmakau about two kilometres from the mine dumping sites. She said that the black dust that resulted from burning materials at the mines often blew into the village and onto the grazing lands. People and animals developed sores on their skin which she attributed to this dust; children suffered from asthma. There were pools of toxic water around the mines which were not sufficiently fenced to exclude wandering livestock. In 2007, 15 of her cattle died over a short space of time having ventured near these lethal ponds. She opened the carcasses and found that they were totally black inside. Laboratory tests indicated that they had been poisoned by waste products from the mines. Many neighbours had also lost stock. Furthermore, she claimed some men who worked at the mines had green tongues and died suddenly, vomiting blood.

Gumada wanted to take legal action but lacked the funds. She felt that the royalties paid to local communities were minimal and failed entirely to compensate for the damage to animal and human health. The state was also, in her view, neglecting its duty to regulate the mining companies and maintain environmental safety. Her plea was for the mines to be closed down so that the 'children and livestock may live'.[165] Such regulatory neglect echoed earlier episodes

[163] Charlie Tshabalala, Phuthaditjhaba, 11 October 2010.
[164] Michale Ncobuga, Lusaka, 3 March 2011.
[165] Francina Gumade, Mmakau, 30 October 2009.

in South African history such as the prevalence of silicosis and phthisis on the Rand and asbestosis in the Northern Cape.[166] Poverty and political impotence made it difficult for rural African communities to challenge the might of the mining companies. In post-apartheid South Africa, some Setswana-speaking communities have been able to benefit greatly from the local mining industry, but the environmental consequences endure.[167]

In Moiletswane in the eastern part of North West Province, stockowners also complained that contaminated water was killing their livestock. The government had failed to maintain the local dams and boreholes so livestock were forced to drink from the rivers and streams into which industries pumped contaminants. Industrialisation was also associated with environmental explanations of diseases. Stockowners from Moiletswane believed that polluted water caused anthrax, abortions and lumpy skin disease.[168] Like Gumada they wanted the government to prosecute and rein in the activities of the industrial companies. They felt that the state was neglecting the interests and needs of the poor and this was having a grave effect on the health of their animals. In Phuthaditjhaba, Charles Tsahabalala vented his anger against the high levels of pollution in the Elands River that flowed through the town. Livestock drank from this contaminated source, full of effluent, sewage, plastics and other types of discarded refuse. He too clamoured for more action to safeguard the environment which was so important for ensuring the good health of both humans and livestock.[169]

Breakages and wounds

In March 2012, we visited the impoverished homestead of a family in Mbotyi who kept seven goats.[170] Unlike many of the kraals in the village, which were carefully constructed of indigenous materials, it was a makeshift pen made of scrap metal and corrugated sheets. Most people tried to cover their goat pens with a roof, unlike the cattle kraals which are open. We were called to the kraal to see if we could help a kid with a broken tibia. A small and inadequate splint had been tied on the leg, which could not immobilise the bone around the break – this is difficult on a small goat. The owner was absent and his wife, busy with young children, seemed at a loss. Roger Davies, the visiting veterinarian, made a more substantial makeshift splint.

[166] Elaine Katz, *The White Death: Silicosis on the Witwatersrand Gold Mines 1886–1910* (Johannesburg: Witwatersrand University Press, 1994); Jock McCulloch, *Asbestos Blues: Labour, Capital, Physicians and the State in South Africa* (Oxford: James Currey, 2002).

[167] Bernard Mbenga and Andrew Manson, *People of the Dew: A History of the Bafokeng of Phokeng-Rustenburg Region, South Africa, from Early Times to 2000* (Auckland Park, Jacana, 2010). There is no similar process of industrialisation yet near Mbotyi although exploitation of the Xolobeni mineral sands on the Wild Coast as well as plans for a coastal toll road present similar possibilities.

[168] Group Interview, Moiletswane Community Centre, 20 January 2010.

[169] Charlie Tshabalala, Phuthaditjhaba, 11 October 2010.

[170] Mbotyi, 21 March 2012, with Roger Davies.

Most of the information gathered from the different sites was based on interviews. However, in the last research visit to Mbotyi, Beinart was accompanied by a veterinary surgeon for two weeks who was able to visit kraals, examine animals, and discuss problems with the owners. This additional exposure to individual kraals opened up a slightly different perspective. Fractures and wounds had certainly arisen in interviews – not least in that with a healer who specialised in this area – but most livestock owners talked more about diseases than such physical damage.[171] The kid had suffered because, unherded, it had trespassed into a garden and been forcibly chased away. This was a particular problem with dogs and goats in the village. But the episode also alerted us to the significance of breakages and wounds more generally. The rough terrain of Mbotyi made for additional environmental dangers. Animals negotiate steep hills, muddy paths and coastal cliffs. We came across a number of breakages such as the cow with a broken tarsal bone, possibly caused by one such hazard. A number of horses in Mbotyi suffered from swollen fetlocks.[172] This was probably the result of damaged ligaments caused by the rough and uneven ground; horse-racing on the veld doubtless exacerbated the problem. One local owner injected Terramycin as treatment.

Another hazard arose from the horns of cattle. Villagers did not cut the horns of cattle, yet the animals were often cooped up in kraals where the dangers of piercing were apparent. As noted in chapter 4, animals left out for long periods in the summer were particularly difficult to control and informants spoke of them as wild. Wounds from cattle horns seemed to be a particular problem in Mbotyi and we saw a cow with a tumour attributed to this cause, as well as a goat with a puncture wound from which its intestines were obtruding. Stockowners treated horn wounds with plant medicines, and the stomach contents of cattle, left in the kraal after slaughter, were said to diminish conflict between animals. In North West Province, by contrast, stockowners regularly dehorned their cattle in order to prevent injuries occurring in the kraal or from fights in the veld. They used special tools to cut the horns when the calves were young and they filled the wounds with Stockholm Tar (pine tar) to prevent flies from laying their eggs in the lesions.[173] Thomas Seemela believed that dehorned cattle were also healthier because 'they grow large and fat' – the sign of optimal health.[174]

[171] Wellington Jikazi, Mbotyi, 24 February 2009.

[172] Cecil Malindi, Mbotyi, 13 March 2012.

[173] Discussion with agricultural science students, North West University, 6 November 2009; Moses Tobosi, Mafikeng, 6 November 2009; Lott Motuang, Bollantlokwe, 10 November 2009.

[174] Thomas Seemela, Madidi, 23 October 2009.

Photo 3.4 Milking dehorned cows at Lokaleng, NWP (KB, November 2009)

Conclusion

This chapter has explored the identification of maladies and diseases in livestock, generally inferred from visible symptoms which are then confirmed, in the case of death or slaughter, by post-mortems. Some names for disease are specific while others describe a range of symptoms. We have demonstrated the importance that African farmers attached to observations and interpretations of environmental conditions and events when explaining the causes of many animal diseases. In their view, the condition of the veld could foster good health, but equally it could bring about diseases like gallsickness and blackquarter. The cycle of seasons was also a harbinger for disease. Changes in the consistency of the grasses, even natural seasonal transformations from green to desiccated veld and vice versa, had a notable effect on digestive functions and thus the overall health of individual animals. In this sense disease was seen as part of the natural cycle, although its intensity could change unpredictably in different years, shaped by environmental events. Diseases such as gallsickness and blackquarter seemed to prevail in the spring and summer. Poisoning from noxious weeds often occurred in spring. Malnutrition and unusual food cravings were hazards in the winter or whenever nutritious grasses were in short supply.

Environmental relationships were at the heart of many discussions of disease and provided African communities with a context for the annual ebb and flow of particular ailments. Yet they sometimes found it more difficult to pinpoint links between environmental changes and the incidence of specific or new diseases. On the one hand, such seasonal and environmental awareness helped livestock owners to explain diseases and take preventative action. On the other, the limits of local knowledge also had implications for effective disease control. Managing the nutrition of livestock remained central for interpreting and coping with disease. Plant medicines are primarily geared to cleansing the blood as well as the gut, in the form of laxatives (chapter 6). They are supplemented by notions about the best environmental conditions for animals. This can include moving livestock between pastures in systems of transhumance which we discuss in the following chapter.

4

Transhumance, Animal Diseases & Environment

'If they keep cattle right through the year here, they will die... the mud is killing the cattle.'[1] (Sidwell Caine, Mbotyi)

The context of transhumance

Transhumance, or trekking in South African settler language, was intrinsic to white and black livestock management up to the early decades of the twentieth century. By transhumance, we mean the movement of people with livestock – a practice common to many societies, especially but not only in regions with large areas of pastureland held in common. The practice is not restricted to pastoralists, or those who specialise in livestock, and it takes a multitude of forms.

In the past, most African societies found it necessary to move their animals during the year, sometimes over considerable distances. Colonisation, private property and veterinary restrictions on movement made this increasingly difficult during the twentieth century. Yet the practice survives on a small scale in particular localities. This chapter explores the rationale and parameters of transhumance historically, and offers two examples of localised movement, in Mbotyi, Mpondoland, and QwaQwa. We suggest that transhumance in South Africa was shaped not least by the imperatives of animal nutrition and health, including the avoidance of specific animal diseases.

At the beginning of the nineteenth century, the German physician Hinrich (Henry) Lichtenstein enjoyed the idea that even he, as a scientist in a large, official touring party, was to some degree subject to the general restless movement of South Africa.

[1]   Sidwell Caine, Mbotyi, 22 March 2008.

109

We were indeed become perfect nomades, sharing the lot of most of the inhabitants of southern Africa, whom nature disposes, or compels, to stated changes of habitation. The colonists are driven by the snow from the mountains down to the Karoo; the Caffre [Xhosa] hordes forsake their valleys when food for their cattle begins to fail, and seek others where grass is more abundant; the Bosjeman is fixed to no single spot of his barren soil, but every night reposes his weary head in a different place from the former. The numerous flocks of light footed deer, the clouds of locusts, the immeasurable trains of wandering caterpillars, these, all instructed by nature, press forward from spot to spot.[2]

Lichtenstein, as in the case of many later authors, saw movement as 'instructed by nature'. Transhumance was often a response to seasonal ecological opportunities and constraints, as well as irregular climatic variation such as drought. Pastures and water are key resources for all ruminants. Wild animals, which move themselves, provide a pointer. The best known example of mass animal movement in southern Africa was the irregular springbok trek from Namaqualand and the southern Kalahari to the wetter eastern Karoo and sometimes in other directions.[3] Millions of animals would trek in some years – giving rise to spectacular descriptions. It was a complex phenomenon, affected both by population dynamics and environmental factors. By the late-nineteenth century this irruption was terminated by hunting, the spread of commercial pastoralism and fencing.

Livestock generally move as they eat and they retain instinctive traits around food and water procurement. If left to themselves, or when they go feral, they can cover significant distances in search of the best pastures. Under human control, livestock were moved in different seasons or from highlands to lowlands or from winter to summer rainfall zones. (In South Africa, it was only the Western Cape which experienced winter rainfall.) Animals were moved less predictably during droughts or to avoid diseases. The regularity of movement depended partly on ecological conditions and the specific terrain. Some environmental changes were seasonal and predictable while others were less so; livestock owners sometimes went 'trekking about after the rain' or 'chasing thunderstorms in the desert'.[4] Different animals could require different strategies. The seasonal dynamics of herd and flock reproduction were also important as ewes, for example, needed adequate food for lambing. Thus multiple local patterns of transhumance evolved. The distances travelled were sometimes, but

[2] M. H. K. Lichtenstein, *Travels in Southern Africa in the Years 1803, 1804, 1805 and 1806*, vol. 1 (Cape Town, Van Riebeeck Society, 1928, first published 1812, 1815), 414.

[3] Christopher Roche, '"Ornaments of the Desert": Springbok Treks in the Cape Colony, 1774–1908', unpublished MA dissertation, University of Cape Town (2004).

[4] R. W. Willcocks (ed.), *The Poor White, Report of the Carnegie Commission on The Poor White Problem in South Africa*, vol. 2 (Stellenbosch, 1932), 9; William Beinart, *The Rise of Conservation in South Africa: Settlers, Livestock and the Environment, 1770–1950* (Oxford, Oxford University Press, 2003), 54.

not always, related to the aridity of the land. Movements were likely to be over a longer distance in the drier western half of the country than the wetter eastern.

Under the conditions of livestock management prevalent in many areas up to the early decades of the nineteenth century, mobility of livestock was required for a more stable supply of food. Transhumance increased the number of animals that could be kept throughout the year because it reduced the number of deaths during the winter dry season when grazing was scarce. The practice was related also to social choices and constraints. To some degree, the scale of mobility was influenced, in different societies, by the extent to which they depended on their herds and flocks for food. Hence it tended to be more pronounced in societies which placed less emphasis on cultivation, such as the Khoisan and Trekboers. Amongst Afrikaners, it was associated with hunting and social freedom during particular periods of the year. There are echoes of such social practices up to the present.

Before discussing disease and transhumance in more detail, it is important to contextualise and elaborate on the changing causes of transhumance. One new factor in the nineteenth century was environmental degradation, such as the deterioration of veld, or drying up of water sources, which was often cited as one of the pressures behind the movement of animals. Second, transhumance was inseparable from power relations as settlers and some more dominant African societies stamped their authority over land, water, and transport corridors. Not only settlers, but African chiefdoms such as the Zulu, Sotho, Mpondo and Pedi were expanding their territories and their potential pastures in the early decades of the nineteenth century. Third, and as a corollary of this point about the political dimensions of transhumance, the management and movement of livestock were sometimes associated with conflict. Every war in the battle-scarred nineteenth century resulted in the death, capture, theft and redistribution of cattle. Fourth, and perhaps most important, transhumance was also intimately connected with trade, commodification and markets. Most settler livestock owners sold a portion of their animals annually, and increasingly Africans participated in the livestock trade. These had to be driven long distances to urban markets and the ports. Farmers sometimes offloaded livestock before the dry season thus using the market to diminish environmental risk. Ox-wagons were essential for the movement of goods. Long-distance transhumance was increasingly related to markets rather than environmental factors in the nineteenth century. The cattle drives in the late-nineteenth-century United States from Texas to the railheads, immortalised in such Westerns as *Red River*, and also those to the southern ports are analogous market-linked patterns of transhumance.[5] In sum, changing patterns of transhumance through the nineteenth century were

[5]   William Cronon, *Nature's Metropolis: Chicago and the Great West* (New York, Norton, 1991); for disease problems, J. F. Smithcors, *The American Veterinary Profession: its Background and Development* (Ames, Iowa State University Press, 1963); Howard Hawks and Arthur Rosson (directors), *Red River* (1948).

only partly related to ecology and disease and must be considered in a more complex analytical framework. Animal health, rainfall, household requirements, commercial considerations, markets, transport riding, conflict and theft avoidance could all be calculations in a particular pattern of movements.

Transhumance and disease

Transhumance, whatever its context, was closely tied up with animal nutrition and – as we have demonstrated – nutrition was imbricated with disease. Animal disease is quite often cited in sources as a factor in explaining the many different localised patterns of movement. Khoikhoi herders north of the Cape peninsula spent part of the year south of Saldanha Bay on the coast, where the grazing was rich. They did so to avoid *lamsiekte* (botulism); this resulted when cattle gnawed on decaying bones.[6] Soils and plants in parts of the interior were deficient in phosphorous, and cattle tried to get the required minerals through eating bone and carcasses. While neither Khoikhoi nor settlers knew the cause of this disease, it was associated with drought and deficiency, and transhumance to particular grazing resources was known to be an effective prophylaxis. Cattle owners in some coastal districts of the Eastern Cape were also threatened with *lamsiekte*. Jeff Peires records that Xhosa chiefs moved their livestock seasonally to avoid what colonial officials called 'stiff sickness' – probably botulism.[7] At that time, as more recently, the Xhosa blamed bad grass for illnesses.

Livestock owners in South Africa have, over a long period, distinguished between sourveld and sweetveld. These are not exact terms but described the general characteristics of the pasturelands. Sourveld, generally found in coastal districts and on some mountain ranges, was characterised by grasses that were highly palatable after the spring and summer rains, but became hard and inedible during the dry winter. Sourveld was usually a denser sward. Sweetveld tended to predominate in hotter inland districts and in the valleys. While the grasses such as *Themeda triandra* (*rooigras*) were more palatable throughout the year, it was seldom as dense and could not sustain continuous heavy grazing. African languages have analogous terms for these grass types. In isiXhosa sourveld is called *ijojo* and sweetveld *isandle*; in Zululand and Mpondoland

---

[6]   Andrew B. Smith, *Pastoralism in Africa: Origins and Development Ecology* (London, Hurst, 1992), 194–5; Nigel Penn, *The Forgotten Frontier: Colonist and Khoisan on the Cape's Northern Frontier in the 18th Century* (Cape Town, Double Storey Books, 2005); Beinart, *Rise of Conservation,* 128–57; Daniel Gilfoyle, 'A Swiss Veterinary Scientist in South Africa: Arnold Theiler and the Explication of "*Lamsiekte*" in Cattle' (unpublished paper presented at the conference on 'Imperial Culture in Countries without Colonies', University of Basel, 23–25 October 2003).

[7]   J. B. Peires, *The House of Phalo: A History of the Xhosa People in the Days of their Independence* (Johannesburg, Ravan Press, 1981), 9.

sourveld is called *ngongoni*. The latter was adopted as the common English name for Transkeian coastal sourveld, and for the dominant species, *Aristida junciformis*, found in it. A classic pattern of transhumance, therefore, was to let livestock graze on sourveld during the spring and summer, moving them to sweetveld during the late summer and winter months. There were, however, very many local practices.

Basil Sansom, an anthropologist, examined in detail the ecological constraints shaping the differences in the settlement and land-use practices of African societies on the east coast and in the interior of South Africa.[8] On the eastern seaboard, where water was widely available from a multitude of perennial rivers and streams, settlement tended to be dispersed and a variety of pastures were generally available close to the homesteads. In the drier, flatter interior, with fewer rivers, especially in Tswana chiefdoms, settlement was concentrated. It was dangerous to keep large numbers of livestock close by and cattle posts were essential. Tswana notables also sent cattle with clients and young men to exploit the Kalahari pastures after rainfall opened brief windows of opportunity on the desert margins. Cattle diseases also played a role in encouraging these strategies. For example, anthrax and *lamsiekte* were problems for the southern Tswana settlement at Dithakong (Kuruman) in the early-nineteenth century when too many livestock were concentrated around the main settlement; the cattle- posts were a means of averting sickness.[9]

Even on the wetter eastern side of the Drakensberg, African communities moved livestock. Mpondo chiefs kept Lambasi, a high-rainfall coastal plain, free of habitation as a seasonal winter grazing ground until the early decades of the twentieth century.[10] The Sotho and Hlubi communities of the northern Transkei moved their animals between the sourveld highlands, where ticks were less prevalent, to lower ground around the valleys and settlements in the cold, dry winter.[11] Zulu livestock owners on the borders of the tsetse belt moved cattle from the lowveld before the summer rainy season to higher land in order to avoid trypanosomosis (*nagana*, akin to human sleeping sickness).[12] Seasonal migration to avoid tsetse was commonplace in Africa. Annual movement from the lowveld to the highveld became a central strategy in settler livestock management in the Eastern Transvaal (Mpumalanga), which persisted into the

---

[8]   Basil Sansom, 'Traditional economic systems' in W. D. Hammond-Tooke (ed.), *The Bantu-speaking Peoples of Southern Africa* (London, Routledge and Kegan Paul, 1974).

[9]   Nancy J. Jacobs, *Environment, Power, and Injustice: A South African History* (Cambridge, Cambridge University Press, 2003), 44, 62–63.

[10]   William Beinart, *The Political Economy of Pondoland, 1860 to 1930* (Cambridge, Cambridge University Press, 1982).

[11]   W. Beinart and C. Bundy (eds), *Hidden Struggles in Rural South Africa* (London, James Currey, 1987), 203.

[12]   John Ford, *The Role of the Trypanosomiasis in African Ecology: A Study of the Tsetse Fly Problem* (Oxford, Clarendon Press, 1973), 481.

1950s. Livestock were also moved in emergencies to avoid infectious diseases during epizootics such as lungsickness in the 1850s and 1860s.[13]

Records of transhumance up to the early-twentieth century suggest that some patterns were intimately related to animal nutrition and health in general and specific diseases in particular. They could be shaped by deficiencies in the grazing, by poisonous plants, or insect-borne diseases. Livestock owners did not necessarily have a clear knowledge of causes or even sometimes the specific character of the diseases they were avoiding, but many learnt by practice and observation. Environmental interpretations of animal disease (chapter 3) could reinforce patterns of transhumance and avoidance. Colonists drew some of this knowledge from Africans.

The demise of transhumance: processes and arguments

Transhumance in its older forms contributed to providing a food supply and managing certain diseases, in the absence of investment into fodder, water supplies and fencing. By the second half of the nineteenth century, however, livestock movements of all kinds were increasingly perceived to facilitate the spread of disease. The Cape authorities and some farmers thought that the epizootics of horsesickness in the 1850s, as well as lungsickness transmitted by imported cattle, were caused in part by animal movements. Lungsickness remained endemic. The new veterinary specialists working for colonial and Union governments saw the shattering epizootics of redwater in the 1870s and 1880s, rinderpest (1896–97) and East Coast fever (1902–13) in a similar light.

Veterinary experts, often supported by wealthier farmers, became increasingly committed to regulating transhumance and the long-distance wagon trade in order to control disease. It is an irony that a major argument against transhumance was also related to animal health. As the state acquired the power to enforce veterinary controls, so scientific prescriptions gained authority and were gradually imposed. It would be wrong to see veterinary measures as the only influence in curtailing transhumance. We should be aware of the multiplicity of processes and causes. For much of the nineteenth century, British missionaries and officials, as well as elite Afrikaners, railed against the social costs of trekking for blacks and whites. They wanted a stable population or, as the surveyor general noted in 1860, 'enterprising landowners, the makers of dams, the builders of houses, and cultivators of soil'.[14] A more settled agriculture was linked to ideas of civilisation, control, progress, commerce, improvement, invest-

[13] Christian B. Andreas, 'The Spread and Impact of the Lungsickness Epizootic of 1853–57 in the Cape Colony and the Xhosa Chiefdoms', *South African Historical Journal*, 53, 1 (2005), 50–72.

[14] Cape of Good Hope, *Copy of Correspondence with the Divisional Councils of Certain Divisions in which "Trekvelden" Exist on the Subject of the Future Disposal of Those Lands* (G.30–1860), 1, Charles Bell.

ment and education.[15] The state and the churches increasingly associated the old pastoral practices with poverty and social retrogression. Matters came to a head in the 1890s when the Cape government tried to introduce compulsory controls over movement, together with dipping, as a means of eliminating scab in sheep. Both frontier Afrikaners and some African communities protested vehemently against these regulations that curtailed their old modes of trekking.[16] They talked of scab as a disease of poverty amongst sheep, a term they related to nutrition. The logic of their position was that transhumance was good for the prevention of scab because it minimised the impact of drought, malnutrition and poverty amongst animals. There are echoes of this approach in our interviews.

In the late-nineteenth and early-twentieth centuries, privatisation of land, the sale of public or crown lands, as well as spatial definition of African reserves stopped some transhumant flows. Fencing with barbed wire cut off routes that livestock owners had formerly used. Equally important, springs were privatised in the process. Officials believed investment in water provision would obviate the need to migrate – which in turn militated against investment.[17] As a senior vet commented in 1928, 'the time of trekking with livestock, especially on a large scale, is past. Most of the land is now densely inhabited, and has so many fenced farms, that trekking with cattle has become impracticable.'[18]

To this social and economic critique of transhumance, botanists and progressive farmers added concerns about the destructive impact of tramping on vegetation and soil.[19] Transhumance was associated with gulleys and soil erosion. It was also judged bad for the animals, weakening them and reducing their weight. Veterinarians argued that transhumance contributed to selective grazing, destroying the most nutritious vegetation, which provided prophylaxis against disease and worms. If animals were kept in one place and allowed to graze freely night and day in a fenced enclosure, they would eat a wider range of species more evenly. This argument echoed down the years and became one of the key justifications for massive interventions in the African rural areas through betterment policies.

It is difficult to periodise the ending of transhumance. There were countervailing processes even in the early decades of the twentieth century, and white

---

[15] D. Nell, '"You can Make the People Scientific by Act of Parliament": Farmers, the State and Livestock Enumeration in the North Western Cape, c.1850–1900', MA dissertation, University of Cape Town (1998).

[16] Mordechai Tamarkin, *Volk and Flock: Ecology, Identity and Politics among Cape Afrikaners in the Late Nineteenth Century* (Pretoria, University of South Africa Press, 2009).

[17] Cape of Good Hope, *Blue Book, 1859*, JJ3–4, Report of the Civil Commissioner, Calvinia.

[18] National Archives Pretoria, Department of Agriculture, Box 2327, file 3435 i, P. R. Viljoen, Deputy Director of Veterinary Services to Secretary of Agriculture, 7.3.1928 – 'die tyd van trek met vee, veral op groot skaal, verby is. Meeste dele van die land is nou dig bewoon, bevat soveel omheinde plase, ens. dat trek met vee onpraktiese geword het'.

[19] Beinart, *Rise of Conservation*, 332–66.

farmers in some districts found ways of moving large numbers of livestock to distant pastures, especially during major droughts. Rail transport provided a route for modernising this old practice in the 1930s. Cattle posts remained a feature of Tswana communities on communal land. But the social features of transhumance amongst the great majority of white and black livestock owners had largely ebbed by mid-twentieth century. The state, the much enhanced veterinary services, and an increasing majority of white livestock owners saw stabilisation and dipping as far more effective than transhumance for controlling disease. Where vaccines were developed, inoculation offered a further defence against animal plagues.

Transhumance and grazing in Mbotyi

In localised areas, such as our research sites at Mbotyi in Mpondoland and QwaQwa in the Free State, forms of transhumance have survived. Within the former homelands, the grazing commonages did not fall into private hands and attempts by the government to introduce fencing during betterment and rehabilitation programmes largely failed over the longer term. Few fences were maintained, many were cut to free up walking, grazing and transport routes and some were removed for other uses, such as homestead boundaries. While the pastures remained as communal land, however, this did not imply that animals could be moved freely over significant distances, because each administrative area, under a headman or chief, controlled their own grazing lands. Since 1994 transhumance may have increased. Controls over movement and state supervision of pastoral activities have declined. Some African communities have acquired access to expanded pastures as a result of successful land claims, government-assisted transfers and purchases. Furthermore in some places the boundaries of farms which had once been conterminous with the former homeland frontiers have become eroded enabling livestock to roam over larger areas.

Mbotyi itself was never subject to a government rehabilitation scheme and the pattern of settlement has evolved in response to local priorities. The homesteads have increased greatly in number over the last few decades. Settlements still snake along the ridges out of the village or penetrate into the forest margins – evidence of the old scattered settlement pattern. But the population has become denser at the heart of Mbotyi, so that the homesteads effectively form an unplanned village. As Mbotyi lies on the coast, surrounded by forest and in the heart of an old area of reserved land, livestock owners have not been able to extend their range. Blessed with a high rainfall and abundant grass, they do not generally have a problem in finding sufficient pastures and water.

Livestock owners at Mbotyi have been practising a form of localised transhumance for as long as anyone in the village remembers. The main grazing ground is Lambasi, a large area of coastal grassland, with very little settlement, to the north-east of the village, and particularly sections of it called Lubala. The Lambasi pastureland differs in its soils and topography from that around Mbotyi village itself. Around the settlement, patches of open grassland are interspersed with forest which reaches right down to the coastal lagoon and even to the beach at points. But beyond the Mvekane stream, about 5km north-east of the village, at the end of what is now called Shelley Beach, the land slopes sharply upwards. The forest stops abruptly on the north-eastern side of this valley giving way to the Lambasi grassland, which stretches to the Msikaba River, about 25km away. Depending on its definition, Lambasi is around 150–200 sq km, or 15,000–20,000 ha. This coastline is well known because it is one of the few parts of South Africa where rock cliffs soar directly up from the sea and a waterfall plunges down off the plain into the ocean.

In pre-colonial and early colonial days, Lambasi was used by the paramount chief of Eastern Mpondoland as a winter grazing ground. The Great Place at Qaukeni is located 30–40 km inland on substantially higher ground where winter temperatures drop more sharply than on the coast. In the winter, the grass in the interior hardens and is less palatable. But the grazing on the wet coastal strip remains relatively fresh throughout the year. At that time, Lambasi was deliberately kept free of settlement. It is also not very suitable for the mixed agriculture that characterised Mpondo subsistence patterns. Much of the land is marshy, cut through with small streams; the soil is shallow and not generally good for cultivation.

Lambasi no longer serves primarily as a grazing ground for livestock from the interior of Mpondoland. This pattern of transhumance was stopped during the period when controls over livestock movement were enforced, and has not re-emerged. Since the 1930s, settlements have spread on the peripheries of the pastureland, such as Cutwini and Lambasi villages, and to a lesser extent within the area itself. The government made some attempt to stop these during the 1940s and 1950s but has been less interventionist since then. Such state intervention was one element in popular discontent that erupted in the Mpondo revolt of 1960.[20] Livestock owners in new villages that have sprung up around this lush pasture run cattle there all year round. Since the early-twentieth century, and possibly on a small scale before, Lambasi has been used as a summer, rather than winter, grazing ground by those in Mbotyi. During the summers there may be over 10,000 cattle on the Lambasi plain.

Our oldest informant in Mbotyi, Zipoyile Mangqukela, was born there around 1918.[21] This is at least roughly confirmed by the fact that he clearly remembered the visitation of locusts (inkhumbi), which was in 1933, and said he

20 William Beinart, 'Environmental Origins of the Pondoland Revolt' in Stephen Dovers, Ruth Edgecome and Bill Guest (eds), *South Africa's Environmental History: Cases and Comparisons* (Cape Town, David Philip, 2002), 76–89.
21 Zipoyile Mangqukela, Mbotyi, 26 March 2008.

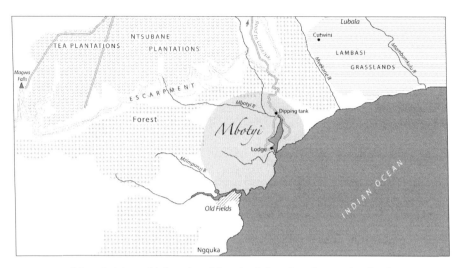

Map 5 Map of Mbotyi and Lambasi showing the grazing lands

was a grown boy at the time.[22] As a herdboy in the 1920s, he drove the cattle to Lambasi, staying with them there through the summer: 'we were not schooling but herding cattle.' They had a shelter, and received supplies of maize and food from home. They also milked the cows and picked wild fruit such as the wild banana and the *numnum* bush (*Carissa bispinosa*). *Numnum* is widespread in the coastal areas and has a red edible fruit, the size of a small plum. The Lubala section of Lambasi was located close enough, within a day's walk, for them to return periodically for supplies and they had to bring the cattle back to Mbotyi for dipping. The seasonal pattern of transhumance was an important element in cultural practice.

Mangqukela recalled that the livestock were so used to going to Lambasi that they would go there themselves. They hardly needed driving at the right time of year: 'you just show them the path.' Sometimes it was difficult to bring them back for dipping, because they wanted to stay there. After dipping, they were keen to return: 'you just dipped them and left them at the tank and they walked straight back to Lambasi.' A number of informants believed that livestock, left to themselves, were adept at finding nutritious pastures.

At the end of the summer, around April, when the rains were largely over, the cattle were fetched back to the village. 'When the colder winds started to blow, the cattle came back themselves', Mangqukela reminisced, 'they got the wind in their nostrils, smelt the aroma of mealies in the air and returned to graze on the mealies.'[23] After the harvest, the cattle were let into the maize fields where they consumed the stalks and dry leaves, while also fertilising the soil with their dung. Throughout South Africa, maize stubble has been a valuable

[22] Beinart, 'Environmental Origins of the Pondoland Revolt'.
[23] Zipoyile Mangqukela, Mbotyi, 26 March 2008.

dry-season winter fodder amongst white and black livestock owners. In September/October, at the beginning of the ploughing season, when the heavy rains came, and when cattle were needed at home for ploughing, the oxen had to be herded because 'when they have finished the mealies and grass here, they hide and go back by themselves'. When ploughing was complete, most of the animals were moved back to Lambasi. Not everyone sent all their livestock away. In earlier years it was the practice to keep some milking cattle at the homesteads throughout the summer. During the first periods of fieldwork in 2008–9, the local sub-headman had about 20 head of cattle at his homestead out of a herd of roughly 90 – probably the largest in the village.

There are a number of reasons for this pattern of transhumance, partly but not wholly related to nutrition and disease. The most commonly cited was the problem of keeping animals off the cultivated fields. Zipoyile Mangqukela associated taking the cattle to Lambasi in his youth, in the 1920s and 1930s, with the planting season. If they were kept at home, livestock would devour growing crops unless they were carefully herded. In those days sufficient labour was generally available for herding. As more boys and youths have gone to school in recent decades, this pattern of transhumance was, if anything, reinforced because it was labour-saving. By the time of this research, livestock were not generally herded at Lambasi on a daily basis – they roamed freely. If the animals were based in the village and kraaled nightly, they had to be watched throughout the day so that they did not damage the crops and gardens.

The spatial layout of the village has also changed in the last few decades. The bulk of maize used to be grown in alluvial fields in the valleys (see below), or on unsettled slopes, rather than next to the homesteads. There was a clearer spatial separation between fields and homesteads with their adjacent small gardens. Cultivation in fields has declined and very largely stopped in recent years and many people have grown the bulk of their maize crop and other food plants in smaller gardens immediately next to their homesteads. Thus the homesteads are not only closer together but the gardens used for cultivation are now interspersed between them and by no means all adequately fenced. If animals are kept in the village during the summer growing season, herding is essential as livestock owners are liable for damages if their animals eat crops and vegetables. Keeping animals away from gardens by moving them out of the village for the summer is one solution to the problem.

We heard of a few arrangements in which livestock owners employed herders. Myalezwa Matwana placed his 12 cattle with a man who lived adjacent to Lambasi and paid a token R100 a month in 2008 for them to be looked after.[24] Two younger men, with good alternative sources of income, employed an older man to herd their 15 cattle daily, paying about R700 a month plus

[24] Myalezwa Matwana, Mbotyi, 22 and 24 March 2008.
[25] Discussion, Mbotyi, March 2012. They requested that this information should not be directly attributed.

accommodation and food in 2012.[25] In cash and kind, this may be roughly equivalent to the pension of R1,140 a month (to which it was added by their employee). It was below the national agricultural minimum wage of about R1,500 a month. But local livestock owners were generally very reluctant to pay for herding at all and would not pay the minimum wage in cash. As it is difficult to find adequate unpaid family labour, older livestock owners often had to watch their animals themselves when kept in the village. In North West Province, older, widowed women as well as men took this responsibility. In Mbotyi it was rarer to see girls or women herding cattle, although some of the women interviewed did recall herding in their youth.[26]

In sum, family labour is limited because boys are at school and youths reluctant to work without payment. 'The youths', one man said, 'are diving and fishing and they get enough money from selling crayfish, so they don't want to work at herding.'[27] Owners have not generally made the transition to wage labour and complain that even if they can find herders, they are unreliable. They don't feel that they generate enough income from their livestock to merit this, and are not prepared to sell more to pay for labour and medicines.[28] Labour shortage in a context of high unemployment is a complex phenomenon, but it tends to reinforce the movement of animals to grazing grounds away from the village.

However, there were and are other important reasons for transhumance. Although the grassland is rich at Mbotyi, the area for grazing around the village is relatively restricted by forests and, of course, the sea. Sikhumba Malelwa said that if they all kept their livestock at Mbotyi in the summer they would exhaust the grass in a few weeks.[29] Sidwell Caine (who owned 23 head of cattle and 2 horses) mentioned: 'if they keep cattle right through the year here, they will die. We have to put them in the kraal every night and the mud is killing the cattle.'[30] He saw kraaling during the rainy summers as unhealthy. Many people talked about mud causing a disease called *umkhondo*, which was manifest in a lack of condition, but also referred to supernatural affliction (chapter 7).

Moreover, the grass types at Mbotyi and Lambasi differ. At Lambasi, *Nkonkone (ngongoni/Aristida junciformis)* is dominant, although in some places the grassland is mixed and even includes the sweetveld *Themeda triandra (rooigras)*. As in the case of coastal grasslands at neighbouring Mkambati, where Thembela Kepe researched, sections of Lambasi were burnt regularly, mostly by people from surrounding settlements rather than those in Mbotyi.[31] Veld burning is a long-established practice designed to clear the old dry grass and encourage fresh

---

[26] Mamcingelwa Mtwana, Mbotyi, 13 December 2011.

[27] Simphiwe Yaphi, Mbotyi, 13 March 2013.

[28] Tata (Alfred) Banjela, Mbotyi, 20 December 2011.

[29] Sikumba Malelwa, Mbotyi, 28 March 2008.

[30] Sidwell Caine, Mbotyi, 22 and 23 March 2008.

[31] Thembela Kepe, 'Grasslands Ablaze: Vegetation Burning by Rural People in Pondoland, South Africa,' *South African Geographical Journal*, 87, 1 (2005), 10–17.

new growth. For much of the year the pastures at Lambasi form a dense low grass sward, kept short by burning and heavy grazing and kept green by the high rainfall. This not only provided what stockowners viewed as excellent summer grazing, but also grasslands that appeared freer from ticks.[32] By contrast, informants saw the pasture in Mbotyi as more varied and to include a wider range of grass types, including thatch grasses, judged to be less nutritious. It is difficult to burn because of the number of homesteads and some areas were left to grow on the peripheries of settlement. The absence of cattle and fire for this period protected supplies of thatch grass. Pastures were usually burnt in and around the village in August and September, towards the end of winter, when the period of grazing in the village was ending, the grass had become hard and the rains were about to arrive. Transhumance to Lambasi from Mbotyi enabled owners to keep more livestock than they could if they used only their local area. They felt that cattle benefited from the richness of grazing there in the summer and its relative freedom from ticks: 'the oxen', an informant said, 'came back fat from Lambasi.'[33] Enough people were still pursuing this option for the droving routes along the coast to Lambasi to be a series of muddy furrows in the coastal grassland, most less than 50 metres from the sand and rocks.

Informants suggested that longer grass in the summer immediately around the village harboured more ticks than the short sward at Lambasi. It is also likely that the livestock grazing freely at Lambasi, where they are not kraaled every night, were less susceptible to worms and infection. Stockowners recognised the dangers of worms (izilo or iskelem), but did not associate them directly with muddy kraals full of dung, or with nightly kraaling. However, informants mentioned two diseases related to grazing in Lambasi: nonkwanyana (stiffness) and uqhonqa (when a cow with a calf becomes hunched up and the backbone appears painful).[34] We are uncertain about the nature of these ailments although nonkwanyana, which is apparently seldom fatal, and tends to occur towards the end of summer around April, may be three-day stiff sickness. It is possible that both diseases, neither of which posed major problems, were associated with continuous grazing on sourveld at Lambasi.

This pattern of transhumance was threatened, in the minds of the cattle owners, by what they perceived as a rising rate of livestock theft in the rural areas. Lambasi was seen as particularly attractive to cattle thieves due to the large concentration of inadequately herded animals. Such complaints were also a major factor in our interviews in QwaQwa. Sidwell Caine decided not to send his cattle to Lambasi in the few years before 2008 because of the problem of theft.[35] Instead he kept some in the village and sent others a few kilometres

---

[32] Myalezwa Matwana, Mbotyi, 22 March 2008, 24 March 2008; Sidwell Caine, Mbotyi 22 and 23 March 2008.

[33] Zipoyile Mangqukela, Mbotyi, 26 March 2008.

[34] Nongede Mkhanywa, Mbotyi, 23 and 29 February 2008.

[35] Sidwell Caine, Mbotyi, 22 and 23 March 2008.

south of the village to the coastal area between Mbotyi and Manteku at Ngquka – an area named after the local stream. It used to be more heavily settled but some families have left because of its remoteness. This isolation made it less accessible to stock thieves and the area has good grazing, with forest and beach nearby. Caine generally left his cattle there without herders and noted that they had a reasonably predictable pattern of grazing.

> Early in the morning, the cattle graze in the forest, then on the pastures. Around lunch they come into the sun to rest. Then they go down to the beach there to get away from the flies that annoy them. They like to go onto the beach and stand in the waves up to their knees as there are no flies there. Horse flies are dangerous to the cattle and there are many in the forests as well as on the pastures at night. If you want to find your cattle, you just go the beach in the afternoon and find them there.[36]

It is common to see livestock on the sands at Mbotyi and neighbouring beaches.

However, in 2008 Caine had to bring his livestock back to Mbotyi in late March, which was earlier than he would have liked, because of disputes and problems at Ngquka. This was not an established area for regular transhumance from Mbotyi and rights of access were therefore contested. In April 2008 he was herding the cattle himself in Mbotyi village. Other villagers also switched grazing grounds from Lambasi to the north-east of the village to Ngquka to the south-west and by 2012 issues of access appeared to be resolved. Mayikalisa Jikijela, one of the most successful women owners, with 23 cattle and 18 goats, left her animals there for most of the year because the pastures were rich and they did not need regular herding.[37]

Richard Msezwa lived inland about seven kilometres outside of Mbotyi village, in a cluster of three rather isolated homesteads at a place called Makwane, tucked under the escarpment and near the bottom of the famous Magwa falls.[38] He had 16 head of cattle and 14 goats in 2008 and he was able to graze them around his homestead and on the fringes of the local forest, throughout the year. The grasses are varied there, and settlement has not increased because of its isolation; people did not like to walk all to the way to Mbotyi for transport, shops and facilities. He said that a few Mbotyi people were now also sending livestock towards his homestead because of theft at Lambasi. However, he felt that his area was not as free of ticks as Lambasi. Msezwa experienced particular problems with bont ticks and reckoned he had lost nine cattle to the *qologqibe* in the few years prior to 2008.

Nonjulumbha Javu, in his sixties, who lived in Mbotyi village along the Nyambala stream, owned 45 head of cattle as well as a horse in 2009; by 2012

---

[36] Sidwell Caine, Mbotyi, 22 and 23 March 2008.

[37] Mayikalisa Jikijela, 12 December 2011.

[38] Richard Msezwa, Mbotyi, 25 March 2008.

Photo 4.1 Cattle on the beach at Mbotyi (David Hemson, 2009)

these were reduced to about 38 because he found it necessary to meet some large expenses through sale.[39] He was one of the most successful and careful managers of livestock although he lived in a very simple homestead, with thatched, mud rondavels. Javu used to send animals to Lambasi but said that livestock were stolen one year by *amakwerekwere* (foreigners). He grazed some of his cattle on the valley slopes immediately above the river, and he took others a few kilometres directly inland, where there is a mosaic of forest patches and grassland. He used to live in this area, about half way to the escarpment, before he moved down to Mbotyi, and knew the terrain well. Nonjulumbha has a marked limp, the result of being shot by a ranger when hunting illegally in the forests as a youth. Despite this disability, he walked frequently to check his animals. We went with him one day to a spot, about four kilometres away, without homesteads on the hills overlooking Mbotyi. This patch of grassland was adjacent to the forest, and difficult to reach except on foot, where few other owners took their animals. He felt it was safe to leave some of his herd there overnight.

In our visit in March 2012, when we visited a number of homesteads of people that we had interviewed in earlier visits, it seemed that more people were keeping their livestock in the village for longer periods than in previous years. Some of the most substantial and successful owners, for example Sidwell Caine (about 23 cattle), Alfred Banjela (28), Cecil Malindi (34) and Nonju-lumbha Javu (38), all spent a good deal of their time watching livestock while we were visiting the village. They were all older men who were no longer

[39] Nonjulumbha Javu, Mbotyi, 11 February 2009, 19 December 2011, 22 March 2012.

working for wages and who could occasionally call on family members to help with herding. Banjela, a key figure in the dipping committee, said that sending his animals to Lambasi without herders was 'like throwing away cattle'.[40] This was also very likely one reason why their livestock were in better condition than most. They kraaled some or all of their animals nightly, or visited them regularly, kept a close eye on them, and treated them more frequently than owners who left their animals out in the grazing lands for long periods. Though these herds were not free of ticks, they were dipped and sprayed more effectively than most. If they had additional problems with worms, because of frequent kraaling, they treated these with Valbazen (albendazole) or similar biomedicines.

By contrast, some large herds, such as the former sub-headman's, were still kept at Lambasi, where they were herded.[41] So were some smaller herds. Simphiwe Yaphi continued to keep his five cattle at Lubala, on the Lambasi plain, for most of the year. He worked full time at the Lodge, had young children, and thus no time to look after them on a regular basis. Generally he could find them easily, because they tended to congregate on a hill above the Mlambomkulu stream and did not stray far or go down to the coast. But, he continued:

> They are wild. It is difficult to get close to them.... I have to whistle, and call them by their names. Although they are wild, they know their names. But you have to run to catch them and round them up. The old men find it difficult to bring them back for dipping and spraying. I helped one old man do this. When I catch them, I can't use Deadline [flumethrin] because they keep moving and it is difficult to spray them.[42]

So there are other costs in leaving animals unattended at Lambasi.

Goats are also sometimes sent out of the village to the grazing grounds, but a higher proportion are kraaled nightly in the village and owners tried to train them to go into the forests or down to the sea where they were further away from gardens. Informants perceived that goat numbers had declined more sharply than cattle. Disease (particularly heartwater) and lameness from ticks may have been major factors here. In 2012 there were losses of goats at Lambasi from a disease that made goats shiver, possibly gallsickness or heartwater. A herd kept in the village suffered a very high rate of late-term abortions. Those who had more goats than cattle seemed (from the evidence of their homesteads) to be poorer and we heard of no large herds of over 30. One exception was a younger man in his thirties who was breeding goats commercially to sell them for sacrifices.

In sum, transhumance had long been practised in Mbotyi and there was a regular seasonal movement of animals to Lambasi by the bulk of livestock

[40] Tata (Alfred) Banjela, Mbotyi, 20 December 2011.
[41] In 2011, the sub-headman, who suffered a stroke, had been replaced by his daughter.
[42] Simphiwe Yaphi, Mbotyi, 13 March 2013.

owners. It was seen as essential for adequate pastures and for animal health. Young boys would be sent with the herds to look after them. While it was possible for some owners to keep some livestock around the village, it was difficult for all Mbotyi owners to do so through the summer months. But the pattern was changing in the early twenty-first century, and becoming more individualised; Lambasi was perceived as more risky and some owners preferred to move animals in other directions. A few of the most successful owners tried to keep their animals in the village and take them out daily, when necessary, to access fresh grazing. Erosion of the old-established pattern of transhumance may not be ideal for animal health, and this is compounded by declining fodder resources during winter (see below). This may be one factor in the decline of milk supplies. As one man put it: 'people are scared of milking their cattle because they think that the calves are dying, all of them, for lack of milk.'[43] Tick infestation and also possibly nutritional problems seemed to have reduced the supply of milk. However, wealthier owners who now keep their livestock based in the village for longer periods and manage them carefully appear to have healthier herds because of the attention and treatment they can give.

The older pattern of transhumance was also affected by the decline of small-holder arable agriculture in the Eastern Cape. In some parts of the former Ciskei, many fields and even gardens have been abandoned for the last couple of decades.[44] Maura Andrew, an anthropologist, noted in the 1990s that in southern Transkeian districts with high rainfall, families were abandoning their main fields and cultivating maize in gardens next to their homesteads.[45] In Mbotyi, cultivation of the main fields seems to have continued till about 2005. These are located in a rich alluvial valley of about 50 ha along the Mzimpunzi stream, immediately to the south of the main village settlement, and many of the old-established village families have rights to fields there. The number of people using it declined in the early-twenty-first century. During a visit in 2009, we saw only one field successfully cultivated, amounting to about 2 ha. Some cannabis bushes were inter-planted with the maize and pumpkins, which probably made cultivation worthwhile. In 2011, even this field was fallow and deserted.

Older villagers in Mbotyi saw the main reasons as being the pension, and other social payments. Nongede Mkhanywa felt that people neglected cultivation because they could afford to buy maize and other food.[46] Agricultural labour did not give a sufficient return. Young people were impatient with agri-

[43] Nongede Mkhanywa, Mbotyi, 23 February 2008.

[44] Paul Hebinck and Peter C. Lent, *Livelihoods and Landscapes: The People of Guquka and Koloni and their Resources* (Leiden, Brill, 2007).

[45] Maura Andrew, 'A Geographical Study of Agricultural Change since the 1930s in Shixini Location, Gatyana (Willowvale) District, Transkei', unpublished MA thesis, Rhodes University (1992).

[46] Nongede Mkhanywa, Mbotyi, 23 and 29 February 2008.

cultural work, and the decline in patriarchal relationships meant that older men could no longer command the labour of women and youths. As in the case of herding, explaining the demise of field cultivation and the shortage of labour in a context of high unemployment is complex. A similar situation pertained in North West Province.

Another important factor in the abandonment of fields, which is unique to Mbotyi and some other parts of coastal Mpondoland, were bush pigs. The forests are no longer rich in wildlife. But bush pigs seemed to be thriving and expanding in numbers.[47] Hunting used to be a popular pastime amongst men in Mbotyi who would go illegally into the protected forests at night with dogs. Ironically, although the forests are no longer policed as tightly as they used to be, young men are less keen on the difficult task of catching bush pigs. The latter had clearly adapted to living partly on the forest margins, feeding off gardens and fields. The decline of hunting may not be the only factor in the expansion of bush pig numbers; similar trends were reported from other parts of the eastern seaboard. But as the number of fields cultivated at Mzimpunzi declined, the damage to the remaining maize crops became intolerable, and this was offered as a key explanation of why the remaining cultivators gave up.

The decline in cultivation removed substantial amounts of winter fodder from maize stalks. In this way also the older form of transhumance was being modified. But it still survived to a degree and people continued to take their livestock to Lambasi and surrounding grasslands. Livestock owners had a detailed knowledge of the local environment and were adapting their strategies to changing local contexts, shaped by labour availability, environmental change, and social problems such as stock theft.

Trekking between mountains and plains in QwaQwa

Whereas Mbotyi livestock owners moved animals between contiguous areas of different coastal grasslands, those in QwaQwa moved theirs from lower-lying villages to mountainous areas in the direction of Lesotho and KwaZulu-Natal. Two reports, one from 1979 and one from 1991, described seasonal patterns of transhumance.[48] In the summer the livestock grazed on predominantly sourveld mountain grasslands. In the winter, when the veld dried out, the animals returned to the warmer sweetveld valleys, which retained some nutritional

---

[47] Sikumba Malelwa, Mbotyi, 28 March 2008; George Mkhanywa, Mbotyi, 11 December, 2011.

[48] J. J. Human, D. P. J. Opperman and C. S. Blignaut, 'Agricultural Potential in Qwaqwa', paper delivered at a conference on the Planning and Development of Qwaqwa, September 1979 (Bloemfontein, Institute of Social and Economic Research, University of the Orange Free State, 1980), 93–106; T. J. Bembridge and J. L. H. Williams, 'An Evaluation of the Qwaqwa Extension Service', Agricultural and Rural Development Research Institute (ARDRI), University of Fort Hare Report no. 2/91, February 1991 (Department of Agriculture library), 18–36.

value. As in Mbotyi, the change in pastures enabled livestock owners to keep more animals than they could if they restricted grazing to a single zone. QwaQwa's 1975 livestock census suggested that there were about 13,000 each of cattle and goats, just under 6,000 sheep and 2,000 equids.[49] The 1991 report showed that livestock populations had risen to about 15,000 each for cattle and goats and 7,000 for sheep.[50]

In their 1991 report, Bembridge and Williams were convinced that the veld was overgrazed and that animals were destroying the fragile mountain ecologies. By the late 1970s, if not before, weed encroachment had become a problem. Toxic species like the *bitterbossie* (*Chrysocoma tenuifolia*) colonised the veld. The *bitterbossie* causes *kaalsiekte* in small stock and the animals lose their pelage. In 2010–11, however, local stockowners were less concerned about overgrazing and did not discuss the problem of *kaalsiekte*. Informants perceived that pastures were being reduced by settlement, rather than degraded by overgrazing and undesirable weeds. They thought that rising human populations reduced access to nutritious vegetation and made it harder for stockowners to accumulate larger flocks and herds.

Sotho men in their 50s and 60s, who reminisced about herding in their youth, invoked positive and idealised memories. In discussing grazing strategies, some corroborated descriptions of transhumant patterns between the mountains and the plains which featured in the official reports. They associated herding with camaraderie between youths and a sense of freedom away from adult scrutiny. Some herders combined husbandry with schooling. Milking the cows and taking them to and from nearby pastures fitted in to the school day. Seasonal herding could also, to some degree, be contained within the school year. Herdboys took the animals to the mountains in the summer when schools broke up. In the village of Lejwaneng, Chadiwick Mbongo spoke nostalgically about the joys of herding and the annual trek to the mountains. He explained that the herdboys operated in groups of 12 to 20 and,

> we would pick valleys high in the mountains which were hard to access as there was only one viable entry point to the grazing lands. This made the area easier to defend against jackals and thieves and also enabled us to control the animals by erecting a makeshift fence across the entrance of the mountain valley.

As a child Mbongo was in charge of about 50 cattle, sheep and goats, and he drove them up to the mountains on horseback. The animals slept overnight in the open, whilst the herdboys lived in caves. This practice occurred in other parts of South Africa where villages were contiguous to mountains, such as Herschel district in the Eastern Cape. In the deepest caves Mbongo and the other herdboys constructed sleeping huts. The main source of food was *lahala*

---

[49] Human, et al., 'Agricultural Potential in Qwaqwa', 93–106.

[50] T. J. Bembridge & J. L. H. Williams, 'An Evaluation of the Qwaqwa Extension Service, 18–36.

made from boiled milk and maize meal. They also had occasional access to meat if an animal died. Otherwise they hunted small wild mammals for protein. 'Sunday was the best day as it was hunting day and we would group together to chase small antelope and hares.' For most of the week though, they herded and played – stick fighting being particularly popular. They milked the cows daily and sent the produce down to the valleys from time to time. Younger children from Lejwaneng were responsible for collecting it.[51]

Daniel Khonkhe from Ha-Sethunya used to drive the cattle to a mountainous area with caves in KwaZulu-Natal each summer for two months at a time. As in Mbotyi, this enabled villagers to keep the animals out of the planted fields in the lowlands. Again herders lived in the caves in groups, and these were big enough to confine their animals at night. Sometimes, when the grasses were poor, they brought mealie husks up the mountains to supplement their diet. Here too, he recalled that the herd boys spent time hunting. Khonkhe recounted how 'we had to deal with lots of jackals which preyed on small stock and calves and we kept vicious dogs to chase away the jackals and to hunt. We trained the dogs to attack and catch jackals, which we killed by clubbing them to death.'

If the animals were sick, the herdboys, rather than the village-based adults, had to treat them with medicinal herbs like *leshogkwa* (for gallsickness), *mositsane* (for worms and digestive complaints) and *mofifi* leaves as a tick repellent. Distance from home gave the boys more responsibility for looking after animals. It also meant that the elders imparted knowledge about medicinal herbs to them at a young age.[52] In Mbotyi, herdboys also knew the plant medicines for gallsickness which they administered themselves at Lambasi. In some Sotho families, girls were sent to look after animals but this was not the norm. Gladice Mokoena was the only women interviewed in QwaQwa who had herded as a girl. In her family, girls could herd until puberty. She recounted: 'I loved herding the animals, especially when we went hunting too. We would run around the veld, singing at the top of our voices.'[53]

The emphasis in these narratives was on the sense of freedom, personal autonomy and the authority that herding gave young people. Sotho herders were responsible for selecting grazing lands, treating sick animals and protecting themselves against hunger and danger. Interestingly, none of our Sotho informants talked about trekking in relation to disease prevention. Respondents did not differentiate between diseases of the valleys and diseases of the mountains. Many associated *nyoko* or gallsickness for example with deficient grasslands in any ecological zone. This could be because even the valley areas of QwaQwa were not so blighted by the presence of ticks or other biting arthropods as were the hotter and more semi-tropical parts of the country. In Mbotyi, by contrast,

[51] Chadiwick Mbongo, Lejwaneng, 12 October 2010.
[52] Daniel Khonkhe, Ha-Sethunya, 13 October 2010.
[53] Gladice Mokoena, Ha-Sethunya, 13 October 2010.

one reason for transhumance was to avoid ticks in the summer. However there were links between transhumance and good health, since Sotho livestock owners, just like their Tswana counterparts believed plentiful and palatable grass cover prevented or mitigated levels of infection (see chapter 3). Trekking in QwaQwa was about optimising the extent of the grazing lands as the valleys had become increasingly subject to denser settlement and diminished veld.

In recent years, fewer Sotho farmers have sent their stock to the mountains. As in Mbotyi, stock theft is blamed for this change. People keep their livestock in the valleys and near to home so they can monitor them more effectively. One man had a search light to try and detect thieves who came at night. Some said that this new concentration of animals was leading to overgrazing around the villages and kraals because people 'were scared of going to the mountains'.[54] High levels of stock theft posed a continuous threat to livelihoods and undermined the belief that animals were a good investment for retirement, or for financial emergencies. Abel Lebitsa who worked in maintenance at the University of the Free State in Phuthaditjaba, explained that he kept 18 cows in his village, Mountain View, and then rented land from the government for a further 11. In this way he hoped to divide the risk. However, over the years he had lost 'scores of animals' to thieves and he had become 'disillusioned with the whole prospect of farming'. He felt, with sadness, that it would 'never provide security for old age'.[55] More generally, the decline of transhumance to the mountain pastures has reduced optimum use of grazing capacity, as animals are now largely concentrated in the valleys throughout the year.

Daniel Khonkhe recounted his numerical impressions of the impact of stock theft. He explained that in the 'olden days', the 1960s, many people had 40 to 50 head of livestock; now most struggled to keep 10. He perceived this to be the result of escalating stock theft since the 1980s, which he attributed to higher levels of unemployment.[56] Other farmers concurred; rising unemployment and retrenchment from the mines exacerbated rural poverty, whilst the relative increase in the value of cattle has provided a strong incentive for some to steal. The demography of livestock ownership is less clear in local narratives. Khonkhe may have underestimated the number of small owners in the past – average holdings of about 10 or less were more common. Neither are smaller herds necessarily an indication of declining overall livestock numbers. The population of QwaQwa, and possibly livestock owners, has increased sharply since the 1960s.

In Mafikeng, some Tswana informants attributed theft to the breakdown in the *mafisa* or cattle loan system. Traditionally, wealthier livestock owners would loan out a few head to younger men and poorer farmers, so that they had access

---

[54] Chadiwick Mbongo, Lejwaneng, 12 October 2010; Maria Khoboko, Lejwaneng, 12 October 2010; Jacob Tshabalala, Phuthaditjhaba, 13 October 2010.

[55] Abel Lebitsa, Mountain View, 14 October 2010.

[56] Daniel Khonkhe, Ha-Sethunya, 13 October 2010.

to milk and draught. This also avoided heavy concentrations of livestock and helped to optimise access to grazing. If managed effectively, the clients' animals could increase and provide them with the nucleus of a herd. However, according to Victor Moagi, a retired schoolteacher from Modimola near Mafikeng, the gradual collapse of this system from the 1960s had resulted in a general decline in community obligations and responsibilities. His impressions coincided with a much more generally held view, discussed in chapter 2, that Setswana speakers had become more individualistic in their approaches to livestock management. In the absence of the *mafisa* system, Moagi felt: 'there is now less of a community safety net against poverty and the poorest families cannot afford to purchase animals. This has resulted in greater social tensions and led to an escalation in stock theft.'[57] The *mafisa* system had also characterised Sotho livestock management, and featured in ethnographies of the Pedi (North Sotho) and the Basotho written in the 1950s and 1960s.[58] However, no QwaQwa respondent referred to a cattle loan system and they did not attribute the increase in stock theft and other social problems with its demise. They situated stock theft within a broader picture of poverty and lack of perceived opportunity for economic progress.

In Koppies we also found that stock theft was a problem especially on the communal grazing lands around the township of Kwakwatsi. Johannes Malinde described how thieves often slaughtered the animals in the veld and ran off with the meat, leaving the carcass to rot.[59] He took us out to see just such a scene. The dead animal's carcass lay bare to the bone, swarming with flies. The head remained relatively intact with the eyes bulging from their sockets. Informants in QwaQwa, by contrast, suggested that thieves were organised and seized significant numbers of live animals, rather than just killing a single beast for food. This was more difficult in areas such as Koppies, where the African owners were surrounded by farms and were able to police the communal grazing lands more effectively. Several farmers in QwaQwa explained that theft involved a network of local African people in South Africa, sometimes organised by whites, who had links with thieves and speculators across the mountains in Lesotho.

Joseph Mapetla, a security guard at the University, who kept cattle near the town of Kestell to the north of QwaQwa, gave the most detailed description of these operations. As he saw it, the main problems were that African owners do not brand their animals, that liaison between South Africa and Lesotho was weak and that some policemen were corrupt. Lesotho has land, he said, but lacked cattle. Hence there was plenty of grazing for stolen animals. Much of this land was rented by white farmers. These farmers, as well as other criminals in

---

[57] Victor Moagi, Mafikeng, 11 February 2010.
[58] V. G. J. Sheddick, 'The Southern Sotho' in Daryll Forde (ed.) *Ethnographic Survey of Africa* (London, International African Institute, 1953), 21–22; H. O. Monnig, *The Pedi* (Pretoria, Van Schaik, 1967), 164–70.
[59] Johannes Malinde, Kwakwatse, 8 October 2010.

Lesotho, made deals with poor South Africans who agreed to drive animals across the border in return for money or *dagga* (marijuana). The ringleaders then branded or rebranded the animals with their own mark and claimed they were legitimately theirs. These brands were then registered in Maseru or Pretoria. Mapetla was convinced that it was 'big men' who were orchestrating all this as they were the only ones with the funds to pay thieves and bribe officials to turn a blind eye. They could also convince butchers not to question the source of the meat. Armed gangs were prepared to use violence and QwaQwa livestock owners sometimes used firearms themselves to protect their assets. Some farmers said they invested not only in firearms but also packs of dogs to fight intruders. In September 2010, the police did recover 200 cows and 400 sheep from Lesotho. However locals viewed this as a rare occurrence and stock-owners remained suspicious of the police, their intentions and their capacity to deal with crime.[60]

As in Mbotyi, established modes of transhumance in QwaQwa were under threat from theft, social dislocation and social change. There were echoes of the nineteenth century when wars, contested boundaries and livestock theft were frequent hazards for livestock owners. The borderlands of Lesotho have become a frontier area again, not only formally, but in the sense of a no man's land where state authority is limited, or weak. As noted in the Introduction, organised long-term transhumance requires a degree of political control, or at least effective negotiation between neighbouring settlements and countries. It may also be that stock thieves can exploit relatively isolated, unpoliced, but under-utilised mountain grazing.

Cattle posts in North West Province

In North West Province the majority of stockowners we interviewed let their animals graze around the village, and there were no surviving seasonal patterns of transhumance. Stockowners had kraals in their back yards, separate ones for cattle and small stock, and they drove their animals between their homes and the pastures on a daily basis. There was little in the way of maize cultivation, and informants did not seem to depasture their animals on harvested grain fields during the autumn and winter months.

The changes that we discussed above, such as the expansion in private property rights and fencing since the nineteenth century, have severely diminished

---

[60] Joseph Mapetla, interviewed in Phuthaditjhaba, 15 October 2010. Similar stories of violence and links between thieves in SA and Lesotho were introduced by Charles Mbongo, Lejwaneng, 12 October 2010; Maria Khoboko, Lejwaneng, 12 October 2010; Gladice Mokoena, Ha-Sethunya, 13 October 2010; Daniel Khonkhe, Ha-Sethunya, 13 October 2010; Jacob Tshabalala, Phuthaditjhaba, 13 October 2010; Ephraim Mafokeng, Phuthaditjhaba, 15 October 2010; Thomas Mahlaba, Tseki, 15 October 2010.

the capacity of Tswana stock owners to drive their animals beyond the environs of their villages. To maintain their animals at cattle posts, stock owners would also have to have access to labour. Agricultural work remains unattractive due to the comparatively low wages and it has become difficult to recruit willing, as well as trustworthy, herders. Nevertheless, some wealthier farmers could afford to rent more distant camps as cattle posts and to hire labour. In this way they hoped to boost access to grazing and thus promote better health.

Nick Lebethe from Bethanie, for example, owned 80 cattle and 35 goats in 2009. He used money he had accumulated from teaching to invest in animals and was one of the wealthiest African stock farmers we met in North West Province. He kept his animals at a fenced camp about 20 km from Bethanie and paid a herder to look after them. The herder was also responsible for carefully monitoring the health of the animals and he was authorised to administer medicines when necessary. Lebethe used biomedical drugs which he kept in a hut at the camp, so he had to employ herders who had some understanding of when and how to give them. He called the farm his 'cattle post' harking back to an older vocabulary. There were two major advantages of keeping animals at the cattle post, he explained. Firstly, there were fences that enabled veld rotation. The farm was 328 hectares in size and divided into two camps. Rotational grazing improved feeding and enabled the regeneration of the veld. Lebethe was aware of issues like carrying capacity and knew it was important to keep cattle numbers down to sustainable levels. He provided food supplements for his animals in the winter when the grazing was less rich, so that he could maintain his sizeable herds. The second advantage was that he could control breeding cycles. Lebethe was able to ensure that other people's bulls did not copulate with his heifers, potentially spreading diseases and producing generations of less-productive animals.[61]

There is conflicting evidence about the purpose and the extent of the cattle post system in North West Province where we interviewed. Observations in Mabeskraal, suggested that many people did keep their cattle and goats in the village – they were clearly visible grazing by the side of the road and many stockowners had kraals in their back yards. However, Lucas Mabe, a farmer in his 70s, explained that in the past livestock owners were not allowed to keep animals in the village lest they invaded the crop fields or people's gardens.[62] Stockowners were expected to keep their animals at a cattle post some distance from the village. Mabe owned six cows and he still sent his animals away to a camp shared by 50 or 60 other people. Each stockowner had their own kraal within the camps and there were huts for the herders. They also kept medicines on hand in the huts. Grazing within the camps was communal and there did not appear to be any planned system of veld rotation. Bosie Morapadi also from Mabeskraal chose to keep his two

---

[61] Nick Lebethe, Bethanie, 26 October 2009.
[62] Lucas Mabe, Mabeskraal, 2 February 2010.

remaining cows at a cattle post because he felt there was greater protection from stock theft.[63]

Our informants suggested that people's decisions were increasingly contingent on their perceptions of security as well as optimal grazing. For example, Lekunutu Ramafoko from Mabeskraal was a large-scale farmer who had retired as a chauffeur in Johannesburg and invested his capital in cattle. Ramafoko now had a herd of around 200 head, kept at a nearby cattle post at Ditaung about 8 km away from Mabeskraal. In theory these community grazing lands were fenced to enable veld rotation. However, he claimed that 'people came with donkey carts and stole the poles for fuel'. In his view 'many poorer people are jealous of farmers who have access to fenced camps and they steal fencing materials to show their anger and to get revenge'. Ramafoko could not trust herders, lest they were involved in stock theft, so he looked after his livestock himself. He suggested: 'many men would rather steal cattle than farm ... and you could trust nobody.' Consequently, Ramafoko felt that the only way to progress was to have a private farm which he could guard and manage. He aspired to developing a viable commercial business, but he was convinced that this was impossible under the existing cattle post system.[64]

By contrast Matthews Moloko from Magogwe near Mafikeng had been able to lease private land. He had access to nearly 300 hectares which he and his uncle rented for R5,000 per year. Moloko had attended a number of agricultural courses, organised by the Department of Agriculture's extension service, so he was knowledgeable about biomedicines and pasture management. He described it as a big farm with mixed veld which he divided into six camps for rotational grazing. His cattle spent six months at a time in each camp. These were not cattle posts but rather subdivided sections of a farm. The costs of private land, as well as the development of a pasture rotation system and the maintenance of boreholes to supply water made the price prohibitive for most livestock owners. There were also labour costs. Purchase of the farm was financially impossible and also politically difficult as there was much competition for land. Moloko felt: 'unless you are a chief, or have political connections, your chances of becoming part of a state-assisted land reform project are very low and probably impossible.'[65]

In North West Province our interviews suggested that the patterns of transhumance had been more radically affected than in Mbotyi, which is probably untypical of much of the Eastern Cape because of the availability of accessible pastureland. Yet evidence from North West Province indicated that there was more scope for livestock accumulation. Those with larger numbers of livestock adopted a very diverse range of grazing strategies. A few people had been lucky enough to find fenced private farms for purchase or rental which enabled them

---

[63] Bosie Morapadi, Mabeskraal, 2 February 2010.

[64] Lekunutu Ramafoko, Mabeskraal, 2 February 2010.

[65] Matthews Motshwrateu Moloko, Magogwe (near Mafikeng), 11 February 2010.

to rotate the grazing camps, rather than having to move animals to alternative sites. All of those whom we interviewed continued to live in villages and most employed herders to mind the animals. The most successful were in fact following the recommendations of officials: stop transhumance and manage livestock on a single farm.

Others had to cope with the problems of communal grazing areas. In one case at Lokaleng, near Mafikeng, communal grazing lands were fenced and a small group of seven livestock owners were able to manage veld rotation.[66] In Mabeskraal there was some collective attempt to use a cattle post system with a significant numbers of villagers involved. But elsewhere strategies tended to be more individualised. Even in Mabeskraal, observations suggested that there was not a single pattern or practice. Our overall impression was that in recent decades older systems of cattle posts have largely been abandoned, and live-stock owners have had to negotiate different forms of tenure and social contin-gencies. In particular, many spoke of the problems of theft as shaping their grazing strategies, resulting in a sub-optimal usage of the pastures. Some mentioned that theft was more likely at cattle posts, whereas others felt more vulnerable in their own home villages.[67]

Conclusion

Across all of our sites, the evidence indicated that while rights to land within existing communal areas remain strong, control over livestock and land by local people was sometimes diminishing and this affected the remnant patterns of transhumance. Theft was cited as the most common factor in the demise of these older systems of pastoralism. In Mbotyi it is likely that this affects the range of nutritious grasses and winter maize fodder to which access can be gained. However, evidence suggested that owners who kept their livestock in the village were also more careful about checking for and treating disease. Some of the best herds were no longer sent away to more distant pastures at Lambasi. In QwaQwa as well, owners were constrained in sending their animals longer distances and this reinforced their perceptions that access to pastures was declining. In North West Province some owners had been able to resolve these problems through adopting rotational grazing on private or municipal land, but there too, transhumance to cattle posts was restricted.

---

[66] Zachariah Tlatsane, Lokaleng (near Mafikeng), 9 February 2010.
[67] Richard Molebalwe, Bethanie, 26 Oct 2009; Peggy Setshedi, Fafung, 27 Oct 2009; Frederick Matlhatsi, Mabopane, 29 Oct 2009; Tollo Josiah Mahlangu, Mabopane, 29 Oct 2009 ; Chris More, Kgabalatsane, 13 Nov 2009; Simon Moeng, Sutelong, 26 January 2010; Martha Rakgomo, Mabeskraal, 1 February 2010; William Komane, Mabeskraal, 4 February; Ernest Medupe, Mafikeng, 10 February 2010; Reuben Ramutloa, Disaneng, 11 February 2010; Victor Moagi, Mafikeng, 11 February 2010; Modisaotsine Taikobong, Magogwe, 11 February 2010.

Transhumance is no longer a significant route to maintaining animal health.

Discussion of transhumance in Africa tends to be critical of colonial and post-colonial policies which have attempted to control or sedentarise pastoralists.[68] A strong strand in the literature emphasises the salience of local knowledge and argues for less external intervention. In part this is a rights-based approach, which echoes claims by pastoralists for control of their land and for their cultural practices. The literature emphasises the political marginality rather than the power of transhumant societies. New range ecology, developed in the 1980s, added a critique of ecological arguments in favour of rotational grazing and the idea of sustainable livestock carrying capacities. These prescriptions underpinned both betterment and the brief introduction of destocking policies in South Africa. Many Africanists were critical of arguments about the tragedy of the commons – the idea that livestock owners would seek maximum advantage from access to common land, without regard for the ecological costs. The counter-argument suggested that common grazing land was often subject to some control and management while livestock numbers were adjusted to climatic and vegetational cycles. In essence these new approaches offered the idea of disequilibrium in rangelands that fits well with non-interventionist approaches and the social defence of pastoralists and transhumance.

Our South African examples point to a different history. Longer-distance transhumance has been difficult for over a century in many areas.[69] Veterinary policy offered an alternative means of prophylaxis and disease management through dipping, inoculation and the control of contagious diseases. The state advocated fenced grazing camps which could be rotated in order to provide fresh seasonal grazing. In some senses this strategy echoed transhumant practices within smaller, fixed areas of land. An attempt was made to introduce controls over grazing during the implementation of betterment. But these state-sponsored systems have now partly broken down and African livestock-owning communities have had to devise alternative ways of managing seasonal access to grazing. In general those we interviewed have not used fodder. Few have control of fenced camps through which grazing can be rotated. Where local transhumance has survived, as in QwaQwa and Mbotyi, informants thought that they had been restricted in recent years by the scale of stock theft and by the lack of labour. African farmers can neither resort to the old practices of

---

[68] Roy H. Behnke, Ian Scoones and Carol Kerven, *Range Ecology at Disequilibrium: New Models of Natural Variability and Pastoral Adaptation in African Savannas* (London, Overseas Development Institute, 1993); Katherine Homewood and W. A. Rodgers, 'Pastoralism, Conservation and the Overgrazing Controversy' in D. Anderson and R. Grove (eds), *Conservation in Africa: People, Policies and Practice* (Cambridge, Cambridge University Press, 1987), 111–28; Ben Cousins, 'Livestock Production and Common Property Struggles in South Africa's Agrarian Reform', *Journal of Peasant Studies*, 23, 2–3 (1996), 166–208.

[69] William Beinart, 'Transhumance, Animal Diseases and Environment in the Cape, South Africa', *South African Historical Journal*, 58, 1 (2007), 17–41.

regular transhumance in order to find the best pastures, nor to a systematic rotational grazing supplemented by biomedicines. While some more successful stockowners in Mbotyi felt that there were benefits in keeping animals close to the village under their surveillance, rather than sending them to more distant pastures, many perceived a worsening disease environment. They felt that they no longer had the means to cope with a broadening range of infections, in a context where state veterinary services were diminishing. Our next chapter looks more closely at the use of other local resources, in particular plants, in attempts to manage animal health.

5

Plants & Drugs

Medicating Livestock

In the past there were no vaccines and animals were never identified as sick. Now animals are like children; they have to have injections and be cared for. (Caroline Serename, Kgabalatsane, North West Province)

Introduction: plants and biomedicines

Caroline Serename and her husband had farmed in Kgabalatsane (near Garankuwa) for over 30 years. Farming was a family business and as children they had been involved in looking after animals. She had worked as a domestic servant in Gauteng and her husband in factories. They had invested their savings, and now their pensions, in cattle. In November 2009 they owned about 36 Brahman cattle. They favoured Brahmans because they thought these were comparatively easy to handle and grew fast, being ready for market in about 18 months. Their main buyers were African customers who needed cows for ceremonies and funerals. The Serenames were committed to modern medicine. Caroline had a good biomedical knowledge of diseases. She knew ticks spread heartwater, insects propagated lumpy skin disease and that blackquarter was endemic in the area. She sprayed her animals against ticks twice per month and also injected them with parasiticides like Ivomec (ivermectin) to deal with worms as well as acarids. She vaccinated for blackquarter and anthrax and always had the antibiotic Terramycin (oxytetracycline) on hand in case her cattle fell sick. Caroline Serename was adamant: 'if you love your livestock, like your children, then you have to invest in vaccines and antibiotics.' Stockowners, she believed, were far more aware of diseases now than in the past. Even if there had been more infections around in recent years, at least there were also good preventatives and cures. She avoided traditional medicines because she believed that modern drugs were so much better.

Serename's definition of modern drugs was quite broad. Like many African livestock owners from North West Province, but unlike those in Mbotyi and QwaQwa, she dosed her animals with salt to improve their appetites and overall health. Informants regarded salt licks as a traditional remedy that pre-dated the arrival of vets. She also administered sorghum beer to cattle that had ingested plastic wrappers and other inedible objects. Salt and beer had laxative properties that flushed out the toxins and foreign matter thereby promoting good digestion. In this respect her views of animal health accorded with those generally held. However, as Serename saw it, the problem was that many farmers did not vaccinate. Some used traditional medicine either because it was cheaper or because of apathy. As a consequence their animals continued to die. She mused:

> People wait for the government to act. And if the government vets and Animal Health Technicians don't appear with their bakkies [small pick-up trucks] and pharmaceutical stuff local stockowners do nothing – for reasons that are totally beyond me.

For Caroline Serename cattle brought money and happiness, but only if you respected them like your own offspring and invested in the best therapeutics to promote their wellbeing.[1]

Zulu Mokoena, a traditional healer from Lusaka in QwaQwa had a totally different attitude towards western medicine and animal husbandry.[2] In March 2010 he and his father owned 12 head of cattle. Most of his clients visited him for human sicknesses but sometimes they consulted him about animal medicines too. Unlike Caroline Serename he was convinced that biomedicines were essentially harmful and toxic with dangerous side effects. Modern drugs might cure some conditions but he also believed that many human infections were a result of people taking biomedicines. The same was true of animal diseases: 'vets never tell people what they are injecting for; the number of livestock diseases is increasing because vets have infected stock by administering all kinds of harmful injections and drugs.' Mokoena might have had a vested financial interest in promoting traditional medicine, but his testimony was striking because it demonstrated a total dedication to Sotho lore and a passionate dismissal of western knowledge and practices.

Mokoena stated that he would never consult a veterinarian or visit a farmers' cooperative for advice or for drugs, because his pharmacopoeia 'lies in the veld – the veld of my ancestors'. For it was his forefathers who had called him to be a healer and taught him the essence of Sotho medicine. They revealed their ancient knowledge to him in dreams and he experimented with local flora and

[1] Caroline Serename, Kgabalatsane, North West Province, 13 November 2009.
[2] Zulu Mokoena, Lusaka, QwaQwa, 2 March 2011.

other substances to make potent brews that he claimed cured sick people and livestock without causing any harmful side effects. Mokoena was proud of his Sotho culture and traditions. In a sense, he served as a relatively rare example of a full commitment to indigenous knowledge, firmly rooted in the land and older ways of knowing. His conception was in part a cultural statement, unusual even for traditional healers. For Mokoena, showing respect for animals did not involve the administration of a vaccine or an antibiotic, but a conversation with the ancestors and the selection of the best natural medicines to promote health and ward off sickness and death. His practices were somewhat different to isiZulu- and isiXhosa-speaking specialist healers who have been flexible in incorporating new ideas and medicines.[3] His ideas stood in stark contrast to Serename's approach and also that of many informants who felt that veterinary responses should evolve and adapt to changing opportunities and conditions.

These two testimonies were at the polarities of the approaches we encountered in our interviews and they were symptomatic of the diverse narratives and practices that stockowners pursued in relation to animal health care in twenty-first-century South Africa. Throughout our regions there were some farmers who prioritised biomedicines and others who held more firmly to older practices. Although, as we noted in the Introduction, informants often articulated a binary view of tradition and modernity many seemed to follow a middle way, which is why we use the term local knowledge. They appreciated particular biomedical treatments, but also believed that some local plants were more effective for certain conditions, or they felt culturally committed to some older practices. Farmers might use a range of vaccines and antibiotics, for example, but believed that herbal concoctions were far better than commercial products for flushing out the bowels or the placentas that refused to drop. While few expressed their commitment to African culture quite as forcefully as Mokoena, informants who professed to using biomedicines might also continue to commune with the ancestors through slaughters, practice circumcision and demonstrate patriarchal values. They held onto these even though the material context in which they lived had clearly changed.

This chapter explores the remedies that informants deployed to tackle animal diseases. We discuss access to, and transfer of, different forms of knowledge about medication at a local level. We explore the contingencies of generation and gender which have influenced the effectiveness of local knowledge transmission within families. This is followed by an examination of the interaction between biomedical practitioners and stockowners to show how lack of education, together with problems within the state veterinary service have hampered the

---

[3] Karen Flint, *Healing Traditions: African Medicine, Cultural Exchange and Competition in South Africa, 1820–1948* (Athens OH, Ohio University Press, 2008); Chris Simon and Masilo Lamia, 'Merging Pharmacopoeia: Understanding the Historical Origins of Incorporative Pharmacopoeial Processes amongst Xhosa Healers in Southern Africa', *Journal of Ethnopharmacology*, 33, 3 (1991), 237–42.

flow of biomedical knowledge to African farmers. Finally we look at the factors that influenced the choices our informants made when deciding which treatments to use on their livestock.

The transfer of local knowledge: specialists, generations and gender

When we began this research we tended to assume that while access to biomedical knowledge would be constrained by lack of education, local knowledge would be freely available. We found, contrary to our expectations, that it too was restricted. Local veterinary knowledge was not transmitted formally, for example in a training process similar to that experienced by a *sangoma* (*igqira*) or *ixhwele* (isiXhosa for herbalist), but there were nonetheless conventions that affected its spread and gave it authenticity. We heard of only one example where written records were kept of medical recipes, even amongst those who could write; knowledge was passed on by word of mouth. In North West Province and QwaQwa, stockowners would normally treat animals themselves rather than consult a traditional healer about livestock diseases. However, in North West Province informants occasionally spoke of individuals they could turn to in an emergency. These were usually older men, respected for their knowledge and experience. Some such as Thomas Seemela were also literate. Seemela lived in Madidi, north of Pretoria and was 85 years old when we interviewed him in 2009. He had worked at a ceramics factory in Johannesburg and invested his money in cattle. He owned about 30 Bonsmaras. Neighbours said they consulted him if they encountered birthing problems in their livestock or needed advice.[4] Seemela used modern drugs and followed the vaccination schedule recommended by Onderstepoort. He was happy to help his neighbours whenever they called on him for assistance.[5]

In Mbotyi there were local specialists such as Myalezwa Matwana, in his mid-seventies, who treated sick livestock.[6] Matwana rarely visited the animals unless he had to intervene physically such as during birthing. He relied on stockowners' descriptions of symptoms when making his diagnoses and prescribing a cure. Unlike Seemela who used modern drugs, Matwana provided his clients with bottles of ready-made *muthi* or else he collected fresh herbs. He did not always make direct charges for his medicines. He took money when he could but often received payment in kind, such as a chicken, which had a monetary value of around R30–40 in 2009. Matwana said that he was not a *sangoma* or herbalist, but that he inherited his knowledge of animal remedies from his parents. While a couple of other men were recognised locally as

[4]   Eva Sesoko, Madidi, 23 October 2009.
[5]   Thomas Seemela, Madidi, 23 October 2009.
[6]   Myalezwa Matwana, Mbotyi, 22 March 2008, 24 March 2008, 7 December 2011.

specialists, and at least one of the *sangoma*s also assisted in the diagnosis and healing of animals, most of those in Mbotyi who mentioned that they had received help with *muthi*, acquired it from him. He and Sidwell Caine were also in demand for slaughtering animals, and Matwana frequently carried out or oversaw sacrifices and post-mortems – especially of cattle. As a result, he was deeply familiar with the internal organs and skilled at diagnosis. Research by Patrick Masika in other parts of the Eastern Cape revealed that about half the stockowners who used medicines made their own; others went to specialists whom they often paid in kind.[7]

Most stockowners learnt about traditional medicines from their parents or elders. However, fathers did not necessarily pass on their knowledge to all their children. In many families and communities secrecy surrounded the transmission of medical knowledge. Informants, especially from Mabeskraal and Mantsa in North West Province, explained that this knowledge could only be handed down to certain men in the family. As children, herders were expected to look out for signs of illness in the animals they tended, but they had to summon the elders if the animals fell sick and they were not allowed to administer treatments themselves.[8] This reflected a patriarchal social hierarchy with regard to the ownership of important and powerful forms of knowledge. Bassie Monngae and John Modisane from Mabeskraal said that their fathers passed on information to a favoured son, but if the father passed away before that knowledge could be conveyed, then the plant remedies died with him. As many young men sought work outside the village in the cities and mines, there were not necessarily the opportunities to pass on information in time. People sensed that knowledge had been lost because of long-term migrancy.[9]

Some informants expressed concerns that the younger generation was no longer interested in inheriting traditional knowledge. Molekwa Mabe from Mabeskraal was an elderly man who had once served as the acting chief. He owned 60 cattle in 2010 and said he came from a family of traditional healers. He used both biomedicines and plants, especially aloes to treat his livestock. Mabe was convinced that aloes could perform miracles 'by the grace of God' and explained how his mother had cured herself of breast cancer and lived for one hundred years by taking aloe infusions. His belief in the healing power of plants meant he was particularly alarmed because:

[7] P. J. Masika, W. van Averbeke and A. Sonardi, 'Use of Herbal Remedies by Small-scale Farmers to Treat Livestock Diseases in central Eastern Cape Province, South Africa', *Journal of the South African Veterinary Association*, 71, 2 (2000), 87–91.

[8] Ernest Medupe, Mafikeng, 10 February 2010; Modisaotsine Taikobong, Magogwe, 11 February 2010.

[9] Bassie Monngae, Mabeskraal, 1 Feb 2010; John Modisane, Mabeskraal, 2 February 2010; Zachariah Tlatsane, Lokaleng, 9 February 2010.

Our community is abandoning our culture as parents don't pass on information to their children. The young are rejecting the knowledge of their elders. Young people see traditional medicines as primitive and backward. They want to appear modern by using modern drugs.[10]

His views were an example, as noted above, of the tendency to see a sharp distinction between tradition and modernity. He felt challenged by young people who did not wish to be part of an older culture, drawing on ancestors for guidance and the veld for its cures. Despite these anxieties, we found that younger informants were not noticeably less interested in traditional medicines. By contrast, some older farmers totally disavowed herbal remedies and could only vouch for biomedical products.[11]

In both North West Province and QwaQwa younger as well as older men provided information about medicinal herbs and talked about a range of ways in which they treated their animals. These more youthful farmers did not feel that they lacked sophistication because they mixed herbs and many young men, especially in QwaQwa, were proud of their medical traditions. One of the most passionate followers of traditional medicine, in the mode of Zulu Mokoena, was 22-year-old Sethunya Mathu from QwaQwa. He narrated a list of plants he commonly used to treat diseases like gallsickness and explained his great distrust of veterinarians and biomedicine. In 2009 many cattle in his village of Ha-Sethunya had aborted and the veterinary department began to vaccinate. He did not have a name for the disease, but it could have been brucellosis which has been a problem in the Free State in recent years. However, despite the injections, the cattle continued to miscarry. Mathu was convinced that the veterinarians 'do not have the means to cure animals and their knowledge is useless'. He told us that he did not need vaccines and could cleanse the cows by mixing *mowa* (ash scraped off thatched roofs, blackened by fires) with water. In his view one litre of this sooty liquid was enough to prevent future miscarriages in his livestock.[12]

Despite Mathu's confidence, the transfer of recipes and methods from one generation to another was uneven, and by no means assured. As we noted, local knowledge was not openly available. Lekunutu Ramafoko from Mabeskraal was quite bitter that his father had told him nothing about traditional healing as local plants were much cheaper than biomedical products, and he felt his medical knowledge was very limited as a consequence of his father's secrecy.[13] The main explanation respondents gave for secrecy in North West Province was witchcraft. Farmers said they were afraid that if witches knew their medical formulae they would use this information to doctor their cattle and bewitch

---

[10] Molelekwa Mabe, Mabeskraal, 1 February 2010.
[11] E.g., Moiletswane group meetings, 20 and 21 January 2010; Piet Hlongwane, Sutelong, 27 and 28 January 2010; Charles Matome, Sutelong, 28 January 2010.
[12] Sethunya Mathu, Ha-Sethunya, 13 October 2010.
[13] Lekunutu Ramafoko, Mabeskraal, 4 February 2010.

their kraals.[14] In the less densely populated more rural and traditionalist areas of North West Province, especially in the west towards Botswana, the barriers to the transmission of knowledge seemed to be greater. In the peri-urban areas of Garankuwa, this seemed less of a problem.

Patrick Masika also found that stockowners in Eastern Cape were unwilling to reveal their knowledge to their neighbours. He suggested that knowledge has a monetary value and people may only divulge it in return for payments.[15] In QwaQwa and North West Province a number of informants also said that they would only share their knowledge in return for cash. Abram Lebesa, for example, from QwaQwa was proud of his family's tradition of treating animals with medicinal herbs. They were not recognised traditional healers, but over the generations they had learnt, through their own experiments, which plants had therapeutic qualities and which did not. They had started to record their recipes in a book, the contents of which they were not prepared to reveal to anybody. He explained that this knowledge had been built up over time and had a monetary value. Despite treating other people's animals for cash, the Lebesas, like several informants from QwaQwa, saw the use of local medicines as a statement of Sotho cultural independence and identity.[16] It is intriguing that Lebesa saw local veterinary knowledge as both economically and culturally important.

This issue of secrecy and payment was alluded to but perhaps less central in Mbotyi. Some remedies were widely known; for example, herdboys were expected to learn the four-plant mixture given as a drench for gallsickness because people felt it was more likely to be successful if applied as soon as symptoms appeared. In the past, when they took livestock some distance to Lambasi, boys mixed and administered this themselves. In QwaQwa also those out with the cattle for long periods were expected to develop some veterinary knowledge. In Mbotyi a few younger men said the old were secretive, and older men claimed the youth were uninterested. Matwana did not seem to have transmitted all of his knowledge to his family. He joked that if he divulged all the *muthis*, his son would simply go and sell the secrets. He was, nevertheless, prepared to tell us about more than 10 plants he used for various purposes and we heard about additional remedies from others. The four interviews with him, and additional conversations, were conducted with Sonwabile Mkhanywa, a local interpreter, so that he was quite aware that such remedies would not be kept secret. Lungelwa Mhlwazi, the herbalist, also told us about a few of the plants that she used, and showed us samples. However, when we asked her to take us on a tour to find the plants, she demurred, and joked that it would cost us two cows.[17] We decided not to commit this amount (about R12,000 in 2009) to the research.

[14] Hendrik Metswamere, Mantsa, 12 February 2010; Kgomotso Goapele, Mantsa, 12 February 2010.
[15] Discussion with Patrick Masika, Fort Hare University, 23 February 2011.
[16] Abram Lebesa, Ha-Sethunya, 2 March 2011.
[17] Lungelwa Mhlwazi, Kunzimbini, 19 February 2009.

Zipoyile Mangqukela told us a number of the *muthi*s he had used when he had cattle, such as his eight-plant remedy for redwater. But he said he would not reveal all and was particularly reluctant to talk about birthing medicines for humans and animals. These he considered particularly valuable and he suggested we might pay him for the information because this knowledge could make us money. He said he would pass his knowledge to his wife (though she too was old). Mangqukela noted that one man, who was a specialist in healing snake bites, passed away without telling people his medicines: 'it is not an easy thing to let someone learn your medicines'.[18] George Mkhanywa, 60-year-old brother to a *sangoma* in Mbotyi, told us that he knew very little about local plant medicines.[19] He explained that he did not learn much about them as a boy, had been away working for a long period and had started with livestock late in life with limited success. At one time, 10 years before, he owned 14 cattle but had lost many from *inyongo* and sold two. In 2011 he had four cows and five goats. As noted, local remedies were not always shared even within families. He used Terramycin when he could afford it.

We do not know how many plants may be in particular specialist repertoires, and whether they do indeed have a much more extensive knowledge. Clearly, if we had stayed longer in one site, or worked with experts over an extended period, as Dold and Cocks have done, it would be possible to record more plants in use. Perhaps we could have bought additional information. In all, we heard of about 60 plants used in various sites for veterinary purposes, a number of which are also recorded in human pharmacopoeias. There is a rich store of written knowledge about medicinal plants in South Africa, collected over a century, and it would be possible to construct a generalised picture for each region that would almost certainly go far beyond the knowledge of specific local healers (see chapter 6).

For all their caution, these specialists were by no means wealthy. Matwana lived in a modest, traditional homestead with a few round, thatched huts. Unlike many of the wealthier villagers, he had not invested in a central square house. He owned about 12 head of cattle – although dipping records suggest that in 1976 (when he was a migrant worker), he had as many as 30.[20] Matwana was quintessentially a man of the village, frequently drinking with his wife and friends at the *spaza* (informal convenience) shop, and regularly present at ceremonies and feasts where he officiated over the slaughter and the carving of meat. He had a strong sense of humour. Generally clothed in a blue overall and gumboots, common village apparel, he sometimes dressed up for slaughters – for example donning a horse-riding helmet on one occasion and a women's

---

[18] Zipoyile Mangqukela, Mbotyi, 29 March 2008.
[19] George Mkhanywa, Mbotyi, 11 December 2011.
[20] Mbotyi Dipping register, 1976. Eight very dilapidated Dipping Register books were found in an old shed near the tank, covering separate, single years in the period from 1952 to 1992. The rest seem to be lost.

Photo 5.1 Slaughtering cattle in Mbotyi. Post-mortems were part of the butchering process (WB, 2012)

skirt on another. In earlier years he had been part of the *amatshawe* beer-brewing and drinking groups, where he acquired the nickname Judge, after one of the positions in the organisation.[21] He was known for his success with women. His knowledge and skills clearly placed him at the centre of village networks, and gave him authority, but he was not in a position of power or wealth. He was part of a group who kept their distance from the sub-headman and his close advisers.

Information did not necessarily come from the parents or local herbalists. Some people believed that knowledge was transmitted by the departed via dreams. Traditional healers attributed much of their wisdom and powers of divining to dreams and the voices of the ancestors. But so do some stock farmers – although we did not find evidence of extended training or visions amongst those who said that they had been influenced by the ancestors. In 2010 Kgomotso Goapele, a 40-year-old farmer from Mantsa near Mafikeng, provided a vivid narration of the linkages between men, cattle, ancestors, knowledge and the land. Compared with his neighbours in this relatively poor village, without running water or electricity, he had accumulated unusually rich hold-

[21] To our knowledge, there have been no publications on the *amatshawe* groups but Beinart has interviewed on them.

ings – 30 cattle, 12 sheep, 11 goats, 10 donkeys and 12 horses. He came across as a deeply spiritual man and felt he was blessed by this wealth. He told us:

> Without cattle the Tswana have no worth. A Tswana man requires cattle to sacrifice to his ancestors; the blood of a slaughtered animal forms an invisible umbilical cord connecting the living with the dead. In return for this blood, the ancestors visit me in my dreams and show me how to cure sick animals and stop diseases. I have succeeded due to the power of these dreams.

Goapele demonstrated a rich knowledge of local flora and his recipes resulted from experiments with plants and substances to which he said his ancestors had directed him.[22]

Goapele also saw this knowledge as an affirmation of a masculine identity which translated into a strict ban on women handling livestock or gaining familiarity with animal medicines. As noted above, he was not alone in regarding this as a male preserve. Majeng Motsisi from Mabeskraal stated that it was only when a father believed his son was 'man enough' to handle the family's secret medical lores that he would impart this information to his offspring.[23] Some women such as Emily Diole, also from Mabeskraal, felt these traditions reinforced the bonds between men and cattle through the deliberate exclusion of women. Given the importance of livestock, especially cattle, in African communities the possession of this knowledge about animal diseases and treatments helped to consolidate men's elevated social position within society. It was not just cattle ownership that conferred status; ownership of veterinary knowledge did too.[24] None of the women livestock owners interviewed in Mbotyi professed to know much about plant *muthi*s for livestock.[25] A few said that when they needed such medicines, they got them from Myalezwa Matwana. Nevertheless, as we discuss in chapter 7, some male stockowners did call on female *sangoma*s and herbalists to protect their kraals against supernatural forces.

The exclusion of women from local veterinary knowledge was certainly evident in our interviews but could not be construed as a blanket prohibition. Women could also own cattle. Many specialised in small stock such as goats, pigs and chickens, both as owners and carers. Betty Gumede from Garankuwa explained that chickens rarely got sick, because she used preventative medicine. She described a sick chicken as one that would not eat or drink and isolated itself from the others. 'A healthy bird is a strong bird. It's energetic and sociable and likes to jump onto fences.' To ensure good health she dosed her birds with

[22] Kgomotso Goapele, Mantsa, 12 February 2010.
[23] Majeng Motsisi, Mabeskraal, 3 February 2010.
[24] Emily Diole, Mabeskraal, 2 February 2010.
[25] Jocelina Mtunyelwa, Mbotyi, 10 December 2011; Mamdenyeshwa Ndube, Mbotyi, 10 December 2011; Mayikalisa Jikijela, Mbotyi, 12 December 2011; Victoria Madikane, Mbotyi, 12 December 2011; Mamcingelwa Matwana, Mbotyi, 13 December 2011; Masamekile Satsha, Mbotyi, 20 December 2011.

Epsom salts as a laxative and added potassium permanganate to their water troughs every month to purify the blood so that 'they would not be weak'. To keep away *matsetse*, a type of tick or mite, she sprayed them with Doom Blue Death powder – a commercial insecticide for controlling crawling insects.[26] Norah Mabena, also from Garankuwa, gave her chickens aloes mixed with potassium permanganate to cleanse the insides and stop sickness. These substances purified the body and promoted good health.[27] We also saw chopped aloes added to chicken feed as a laxative in Mbotyi – apparently the only local use of these plants.

Transmission of biomedical knowledge and medicine

Alongside the uneven process of knowledge transfer within communities and families, there is an equally fragmented parallel process of transmission of biomedical ideas and treatments. Rural livestock owners come into contact with these networks of information but few receive any sustained exposure to them. As we noted in chapter 2, a century of dipping did not cement new knowledge of tick-borne diseases. In the last couple of decades, informants have connected irregularly with various experts with different levels of training in veterinary medicine. Stock inspectors and dipping foremen survived with the remnants of the dipping system till the 1990s in some areas, but are no longer employed. Qualified state veterinarians largely remain office bound, running the administration of services in large districts with tens of thousands of animals. In 2008–9, one government veterinarian in Lusikisiki was responsible for probably over 200,000 head of cattle – as well as all the other livestock. Although he had worked in this position since 1993, and was well-versed in the veterinary problems of the area, he had lost interest in his post, was pursuing other activities, and few of his staff were greatly motivated. Whilst some of the state veterinarians that we met, such as the officer based at Bethlehem who oversaw QwaQwa, appeared highly engaged, finding state veterinarians with adequate commitment remains a problem in the provincial departments of agriculture. Private veterinary services are very rare in these African-occupied districts because of the cost.

In recent years, the specialist officials whom most farmers have encountered are the Animal Health Technicians (AHTs). Day to day responsibility for communication with livestock owners is essentially devolved to them; they have access to motorised transport and play a wide range of roles. For example, we accompanied the young and enthusiastic AHT based at Fort Beaufort to inoculate cattle against lumpy skin disease in the Kat River valley. She said that some

26 Betty Gumede, Garankuwa, 21 October 2009.
27 Norah Mabena, Garankuwa, 21 October 2009.

owners had lost up to 30 per cent of their herds in recent years. She brought the vaccine with her, and administered it herself as local stockowners were not familiar with its usage. She used the opportunity to transmit knowledge about the disease as well as providing treatment. In the Eastern Cape, the AHTs oversee the annual anthrax vaccination, in areas where it is applied. They were also responsible for overseeing the slaughter of pigs in 2007 when African Swine fever, a notifiable disease, swept through the herds.

In QwaQwa the AHTs spoke ardently about their intermediary position and were constantly on the road. Mothupi Molefe said he tried to see stockowners the day they contacted him for help as they could not afford the services of a private veterinarian. AHT visits, by contrast, were free and farmers only had to pay for the medicines. When a stockowner called him, he asked for details of the condition so he could prioritise cases. He dealt with cows with dystocia first as that condition could lead to immediate fatalities. If an animal was sick he estimated how long the animal was likely to survive according to the symptoms. Livestock that had probably ingested poisonous plants received the swiftest attention. When he was not treating stock, he did his best to visit farms as often as possible to test for diseases such as anthrax, blackquarter, redwater, gallsickness and swine fever. He aimed to take blood smears every six months so he could advise and educate farmers about treatments and inform the state veterinarian of any dangerous epizootics in the area. He estimated that he visited about 300 people a month who owned on average about six head of cattle each.[28] Molefe clearly acted as a hub of knowledge transfer.

In our field sites as a whole, however, the extent to which AHTs connected with stockowners varied. The Lusikisiki veterinary department found it difficult to fill all its positions. We interviewed four AHTs operating out of the office and had a follow up meeting with an additional small group. Two were recently trained and seemed engaged with their posts, but one said he had become demoralised, partly by the lack of leadership and organisation in the veterinary office and partly because of the hostility and traditionalism of many of the communities.[29] Villagers believed that the officer for the area that included Mbotyi had little commitment to his post. Responsible for four settlements, each with a dipping tank, he seldom visited and this was one reason why dipping had almost stopped in Mbotyi. He had settled in another village and seemed to spend a good deal of time on his own farming operations. To some extent he had become a receiver rather than a transmitter of knowledge and spoke of his interest in local practices. He knew Myalezwa Matwana, the Mbotyi village specialist, from whom he got traditional medicines for livestock.[30]

---

[28] Mothupi Molefe, Phuthaditjhaba, 14 October 2010.
[29] Mzwandile Ndlela, 19 March 2012. Other AHTs were interviewed in 2009.
[30] Mr Fusile, AHT, Lusikisiki, 19 March 2012.

Few of the people interviewed in Mbotyi visited the veterinary offices to see officials about livestock diseases. Victoria Madikane said she was particularly dependent on assistance from the Department of Agriculture since local live-stock knowledge tended to be the preserve of men, and her husband had died. She had gone to 'the College' (the site of the Veterinary Offices) and received visits from the AHT.[31] Unusually, she concentrated on sheep which she managed to keep alive in this rather hostile sub-tropical environment. Following office advice, she bought Hi-Tet (oxytetracycline hydrochloride) for ticks, injections for scab, and Valbazen (albendazole) for worms; the medicines were kept in a plastic bag under her bed. She also used the plant *intolwane* (*Elephantorrhiza elephantina*) which she fetched personally from Tabankulu, nearly 100 km away.

AHTs participated in organising stock days in their localities. We attended one at Hertzog in the Kat River valley in 2008. About 50 livestock owners participated in this event, addressed initially by a couple of AHTs who spoke about dipping and branding.[32] The main speaker of the day was an African representative of the Pfizer pharmaceutical company who discussed in detail the appropriate Pfizer drugs for a range of animal diseases. Similarly, in Kgabalet-sane, in 2009, proceedings were dominated by a key speaker from the same company.[33] At both venues there was opportunity for lively discussion of partic-ular problems. In both Hertzog and Kgabaletsane, Pfizer provided a free lunch for participants, but no free drug samples which disappointed some villagers. These are important and popular events in some areas, where there is oppor-tunity both for marketing pharmaceuticals and discussing their use. However, there were clearly gaps in understanding as many people in the audience lacked background knowledge of some of the concepts and prescriptions, even though proceedings were conducted in the vernacular.

Private sector involvement in drug supply to the African rural areas is clearly expanding and is a significant element in gradually shifting ideas about biomed-icines. We researched this issue from the vantage point of livestock owners rather than the companies. Our impression was that the demand is not simply being created from above and drugs being pushed onto rural communities. There is a huge demand from below, and many interviewees felt that they could not get sufficient biomedicines for their needs. Farmers learnt from staff at stores where they bought drugs. Livestock owners in Mbotyi visited retail outlets and chemists in Lusikisiki, which did a roaring trade in pharmaceuticals at a high mark-up. Two of those interviewed travelled as far as Kokstad, about 200km away, where they could buy drugs more cheaply at the farmers' coop-erative. For example, in 2012 a 100 ml bottle of Terramycin cost R160 in Kokstad but R250 in Lusikisiki. In a number of cases, informants mentioned that retailers recommended particular drugs. In this sense the shop staff, who

---

[31] Victoria Madikane, Mbotyi, 11 December 2011.
[32] Attended by William Beinart and Vimbai Jenjezwa, Hertzog, 19 November 2008.
[33] Attended by Karen Brown, 12 November 2009.

were not specifically trained in diagnosis, operated in a similar way to local specialists such as Myalezwa Matwana in Mbotyi. They recommended wide spectrum, general purpose drugs on the basis of descriptions of symptoms, or in response to a disease named in isiXhosa – with opportunities both for mis-diagnosis and the sale of inappropriate medicines. Our impression was that farmers bought a relatively narrow range of treatments, especially Terramycin and Ivomec.

In Koppies, in the Free State, all the African smallholders we interviewed had worked for white farmers and learnt from them how to treat livestock with biomedicines. They had a stronger understanding of biomedical explanations of diseases and a stronger commitment to exclusive use of drugs. They were well-informed about the role of ticks in the transmission of key diseases such as gall-sickness and they discussed acaricides and antibiotics such as Terramycin. Johannes Malinde, for example, lived in Kwakwatse, Koppies' township, and grazed his nine cattle on nearby communal land. He affirmed that he had learnt about diseases and medicines whilst working for several Afrikaner farmers and now, in his retirement, he bought dips and medicines from the private vet in town.[34] There is little doubt that the use of biomedicines is expanding signifi-cantly in the African rural areas. Information about them passes through a variety of networks but none of them are very satisfactory or systematic. Farmers also learnt about drugs and dosing from their neighbours, with the inevitable lack of accuracy in administration. Even where they did know about appropriate drugs, the costs were prohibitive for any consistent use by the great majority of smallholders and they could not procure, inject or store some vacci-nations themselves. The task of expanding knowledge about biomedical treat-ments remains a challenge both for stockowners and the state.

Choices of medicine: local medicines, biomedicines and other forms of treatment

Knowledge about medicines was one factor, but not the only one, that informed choices of medication. Notions of cost, as well as availability and effi-cacy, all came into the equation. Choice was based on a complex mixture of ideas and experiences that was not a direct reflection of education, gender, age or wealth. Every community we visited had some exposure to western medi-cines, not only through dipping campaigns but also through state-run vacci-nation programmes to quell diseases such as anthrax, blackquarter, brucellosis and rabies. As with dipping there has been a decline in state investment in live-stock vaccination campaigns in recent years, but many stockowners were aware that there were injections for diseases. Their decisions whether to purchase

[34] Johannes Maline, Kwakwatse, 8 October 2010.

vaccines and other pharmaceutical products were to some degree dependent on their view of the efficacy of inoculations and antibiotics to prevent and control infection.

At the extremes, social position and background had some influence on the choice of medication. But differences were not always a consequence of age, education or wealth. On the one hand, only a few African commercial farmers professed to use biomedicines exclusively or to summon a veterinarian whenever they had a problem. On the other, some of the poorer stockowners, including women, would save up for tick dips and antibiotics because they felt they were far more efficacious than traditional medicines. Women were possibly more dependent on biomedicines as they lacked knowledge about medicinal plants for livestock diseases. In Mbotyi, knowledge of local remedies seemed to reside more with the older generation than the younger. However, that was not always the case. In North West Province and QwaQwa there were informants in their 20s and 30s who knew about local medicines and advocated their usage over biomedical products. In other cases, younger informants were committed to sacrificing to the ancestors as part of a more general strategy for promoting medical and social health.

In 2000 Patrick Masika and his team looked at the use of herbal remedies amongst African small-scale farmers in seven districts in the central Eastern Cape Province. They noted that 73 per cent of those interviewed said they used local plants to treat their animals.[35] More recently one of Masika's students, Oluseyi Soyelu, examined how Eastern Cape farmers dealt with wounds and myiasis (fly strike) in cattle. He discovered that 68 per cent of informants used traditional remedies. Some combined local plants with biomedical products, believing that this increased their potency.[36] The idea of potency and what makes medicines strong or effective was an important issue for the stock farmers we interviewed and is dealt with in greater detail below.

We are cautious about taking these percentages as entirely accurate. The full range of treatments used by individuals is sometimes only revealed through a sequence of interviews. Our informants also gave quite diverse responses in relation to different diseases so that interviews dealing with a narrow range of ailments may not reveal the full extent of local medical practices. However, Soyelu's finding that about 70 per cent of livestock owners used traditional medicines does not sound surprising, and it could be more. Vimbai Jenjezwa, researching for our project in the Kat River valley found that 21 of the 30 interviewed (also 70 per cent) used *amayeza esiXhosa* (Xhosa medicines) or

---

[35] Masika, et al., 'Use of Herbal Remedies'.

[36] O. T. Soyelu, 'Assessment of Plants used for the Treatment of Cattle Wounds and Myiasis in Amatola Basin, Eastern Cape Province, South Africa', MSc thesis, University of Fort Hare (2010); O. T. Soyelu and P. J. Masika, 'Traditional Remedies used for the Treatment of Cattle Wounds and Myiasis in Amatola Basin, Eastern Cape Province, South Africa', *Onderstepoort Journal of Veterinary Research*, 76, 4 (2009), 393–7.

bosmedisyne (bush medicines) – as Afrikaans-speakers called plant remedies.[37] Most stockowners preferred to use western medicines if they could afford them because they were perceived to work more quickly and effectively. In Mbotyi all those interviewed used plants for some purposes and specialists clearly knew about, and said they used, a wide range. In QwaQwa only two out of 38 informants said they never used traditional remedies to treat diseases which they attributed to natural causes, whilst 14 claimed they did not use biomedicines at all, and relied solely on locally available plants to treat sick animals and dispel ticks. In North West Province, the picture was more varied. Whereas about 25 per cent of stockowners from the Odi/Moretele area stated they would not invest in modern drugs, in the village of Mabeskraal over half asserted that they only administered biomedicines to curtail diseases. Overall, our impression is that the great majority of farmers who used traditional medicines also used biomedicines. Nevertheless, usage of local techniques and herbs is very high for birthing problems, retained placentas, constipation and diseases such as gallsickness. Stockowners who relied on biomedicines might also apply 'magical' herbal potions to protect their animals from witchcraft and misfortune (chapters 7 and 8).

Our research does not enable us to accurately quantify what percentage of livestock owners used traditional or local veterinary medicines in each site, or what proportion of their treatments were based on plants. Responses in Masakhane in the Eastern Cape as well as in the Kat River valley, suggested more frequent use of biomedicines because they were perceived to work more quickly and effectively. Both of these places, unlike Mbotyi, were close to areas with more educational institutions, towns and better veterinary services. Interviewees in these communities were generally less traditionalist and had more strongly incorporated biomedical treatments – although not to the same extent as in Koppies. However, when stockowners lacked the means to purchase medicines, or these drugs failed, they would resort to local methods of healing.[38] Sometimes farmers mixed different plants together; at other times they used them singly. In North West Province stockowners tended to combine plants to make *muthi* to protect the kraal from misfortune or evil (see chapter 7), but most dosed their animals with individual herbs to treat diseases they ascribed to environmental causes. In QwaQwa and Eastern Cape, by contrast, a combination of plants was also common for the treatment of specific ailments such as *nyoko* and *inyongo* (gallsickness). The types of flora farmers adopted varied from place to place, depending on what grew locally in the environment. Aloes were probably the most widespread in North West Province and respondents

[37] Vimbai Jenjezwa, 'Stock Farmers and the State: A Case Study of Animal Healthcare Practices in Hertzog, Eastern Cape Province, South Africa', unpublished MA, University of Fort Hare (2010).

[38] Jenjezwa, 'Stock Farmers and the State'; Mike Kenyon, 'Approaches to Livestock Health and Sickness at Masakhane' (unpublished report, July 2009).

regarded them as a particularly versatile plant, with excellent laxative properties to cleanse the system. Their sap-containing mucilaginous leaves produce a soothing unguent that is ideal for treating skin conditions. Stockowners also used aloes as a drench to combat ticks. Only in Mbotyi were they rare; leaves from a small coastal species were occasionally added to chicken feed as a laxative.[39]

Local knowledge is adaptable and exotic alien plants have been widely incorporated as traditional remedies. In the Eastern Cape for example peach leaves (*Prunus persika*) – originally from China but adopted in the Mediterranean and brought to South Africa in the early period of Dutch settlement – featured in local recipes.[40] It is one of the commonest garden fruit trees. In Mbotyi farmers made an infusion of peach leaves (*impitchi*) with an indigenous plant to kill maggots laid by flies and in the Kat River they were used to expel the afterbirth.[41] Prickly pear (*Opuntia ficus-indica*), a widespread alien in the Eastern Cape, was used for a number of medicinal purposes. Its cladodes generally had a laxative effect on animals.[42] Dold and Cocks recorded many introduced plants that were absorbed in to African medical repertoires in the districts around Grahamstown both for human and animal health.[43] More broadly, specialist healers in isiXhosa-speaking areas have incorporated many new plants and substances.[44]

Sometimes plants can disappear from an individual's pharmacopoeia. In QwaQwa a number of stockowners were finding it increasingly difficult to collect from the mountains a plant called *mositsane* (*Elephantorrhiza elephantina*) because it was becoming far scarcer. Former users, who found it an excellent remedy for a wide range of digestive ailments, had to alter their medical recipes as a consequence.[45] In Mbotyi, this plant – known as *intolwane* – had to be

---

[39] Mayikalisa Jikijela, Mbotyi, 18 March 2012.

[40] Andrew Smith, *A Contribution to South African Materia Medica*, Preface to the 2011 edition by A. P. Dold and M. L. Cocks (x–xiv).

[41] Nondege Mkhanywa, Mbotyi, 23 February 2008; peaches were mixed with *uqangazana* (probably *Clerodendrum glabrum*, an indigenous small tree), salt and paraffin for this purpose. The latter has a number of human medicinal uses: M. A. Bisi-Johnson, C. L. Obi, L. Kambizi and M. Nkomo, 'A Survey of Indigenous Herbal Diarrhoeal Remedies of O. R. Tambo District, Eastern Cape Province, South Africa', *African Journal of Biotechnology*, 9, 8 (2010), 1245–54; Jenjezwa, 'Stock Farmers and the State', 118.

[42] William Beinart and Luvuyo Wotshela, *Prickly Pear: the Social History of a Plant in the Eastern Cape* (Wits University Press, Johannesburg, 2011).

[43] A. P. Dold and M. L. Cocks, 'The Medicinal Use of Some Weeds, Problem and Alien Plants in the Grahamstown and Peddie Districts of the Eastern Cape, South Africa', *South African Journal of Science*, 96, 9–10 (2000), 467–73; A. P. Dold and M. L. Cocks, 'Traditional Veterinary Medicine in the Alice District of the Eastern Cape Province, South Africa', *South African Journal of Science*, 97 (2001), 375–9.

[44] Simon & Lamla, 'Merging Pharmacopoeia', 237–42.

[45] Gladice Mokoena, Ha-Sethunya, 13 October 2010.

purchased from some distance away.[46] Stockowners not only administered plants to heal wounds and diseases; minerals such as salt, as well as detergents, motor oil, paraffin, copper wire, tar, Coca-Cola, millipedes and animal dung all formed part of the medical repertoire.

Stock owners believed that many of their medicines worked. Even some of those who used biomedicines often stated that traditional plant extracts were just as effective but they no longer used them because they were not 'modern', they were more inconvenient to prepare, or they were becoming more difficult to obtain due to growing local scarcities. Of course efficacy was partly contingent on what stockowners believed they were preventing or curing and their analysis of particular disease situations might be very different from a diagnosis made by a veterinary surgeon. Phytochemical tests have shown that some plants do have therapeutic qualities. Viola Maphosa, for example, has analysed the properties of *Elephantorrhiza elephantina* (*intolwane*), commonly administered in Eastern Cape for constipation and expelling worms. She found that a decoction made from the roots was indeed a powerful vermifuge.[47] In QwaQwa farmers also administered the plant for digestive complaints. In Mbotyi informants used it as a preventative for the supernatural condition of *umkhondo* (chapter 6) and a sheep owner found it to be a multipurpose cure-all for diarrhoea, weak knees and to prevent late-term abortions.[48] The soothing qualities of aloes to treat wounds and skin conditions were well known and extracts from this species were readily available in many cosmetic preparations. Like aloes, *Senna italica,* known as *sebete* in North West Province, is renowned for its laxative properties. As discussed in chapter 3, stockowners regarded free-flowing bowels as an indicator of good health, and plants with aperient chemicals played an important role in treating diseases.

The importance of diet and grasses in determining sickness and health helps to explain the salience of drugs that informants believed improved appetites and cleansed the digestive system. Plants and substances like salt that have a laxative effect resonate with the cultural understandings of the aetiology of diseases and notions of how sickness affects the body. Respondents often believed that to cure diseases, they literally had to flush them out of the system. The same logic applied to human health and there was some overlap in the plant mixtures stockowners prescribed for themselves and those that they administered to their livestock. Similarly, as we noted in previous chapters on causation, informants also emphasised the importance of 'clean' blood. They associated dirty blood with poor appetites and low reproductive capacity. In North West Province *sekaname*

[46] Victoria Madikane, Mbotyi, 12 December 2011.

[47] Viola Maphosa, 'Determination and Validation of Plants used by Resource-limited Farmers In the Ethno-veterinary Control of Gastro-intestinal Parasites of Goats in the Eastern Province, South Africa', PhD thesis, University of Fort Hare (2009), 37–52.

[48] Lungelwa Mhlwazi, Kunzimbini, 19 February 2009; Victoria Madikane, Mbotyi, 12 December, 2011.

(also spelled *sekanama*; *slangkop*; *Drimia spp*) was a very popular plant, partly because stockowners thought it cleansed the blood of animals (and humans), making them vigorous and healthy. It has large red or white bulbs and its long spindly flower head has a snake-like appearance, giving rise to the Afrikaans name, *slangkop*. In the Eastern Cape, a mix of aloe and prickly pear leaves, boiled and strained, was used as a blood purifier for humans.[49]

We are not in a position to judge the efficacy of medicinal plants. Kassim Kasule, the veterinarian at Lusikisiki, noted that animals have a capacity to recover themselves.[50] Though herbal remedies may not be specific cures for particular diseases, they may have general properties as analgesics, antibacterials or purgatives and could thus have some effect on pain, on secondary bacterial infections, or on worms. By treating secondary infection, the animal's own antibodies could be enhanced. If worms were successfully reduced, for example, then the protein level in the blood could increase and help improve an immune response to other diseases. In this light, laboratory-based assessments are certainly important in testing efficacy but these are based on chemical analysis and the action of plant solutions, or extracts, on particular bacteria. To our knowledge, there have not yet been scientific tests of plant remedies over the long term in the field, given in the combinations that they are commonly administered, with repeated doses or application amongst animals that were accustomed to these treatments.

Advocates of local remedies also praised their qualities because of the apparent lack of side effects, which purportedly distinguished them from biomedicines. Concerns about biomedicines seem to be embedded in popular observations, or at least we heard rumours that modern manufactured drugs were harmful because they were not natural. These ideas arose in testimonies such as that of Zulu Mokoena, discussed above, and in other accounts from North West Province. Samuel Mothapo from Slagboom expressed his anxieties about biomedical products in political as well as medical terms. He used vaccines himself but suggested that the attitude and policies of the apartheid government generated a general hostility to 'white man's medicine' that had not disappeared. During apartheid, people distrusted veterinarians as they allegedly never told stockowners what they were vaccinating for. Suspicions arose that the state was deliberately trying to spread diseases to kill African animals and keep their livestock numbers down.[51]

Similarly in Kgomo Kgomo informants not only accused drug manufacturers of dropping infective ticks from planes (see chapter 2), but also of creating vaccines that introduced new infections such as *lamsiekte* and lumpy skin disease.[52] Noah Motaung and Philip Mokwena from Fafung explained that

---

[49] Beinart & Wotshela, *Prickly Pear*.
[50] Kassim Kasule, Lusikisiki, 18 Feb 2009.
[51] Samuel Mothapo, Slagboom, 10 Nov 2009.
[52] Simon Mokonyana, Kgomo Kgomo, 29 January 2010.

many farmers were cautious about vaccinations, because they believed that veterinarians spread diseases by using only one needle to inject many different animals. In this specific case, their mistrust generated a strong preference for their own tried-and-trusted medicinal herbs.[53] More generally such unease about biomedicines led to uncertainty about the consequences rather than any outright rejection. Many saw them as more effective or acting more quickly, but inadequate for a full treatment. In Mbotyi a few people associated susceptibility to supernatural forms of ill health with the use of western medicines.[54] Zachariah Maubane, a school teacher from Mafikeng who kept livestock near Bela Bela, offered a different perspective. He reasoned that many Africans, of all ages and educational backgrounds, disliked injecting livestock because of concerns that the vaccines themselves could spread diseases to humans. He accepted that animal vaccines stayed in the blood for about 14 days and that it was not advisable to slaughter during this period. However, he explained that some people remained sceptical and were convinced that the meat was permanently tainted and unsafe for human consumption. Farmers entertaining those ideas also rejected inoculation as a veterinary health policy and preferred to use traditional plants.[55]

Unlike biomedicines, the availability of plant medicines can be erratic. Informants mentioned the seasonality of some species as a disadvantage and there was generally very limited capacity to store medicinal flora during the winter months. Farmers lacked the means to dry and preserve plants and many believed that they had to be freshly plucked to be effective. Although they may not have always been aware of this problem, the strength of a plant remedy is probably contingent upon how much of the plant they used, its state, whether lush or withered, and the amount of time it had been steeped in a solvent. Most farmers prepared their medicines using water and had no other chemical means of extracting the active ingredients. Especially in the densely populated area of QwaQwa, informants expressed concerns about the disappearance of potentially valuable flora as a result of unsustainable collecting, not only by rural people, but by those who gathered and sold traditional medicines in city street markets.[56]

---

[53] Interview with Noah Motaung and Philip Mokwena, Fafung, 27 Oct 2009. One veterinarian noted that while it was preferable to use different needles for each animal, this was expensive and there was probably little danger in using one for a single herd as the animals shared infections.

[54] Nonjulumbha Javu, Mbotyi, 19 December 2011; see chapter 6.

[55] Zachariah Maubane, Mafikeng, 2 Nov 2009.

[56] L. J. McGaw and J. N. Eloff, 'Ethnoveterinary use of Southern African Plants and Scientific Evaluation of their Medicinal Properties', *Journal of Ethnopharmacology*, 119, 3 (2008), 559–74; Phatlane William Mokwala, 'Antibacterial Activity of Plants that are used in the Treatment of Heartwater in Livestock and the Isolation and Identification of Bioactive Compounds from *Petalidium Oblongifolium* and *Ipomoea Adenioidides*', DPhil, University of Pretoria (2007); W. Wanzala et al., 'Ethnoveterinary Medicine: A Critical Review of its Evolution, Perception, Understanding and the Way Forward', *Livestock Research for Rural Development* 17, 11 (2005); D. van der Merwe,

Despite ambivalence about biomedicines, many Tswana stockowners did use them. Some invested in vaccines such as Blanthrax – a combined formula for anthrax and blackquarter.[57] The antibiotic Terramycin featured in many testimonies in North West Province and Mbotyi. Terramycin LA (long-acting) is available as a solution for injection and because of this some stockowners treated it like a vaccine, believing they could use it to prevent as well as cure most diseases. Some respondents administered Terramycin on its own, others with a combination of therapeutics, including plants and other substances. Tata Banjela from Mbotyi said that 'Terramycin LA helps for everything'.[58] This long-acting formula, marketed by Pfizer, is widely available in 100 ml bottles. Ivermectin was also a popular biomedical product for worms in North West Province. Our impressions were that knowledge of, and demand for, injectables, particularly Terramycin and Dectomax (doramectin), effective against ticks as well as worms, had grown in Mbotyi over the four years of the project, from 2008 to 2012. During Beinart's visit there with a veterinary surgeon in 2012, a number of stockowners requested *isitofu* (generally pronounced '*stof*' from the Afrikaans original), as injections or vaccines are called in isiXhosa, of these two drugs.[59]

In Koppies all of those interviewed were committed to biomedicines. Isaac Mzima for example rented 60 hectares of land from the Sibongile Trust Settlement Scheme set up in the early 1990s to provide grazing land for African smallholders. In October 2010, Mzima owned about 50 cows and 100 goats, more he said than anyone else in the settlement. He attributed his ability to accumulate animals to the income he had accrued from taxi driving and to his use of modern medicines, such as dips and antibiotics. He called out the state veterinary surgeon from Vredefort to inoculate for diseases like brucellosis and lungsickness. His main problem was poisonous plants such as tulp and *dubbeltjes* (*Tribulus terrestris*) for which there are no vaccines. Mzima claimed he had never used traditional medicines and did not believe many people, if anyone, in the Sibongile Settlement did either. Unlike many Tswana informants he found the idea that many Tswana men allegedly banned women from the kraal because they caused abortions in livestock laughable and unheard of in his neighbourhood.[60] Johannes Malinde, mentioned above, born in 1942, remembered the 1950s as a decade of real change in the area – the time 'when modern medicines came to Koppies'.[61] Before then African stockowners inserted wire in the dewlap to ward off blackquarter and dosed their animals with salt to

(cont) G. E. Swan and C. J. Botha, 'Use of Ethnoveterinary Medicinal Plants by Setswana-speaking People in the Madikwe area of the North West Province of South Africa', *Journal of the South African Veterinary Medical Association*, 72, 4 (2001), 189–96.

57 www.msd–animal–health.co.za – includes useful listings of vaccines in SA.

58 Tata Alfred Banjela, Mbotyi, 20 December 2011.

59 11–25 March 2012, accompanied by Dr Roger Davies.

60 Isaac Mzima, Kwakwatse, 8 October 2010.

61 Johannes Maline, Kwakwatse, 8 October 2010.

improve appetites. The most popular plant was *mositsane,* the leaves of which produced an infusion that aided digestion. By 2010, Malinde claimed, no-one used plant medicines, although salt was still popular as an appetite stimulant.

Nevertheless the administration of biomedicines was often haphazard. We heard of the frequent use of Terramycin for all sorts of complaints, including those for which the antibiotic was not suitable. Many of the stockowners who procured vaccines and antibiotics did not have a biomedical explanation for diseases and any mistaken diagnoses could lead to the expensive, and potentially harmful, applications of the wrong types of medicine, and antibiotic resistance in the longer term. Farmers often administered vaccines with syringes or dosing guns, but they lacked clean needles to inoculate each animal, and ironically, some might potentially have spread infection in the way that they attributed to government veterinary officers. Many vaccines needed to be stored in a refrigerator to remain effective. Few had this facility. In North West Province and Mbotyi we saw medicines stored in outhouses, in cupboards, or under beds where the temperatures could reach over 30 degrees in summer. We also noticed numerous phials of vaccine that had long passed their expiry dates. Vaccine failure could generate scepticism about the efficacy of modern medicines.[62]

In QwaQwa, stockowners placed far less trust in Terramycin as a cure-all and possibly had a greater reliance on herbal concoctions than their counterparts in North West Province. Nevertheless, some did vaccinate for blackquarter, which many informants had identified as the most troublesome disease they faced. According to Mothupi Molefe, an AHT based in Phuthaditjhaba, about 80 per cent of farmers injected for blackquarter and 20 per cent refused. Molefe did not think money was the issue as blackquarter vaccine retailed at about R2 (£0.14 in May 2013) per animal and, in his view, all the local stockowners could afford that. He was sure that there was a latent dismissal or even distrust of modern medicines, and stockowners adhered to older ways of treating animals as an assertion of Sotho identity. *Serotswana* or blackquarter was endemic to the region and stockowners had their own methods of preventing the disease, which they felt had stood the test of time. They claimed that traditional medicines were far more effective and safer to use than the biomedical products that the Department of Agriculture advocated.[63]

In the Eastern Cape, by contrast, the Provincial Government had provided a free anthrax and blackquarter vaccination over many years. Its application was always a little uneven and this might have resulted in sporadic local outbreaks of anthrax. Informants complained that the inoculation campaigns had become increasingly haphazard in recent year. However, neither of these diseases featured significantly in the interviews and it seems that they are less prevalent in Mbotyi

---

[62] Ronette Gehring, 'Veterinary Drug Supply to Subsistence and Emerging Farming Communities in the Madikwe District, North West Province, South Africa', MMedVet (Pharmacology), University of Pretoria (2001).

[63] Mothupi Molefe, Phuthaditjhaba, 14 October 2010.

than in QwaQwa and North West Province. In North West Province the annual inoculation for anthrax seems to have become more sporadic and some farmers suggested that anthrax was on the rise. Molefe, the AHT cited above, was concerned that this rejection of vaccination ensured that diseases like black-quarter, which were controllable, remained a local scourge.

A number of respondents explained that they liked to use a mixture of plants or combine traditional medicines with pharmaceutical drugs to enhance the potency and speed up recovery. Stockowners based their perceptions of potency on their own experiments with different types of drugs, which reflected a will-ingness to mix medicines from a range of sources. An example of this approach appeared in the testimony of Jacob Tshabalala from QwaQwa. Tshabalala used revenues accrued from taxi driving to invest in commercial livestock produc-tion and he grazed about 130 cattle and a similar number of goats on a rented private farm outside Phuthaditjhaba. He administered modern medicines for everything apart from expelling placentas. For the latter he employed *qobo* which we think is *Gunnera perpensa,* or river pumpkin, which is widely used medicinally. He described it as a bulbous plant that looks like a pumpkin and grows in the mountains. He cut the bulb into small pieces, boiled it for 30 minutes and gave the animals a drench in the morning and evening until the afterbirth emerged. Normally this was far more effective, in his view, than the pessaries the veterinarians recommended. However, if the placenta was partic-ularly stubborn, he might use both pessaries and *qobo* as he has found that one intensified the effect of the other.[64] Michael Ncobuga, also from QwaQwa, owned three head of cattle and showed that smaller-scale livestock owners adopted similar practices. Ncobuga had a problem with blackquarter so he bought vaccines from the cooperative and used them jointly with extracts derived from the boiled bulb of *phate ya ngaka (Hermannia depressa)*, which he administered as a drench. He believed that the old and new medicines had together saved his cattle.[65] Evidence suggests that in the Eastern Cape and North West Province there was also the conviction that a mélange of plants, vaccines and modern drugs could improve some treatments.[66] To cure *gala* (gall-sickness) Abraham Meno from Mareetsane mixed salt with fish oil to relieve constipation, but he also injected his cattle with Terramycin for good measure.[67]

Some traditional healers also admitted prescribing biomedical drugs for particular conditions, revealing how adaptive their knowledge was and how local medicine was continuously evolving.[68] Daniel Gambo, a healer and stock-owner from Lusaka in QwaQwa, believed that vaccines were important for controlling contagious and infectious diseases in both humans and animals. In

[64] Jacob Tsahbalala, Phuthaditjhaba, 13 October 2010.
[65] Michael Ncobuqa, Lusaka, 3 March 2011.
[66] Masika, et al., 'Use of Herbal Remedies'.
[67] Abraham Meno, Mareetsane, 4 November 2009.
[68] Simon & Lamla, 'Merging Pharmacopoeia'.

his opinion, biomedicines were less effective for treating chronic conditions and surgical wounds. He had inoculated his 37 cattle against blackquarter and stated that the vaccines worked much better than traditional methods such as placing wires through the dewlap (see chapter 6). However, he used plants such as *qobo* to expel retained placentas, *thobeha (Seddera sufruticosa)* to accelerate the healing of broken bones and *mositsane (Elephantorrhiza elephantina)* for a range of complaints, such as cleaning out the reproductive system and treating inflamed udders (which may be the result of mastitis). Sometimes farmers did consult him about animal medicines and he would recommend that they saw a veterinarian or AHT if he felt that his *muthi* would not be safe or effective. He was not precious about Sotho medicine and thought that traditional healers could learn from other medical systems.[69] Other healers in QwaQwa demonstrated an equal openness. Paulus Motosoeneng, whose wife was an *inyanga* (traditional healer), used a mixture of therapeutics as directed by his spouse to treat his livestock, and Rosalina Johanne would call a vet if she felt her medicines were not working.[70]

When cattle had *umhlaza*, a word which could describe both warts and forms of pox, and probably lumpy skin disease, Sonwabile Mkhanywa in Mbotyi described people using both an injection of penicillin (Peni LA) and an infusion of euphorbia given as a drench.[71] Peni LA could be bought from chemists in Lusikisiki without a prescription. They thought that the euphorbia would kill the foreign matter in the blood that caused the external warts. Some cut off the warts and put a euphorbia poultice on the wound. It was widely believed to stop bleeding. Another man mentioned mixing Epsom salts with the plant remedy for *inyongo* to facilitate more rapid 'toilet' and 'cleanse it inside'.[72] At least one animal had died when it was cut in this way, and the bleeding could not be stopped.

We gained fascinating insight into the possible dangers of local treatments, and the problems of diagnosing, when we visited the homestead of Sipho Madiya in Mbotyi where 11 goats out of a herd of 22 had recently died.[73] Madiya was an absent migrant mine-worker who – judging by his homestead and herds – was prospering. There were substantial new buildings, both square and round, and he had accumulated a herd of 32 cattle as well as goats and sheep. His wife and children looked after the animals. They did not know what had killed the goats, nor was it apparent to others they consulted. Three possibilities arose.

First, the deaths followed an attempted treatment for external ticks and flies using motor oil. Some people use a small amount of motor oil mixed with

[69] Daniel Gambo, Lusaka, 28 February 2011.
[70] Paulus Motsoeneng, Lusaka, 3 March 2011; Rosalina Johanne, Lusaka, 1 March 2011.
[71] Sonwabile Mkhanywa, Mbotyi, 22 March 2012.
[72] 'Rice', Mbotyi, 23 February 2012.
[73] Numzana Madiya, Mbotyi, 23 March 2012.

liquid dip to spray or smear on goats, especially round the horns and hooves, so that the effect is prolonged. Ideally, the oil should be clean and thin. [74] At Madiya's homestead they had used old, dirty oil, salvaged from the workshop at the Magwa tea factory, without dip. They usually used dip but they had run out so they applied the oil alone. The children looking after the goats brought a battered Coca-Cola bottle with murky oil to show us. It seemed likely that in addition to using suspect materials, and getting the mixture wrong, they had put too much of it onto the goats' heads in hot summer weather. The goats apparently started dying only a couple of days after the application of motor oil and they thought that the area around the ears had shed hair and turned reddish. It is possible that too much thick, dirty oil killed the goats by stopping perspiration and damaging the skin.

Second, the goats may have died of an infection. Numzana Madiya thought that they had not put oil on two of the goats which died. They also cut the goats open after death and described lungs and intestines that broke up easily; it was also suggested that the lungs were stuck to the thorax. This may have indicated a longer-established disease, such as a form of pneumonia, or heartwater. The temperatures of the surviving goats were very high and they may have been diseased. The children described some of the goats coughing and making an odd noise, and foam coming from their mouths.

Third, around the time of the deaths a team employed by Working for Water had been cutting and poisoning invasive species, especially paraffin bush – *Chromolaena odorata*, nearby. We saw some of the plants within a couple of hundred metres of the homestead. The branches were cut away and a systemic weed killer painted on the remaining stems. The team was using a blue-coloured poison called Lumberjack 360 SL (triclopyr) made by Arysta Life-sciences. During training, they had not been told that it could be dangerous to livestock, nor had they experienced any problems in the past. It is unlikely, but not impossible, that the goats had grazed on the leafless blue-painted bush stumps. The goats were not herded. They just roamed around in the day, so that they would have the opportunity to imbibe poison without anyone noticing. This conjuncture of possible causes may be unusual. The symptoms match those of an infection that has recently affected a number of animals in Mbotyi. But carelessness in therapeutics, administered by children who were looking after the animals, may have been a cause of death or a contributory factor. The episode pointed to uncertainty of diagnosis and the difficulty of confirming the causes of death without post-mortems and tests.

---

[74] Rice, Mbotyi, 23 March 2012.

Conclusion

In sum, informants from all of our sites used a mix of traditional plant remedies and biomedicines. A growing number saw biomedical treatments as quicker
or more effective, but also believed that some local plants were more suitable
for certain conditions or they felt culturally committed to older practices.
Biomedical products were often used in a generalised way, to combat both
named specific diseases and rather non-specific ailments. Terramycin, for
example, has become something of a cure-all in the rural areas. Dectomax,
while specified for worms, is also seen to help with ticks as well as a general lack
of condition.

Overall our impression is of a gradual shift towards biomedicine, hastened by
stockowners' incapacity to deal with the chronic problem of ticks as well as
some new scourges such as lumpy skin disease. While commitment to plant
knowledge is striking in some areas, informants also expressed uncertainty over
its effectiveness. Transmission is often restricted, the range of remedies probably declining, and some informants who wished to know more, could not
easily gain access to this knowledge. One of the main reasons why many respondents claimed they would always have to rely partly on traditional medicines
was because of witchcraft and stock theft. In this conceptualisation of disease,
sickness referenced environmental and spiritual as well as medical worlds. Good
health amongst animals was partly predicated on managing the supernatural
realm.[75] We turn to this territory in chapter 7.

[75] Elphars Dinne, Mabeskraal, 4 February 2010; George Malatsi, Mabeskraal, 5 February 2010;
Laurence Bodibe, Ramatlabama, 10 February 2010; Matthews Motshwrateu Moloko, Magogwe
(Nr Mafikeng), 11 February 2010; Kgomotso Goapele, Mantsa, 12 February 2010; Hendrik
Metswamere, Mantsa, 12 February 2010; April Nhlapo, Lusaka, 28 February 2011.

6

Medicinal Plants

Their Selection & their Properties

Our plants are powerful and they work well. Vets cannot identify and treat all diseases so we need our own medicines. Cows are like humans they get the same diseases and need the same cures. Plants that are good for humans are good for our cattle as well. (April Nhlapo, Lusaka, QwaQwa).

April Nhlapo kept his two cows and one calf in his kraal in the backyard of his house in the densely populated village of Lusaka. He explained how he loved his cows 'they are my bank and my source of milk'. He sold milk to his neighbours and traded calves when he needed cash. Like many stockowners we interviewed in Lusaka he had a good knowledge of locally available plants and treated himself and his cattle with 'medicines I find in the veld'.[1]

Pre-industrial societies worked with local natural resources on a daily basis and often achieved a rich knowledge of the plant species around them. African societies used plants for a wide range of purposes, from food and building, to furniture and utensils. Plants provided countless elements of material culture and in this way they were part of a deep indigenous knowledge; people knew where they grew, which were edible, which could be cultivated, and how they could be used. Some species had specific cultural connotations or ritual functions. Plants were equally important for medicinal and veterinary uses and, as in many pre-industrial societies, the boundary between food and medicine was not always tight. Medicines were edible plants, or plants that were not poisonous in at least small quantities, and which had – or were thought to have – specific effects on people or animals. These effects were often more extreme or precise than plants consumed in larger quantities for nutrition. Animal products could also be part of the pharmacopoeia, sometimes mixed with plants.

[1] April Nhlapo, Lusaka, QwaQwa, 28 February 2011.

163

Plant medicines have been widely recorded in South Africa over a long period of time. Some were mentioned by nineteenth-century botanists and physicians, at a time when medical training included considerable botanical input. A few South African plants, such as *buchu* (*Agathosma*), achieved global recognition for their properties.[2] Andrew Smith, a teacher at Lovedale School, published a compilation of eastern Cape medicinal species in 1895. Many were also recorded in dictionaries of African languages, produced largely by missionaries, in the late-nineteenth and early-twentieth centuries.[3] Settlers, both Afrikaners and to a lesser extent English-speakers, borrowed and adapted plant remedies from African people; both also experimented and incorporated exotic species that poured into South Africa through the global shipping routes of empire. John Mitchell Watt and Maria Breyer-Brandwijk published two editions of an encyclopaedic collection of medicinal plants, with comments on their use, in 1932 and 1962.[4] Recent popular books such as Ben-Erik van Wyk and Nigel Gericke's *People's Plants* have added to the store of knowledge.[5] There has been a perhaps surprising openness by scientists to these elements of African knowledge over a long period of time. Such compilations mention, and do not generally dismiss, the specific medicinal and veterinary properties that Africans attributed to various plants. Botany has been a particularly strong field in South African science more generally and thousands of species must have been mentioned in these various volumes.

Van Wyk and Gericke's text is an example of a more celebratory approach to local knowledge that increasingly characterises post-apartheid South Africa. It represents an acceptance of more relativist approaches to knowledge, aims to salvage information 'before it is irretrievably lost to future generations', and to 'raise awareness of the role plants play in people's daily lives'.[6] Their publication is part of a more general effort to promote a shared post-apartheid cultural context. Tony Dold and Michelle Cocks offer the most detailed regional discussion of African plant knowledge, which is explicitly celebratory in tone. They go further than recording, and argue both for the centrality of plants in Xhosa culture and the importance of African knowledge in preserving biodiversity in the long term.[7]

---

[2]   Christopher H. Low, 'Different Histories of Buchu: Euro-American Appropriation of San and Khoekhoe Knowledge of Buchu Plants', *Environment and History*, 13, 3 (2007), 333–61.
[3]   A. Kropf, *A Kafir-English Dictionary*, second edn, ed. R. Godfrey (Alice, Lovedale Press, 1915).
[4]   John Mitchell Watt and Maria Gerdina Breyer-Brandwijk: *The Medicinal and Poisonous Plants of Southern Africa* (Edinburgh, E & S Livingstone, 1932); *The Medicinal and Poisonous Plants of Southern and Eastern Africa*, second edn (Edinburgh, E & S Livingstone, 1962).
[5]   Ben-Erik van Wyk and Nigel Gericke, *People's Plants: A Guide to Useful Plants of Southern Africa* (Pretoria, Briza, 2007).
[6]   van Wyk & Gericke, *People's Plants*, 7.
[7]   A. P. Dold and M. L. Cocks, *Voices from the Forest: Celebrating Nature and Culture in Xhosaland* (Auckland Park, Jacana, 2012).

As deputy President in Nelson Mandela's government (1994–99), and as President (1999–2008), Thabo Mbeki promoted an 'African Renaissance' as a vehicle for instilling national and continental pride and as an ideological tool for development. He called for the appreciation and preservation of African knowledge and cultural practices, and these included medical traditions. As a consequence, state funding has been available for scientific investigations into local African plant usage. The Onderstepoort Veterinary Institute, long at the heart of scientific research, is now home to the Jotello Soga Ethnoveterinary Garden of African medicinal plants named in honour of South Africa's first – and for many years only – black veterinary surgeon who worked in the Cape in the late-nineteenth century. Onderstepoort, as well as the Veterinary Faculty of the University of Pretoria, have researched a number of African plant remedies in search of effective chemical compounds.[8] Patrick Masika and his team at the University of Fort Hare in Eastern Cape have focused on local practices, including medicinal plants, and have noted the efficacy of certain therapies. His students, such as Viola Maphosa, have advocated more research into local remedies and the dissemination of useful knowledge amongst African farmers.[9] Proponents of traditional medicines argue that their cheapness and easy availability means they could be ideal solutions for treating some common, endemic diseases, as long as they are not overexploited. They are beginning to develop a scientific and experimental, rather than just a local and practical, dimension to ethnoveterinary research. For example, the Agricultural Research Council unit at Roodeplaat, north of Pretoria, is starting to cultivate medicinal plants to promote their sustainability and they plan to include ethnoveterinary species. In this way, sustainability could be ensured and popular plant medicines made available to more people. It may be possible to translocate useful species to new geographical areas, enabling stockowners to extend their pharmacopoeia.[10]

In view of these many investigations into local medicines, it is unlikely that we have recorded, in our interviews, plant species that have not been noted before. We did not attempt exhaustive research in specific sites and, as mentioned in chapter 5, informants were cautious about detailing all of the plants they knew. Our contribution is to discuss plants in common use in

[8] Discussions with Kobus Eloff, University of Pretoria, 25 February 2010, 28 November 2012.

[9] Viola Maphosa, 'Determination and Validation of Plants used by Resource-limited Farmers in the Ethno-veterinary Control of Gastro-intestinal Parasites of Goats in the Eastern Cape Province, South Africa', PhD thesis, University of Fort Hare (2009); O. T. Soyelu and P. J. Masika, 'Traditional Remedies used for the Treatment of Cattle Wounds and Myiasis in Amatola Basin, Eastern Cape Province, South Africa', *Onderstepoort Journal of Veterinary Research*, 76, 4 (2009), 393–7'; Sipho Moyo, 'Alternative Practices used by Resource-limited Farmers to Control Fleas in Free-Range Chickens in the Eastern Cape Province, South Africa', MSc thesis, University of Fort Hare (2009).

[10] Discussion with Riana Kleynhans and colleagues at Roodeplaat, 27 November 2012.

different sites, the combinations of plants – which have less often been documented – and to think about the relationship of plants to other strategies around animal health.[11]

Choice of plants

Choice of plants depended to some degree on received knowledge and partially on trial and error. Stockowners selected materials that appeared to be effective at quelling or removing the symptomatic manifestations of infection. They used plants to prevent as well as cure diseases, although there was a greater emphasis on remedies than prophylaxes. Stockowners relied on plants with purgative properties but they also rated herbs that they administered to themselves, or those that they had observed wild animals digesting in the veld. Informants had noted that wild fauna, especially species of antelope, seemed to be less susceptible to many livestock diseases, suggesting that they had a particularly healthy diet, which livestock could emulate. There are probably other reasons for these immunities, although Nguni cattle, and goats, which browse as well as graze, might benefit from the wider range of plants that they eat. Other practical reasons for selecting particular species included the availability of particular plants and ease of use. Some flora also bore notable colours or textures which informants believed pointed to particular curative properties.

Anna Pooe, a traditional healer from Shakung in North West Province, owned 20 white goats which she kept for healing rituals as well as for social ceremonies and funerals. She explained that she treated humans and animals symptomatically. She did not look for a collection of symptoms which pointed to a particular disease, but rather sought individual indicators of ill health which she treated separately. Pooe maintained that she cared for her goats in the same way as she looked after her human patients. For people and animals, she prescribed *sekaname* (*slangkop, Drimia spp*) and/or aloes when she diagnosed 'dirty blood' and general ill health. For coughs and runny noses she administered *lengana* (*Artemisia afra),* a popular decongestant and fever suppressant in South Africa.[12] For diarrhoea she recommended *ting* (a kind of porridge made of sorghum) and laced it with methylated spirits, at least for her goats.[13]

Other farmers in Shakung also drew parallels between human and animal medicines. For example, they gave their livestock *Monna maledu* (African potato; *Hypoxis hemerocallidea*) to cleanse the blood and cure diarrhoea, in the same way that they dosed themselves. In Fafung women took an infusion made of *lepate*

---

[11] Spellings of African plant names are phonetic, approximately as an English speaker would write them, and may vary from region to region.

[12] Ben-Erik van Wyk, *Medicinal Plants of South Africa* (Pretoria, Briza, 1997), 44–45.

[13] Anna Pooe, Shakung, 22 January 2010.

(*seepbos*; *Dicerocaryum eriocarpum*), a long creeper with pretty pinky-purple bell flowers, to ease labour pains and expel the afterbirth. Men used the same plant to force retained placentas out of cows.[14] In Mabeskraal people would treat human and animal fractures by scarifying the skin and rubbing ash from burnt pieces of *thobega* bulb (*Seddera sufruticosa*) into the cuts. Some people also buried *thobega* at the spot where the injury had occurred, believing this led to more rapid healing.[15] Sesotho speakers in QwaQwa had a similar name, *thobeha*, and they performed analogous rituals to mend bones.[16] There was thus some overlap in the naming and use of medicinal herbs across Sesotho and Setswana language groups. One specialist practitioner near Mbotyi buried herbal material in the ground to heal bones but gave it a supernatural meaning (chapter 8). A local *sangoma* called this a Sotho practice. Underlining the administration of these plants was the assumption that what worked for humans also worked for animals.

Stock owners in North West Province not only considered the drugs humans took when compiling their pharmacopoeia, but also looked to wildlife for clues about useful plants. Scientific studies of zoopharmacognosy, the scientific term for self-medication in animals, suggest that some species not only instinctively know which plants are good for them, but actively seek them out to improve nutrition and to flush out worms and toxic flora.[17] In North West Province some stockowners claimed that they had learnt about the therapeutic qualities of certain plants by watching their animals graze in the veld. If livestock spontaneously headed towards particular grasses, then it seemed possible that they were attracted to them because of their therapeutic properties. Elias Pooe from Shakung, for instance, had noticed that his cattle liked to eat *rramburo* – a large bulbous plant with long leaves. His own experiments showed that it worked well for cleansing out the reproductive tract following births and abortions.[18] Majeng Motsisi from Mabeskraal had resorted to following animals around the veld because his father had died before handing down his knowledge about medicinal plants. He studied wild animals as well as his cattle and noted how kudu, springbok and impala all feasted on aloes. He consequently transplanted some aloes to his backyard and has been dosing them to his stock ever since.[19]

Some farmers observed that antelopes like kudu, which are largely browsers, as well as animals like porcupines, ate a very varied diet. They surmised that

[14] Fafung group meeting, 25 January 2010.

[15] Henry Ramafoko and Elsie Ramafoko, Mabeskraal, 2 February 2010; Rebecca Molekwa, Mabeskraal, 3 February 2010.

[16] Daniel Gambo, Lusaka, 28 February 2011.

[17] Phatlane William Mokwala, 'Antibacterial Activity of Plants that are used in the Treatment of Heartwater in Livestock and the Isolation and Identification of Bioactive Compounds from *Petalidium Oblongifolium* and *Ipomoea Adeniodides*', DPhil, University of Pretoria (2007), 10–11.

[18] Elias Pooe, Shakung, 19 January 2010.

[19] Majeng Motsisi, Mabeskraal, 3 February 2010.

their dung was therefore rich in nutrients that could protect livestock from toxic weeds and also cure diseases. Samuel Mothapo from Slagboom believed that blackquarter arose from a lack of salt in the diet, too much bile in the intestines, or because people had urinated near the kraal after eating meat from a cow that had died of this disease (chapter 3). Before there were blackquarter vaccines, he explained, stock owners used to mix kudu or porcupine faeces with water and then add the rehydrated dung to the ground-up bulb of the *sekaname* plant and the dregs that remained after processing sorghum beer. Mothapo claimed that some people still used this method to prevent and cure blackquarter as they were convinced that it worked. However, it had become increasingly hard to find kudu dung as these antelopes were no longer so common in North West Province. Consequently, farmers now had to go to the cooperative to buy their blackquarter medicine.[20] Kudu dung also featured as a popular cure for blackquarter in Sutelong.[21] We did not come across dung from wildlife in *muthi*s in Mbotyi but Delinkosi Soyipha swore by pig dung, mixed with crayfish eggs, as a cure for *umkhondo*.[22]

In Fafung informants said that they gave their livestock antelope dung mixed with water to protect them from *mohau* (*gifblaar; Dichapetalum cymosum*) the toxic weed which periodically wiped out large numbers of livestock in that area.[23] In Kgomo Kgomo, supporters of kudu dung said it stopped livestock contracting blackquarter and *lamsiekte* as well as preventing *mohau* and *sekaname* poisoning. Sello Moiosane explained that the dung contained all the poisonous grasses and plants that livestock could possibly ingest and by feeding antelope faeces to domestic animals, he was exposing them to all the harmful agents they could come across in the veld. Drenching livestock with antelope dung, which in his opinion contained an adulterated mix of toxins, enabled his cattle to develop some resistance to diseases.[24] Such explanations suggest a local belief in a type of resistance or immunity derived from administering a substance believed to contain the poisonous or contagious material; this could be found in the veld rather than from vaccines. We do not know whether such beliefs are derived from old medical practice or are influenced by newer explanations associated with vaccines.

Sometimes stockowners looked not to wild fauna but to flora for clues about their medicinal properties. Informants said they sought similarities between plants and the disease they wished to treat. Some species of *sekaname,* for example, have a dark red bulb reminiscent of blood. In North West Province, people chopped up the bulb and added water to make a rich red liquid to treat

---

[20] Samuel Mothapo, Slagboom, 10 November 2009.
[21] Sutelong group meeting, 27 January 2010; Piet Hlongwane, Sutelong, 28 January 2010.
[22] Delinkosi Soyipha, Mbotyi, 18 December 2012.
[23] Fafung group meeting, 25 January 2010.
[24] Sello Moiosane, Kgomo Kgomo, 29 January 2010.

'blood diseases' such as anthrax or blackquarter.[25] Plants such as *lepate* that became viscous when added to water were popular for dealing with retained placentas. They were deemed to be effective as the slimy consistency of the *lepate* infusion resembled vaginal fluids.[26] Philani Mkhanywa in Mbotyi found that *umkhomakhoma* was effective for washing out retained placentas.[27] He classified flora with such properties as 'wet' plants and the idea seemed to include both varieties which exuded moisture when cut and species that grew in wet places. Medicines that were rich in tannins like *sebete (Senna italica)* were commonly used for diarrhoea and constipation, especially in the village of Mabeskraal.[28]

These observations were not unique to North West Province and appeared in other accounts of traditional healing. In his thesis on the ethnobotany of the Venda, Dowelani Mabogo described similar correlations between diseases and certain plants in the north-eastern part of the Limpopo Province. The Venda too looked for red plants for blood diseases. Flowers with a large number of petals conferred fertility; branches full of milky latex encouraged lactation; trees endowed with bark that repaired itself quickly after being cut were ideal for treating wounds. Mabogo drew parallels with the medieval idea of the 'doctrine of signatures'.[29] This was the belief that God had marked medicinal plants so that they resembled the disease or affected parts of the body they were designed to heal. Stockowners that we interviewed did not conceptualise the curative power of plants in religious terms. But over the generations some had looked for physical indicators in their search for medicinal herbs that were appropriate for treating particular conditions.[30] These connections were, however, not omnipresent or even dominant. There were many examples of mixtures of plants used for conditions where a resemblance was not obvious or articulated.

Some stockowners searched for new medicinal herbs and new types of medicine not least because they were aware that certain plants were becoming harder to find in some areas. Thembela Kepe has noted the growing scarcity of some

[25] Anna Pooe, Shakung, 22 January 2010.

[26] Fafung group meeting, Shakung, 25 January 2010; Group meeting, Sutelong, 27 January 2010; Phillimon Mothaba, Kgomo Kgomo, 29 January 2010.

[27] Lenox Philani Mkhanywa, 4 December 2012. This is given as *Nephrodium athamanticum* in Kropf. It is a large fern like plant, common in the forests around Mbotyi.

[28] Mokoboro Rakgomo, Mabeskraal, 1 February 2010; Henry Ramafoko and Elsie Ramafoko, Mabeskraal, 2 February 2010 ; Emily Diole, Mabeskraal, 2 February 2010 ; Philemon Ramafoko, Mabeskraal, 3 February 2010; George Malatsi, Mabeskraal, 5 February 2010; Jeremiah Ramakopeloa, Mabeskraal, 5 February 2010; Mokoboro Rakgomo, Mabeskraal, 1 February 2010; Majeng Motsisi, Mabeskraal, 3 February 2010.

[29] Dowelani Edward Ndivhudzannyi Mabogo, 'The Ethnobotany of the Vhavenda', MSc Thesis, University of Pretoria (1990), 166–8.

[30] See also Christopher Low, 'Khoisan Healing: Understandings, Ideas and Practices', DPhil, University of Oxford, (2004).

medicinal plants for human use in Mpondoland.[31] This is a result of urban demand
and commercialisation of the traditional medicine market. People no longer pluck
or dig up just enough plants for immediate preparation or to store in sustainable
quantities. Rather, the rural poor, especially women, have tried to make a living
from collecting plants and selling them in urban centres like Durban. A few species
that are popular with traditional healers have been depleted.[32]

Kepe's account resonated with a number of testimonies from QwaQwa
where urbanisation and over-exploitation of plants appeared to be a real
concern for many stockowners. April Nhlapo regarded cattle ownership and the
use of traditional medicines as essential for preserving Sotho identity in a
changing world. He owned two cows and was committed to his herbal recipes.
However, closer settlement had led to the disappearance of plants from the
lowlands and now stockowners had to trek into the mountains to collect
medicinal herbs.[33] Gladice Mokoena explained that the need to hike into the
mountains to find *mositsane* (*Elephantorrhiza elephantina*), her preferred herb for
gallsickness and other diseases, had forced her to consider buying biomedicines,
as she was too old to trawl the highlands for medicine. In her opinion biomedicines were no better than *mositsane*, but necessity had forced her to reappraise
her approach to animal health.[34] Daniel Khonkhe was also worried because
mositsane was becoming harder to find even in the mountains. He blamed traditional healers and the collectors they worked with for depleting the veld of
valuable flora in order to sell it on the streets of Phuthaditjhaba at high prices.[35]
Jacob Tshabalala believed that the only reason he could still locate *qobo* (*Gunnera
perpensa*) to treat retained placentas was because the local traditional healers did
not use it for their *muthi*, as far as he was aware.[36] There was thus a sense that
people who should be preserving local knowledge – the traditional healers –
were actually jeopardising its survival by exhausting the veld of Sotho medicines
for their own financial ends.

However, we did not hear of similar problems in Mbotyi. It is surrounded
by some of the biggest indigenous forests in the country. Informants did not
mention a shortage of medicinal plants, except where these were obtained from
outside the village, such as *intolwane* (*Elephantorrhiza elephantina*). When we
raised the issue at the report-back meeting in 2012, people agreed that 'it is

---

[31]   Thembela Vincent Kepe, 'Grassland Vegetation and Rural Livelihoods: A Case Study of
Resource Value and Social Dynamics on the Wild Coast, South Africa', unpublished Ph.D.,
University of the Western Cape (2002); Dold & Cocks, *Voices from the Forest*.

[32]   Thembela Kepe, 'Medicinal Plants and Rural Livelihoods in Pondoland South Africa: Towards
an Understanding of Resource Value', *International Journal of Biodiversity Science and Management*,
3, 3 (2007), 170–83.

[33]   April Nhlapo, Lusaka, 28 February 2011.

[34]   Gladice Mokoena, Ha-Sethunya, 13 October 2010.

[35]   Daniel Khonkhe, Ha-Sethunya, 13 October 2010.

[36]   Jacob Tsahbalala, Phuthaditjhaba, 13 October 2010.

easy to find the plants we need for animals'.[37] We walked around the locality to pick the four ingredients for gallsickness/*inyongo* medicine and Sonwabile Mkhanywa, who had a close knowledge of the local environment, found these in little over half an hour. However, a few of the trees situated closest to the village had been used so often for their bark, that they were quite badly damaged and were unlikely to survive.[38] A couple of people – such as the healer Lungelwa Mhlwazi – did mention that in recent years they had to walk further for some rarer species.[39] It is sometimes claimed that the bark of trees used for medicinal purposes is only cut or scraped on one side of the tree by collectors for a period and then left. We certainly saw evidence of such practices but also of more careless exploitation.

Once they had gathered their plants, there were a number of ways in which stockowners could transform them into medicine. Bulbous plants were especially popular in North West Province as stockowners believed that the bulbs and roots had the highest concentration of healing agents.[40] Farmers also made use of the bark and leaves of certain species, but rarely the flowers. Sometimes they would uproot the whole plant, especially if it were a creeper or bulbous species. They often chopped the bulbs into slices and pounded the bark into a powder before placing them in water to make a medicinal drink. This beverage might be an infusion in which the active principals from the plant gradually steeped into the water, or else they might boil the mixture to create a decoction. Choice of methods depended on convenience and the most effective way of reaching the desired colour or consistency. Dosing was normally through the mouth, but sometimes through the nose, using a plastic bottle. For skin diseases, mucilaginous plants like aloes which contain a rich liquid sap were ideal. Prickly pear was used for similar reasons.[41] Some stockowners also took ash from burnt plants to make a drench or poultice and they incinerated medicinal herbs in the kraal to ward off disease and witchcraft.

Most of our informants collected plants from the wild as and when they needed them. Michael Ncobuga from QwaQwa believed that 'plants are far more potent if they are fresh'. He asserted that he would never buy medicine from the streets of Phuthaditjhaba because 'widows, menstruating women and men who had just had sex and were unclean might have walked past the *muthi* and made it powerless (chapter 7)'.[42] Rosalina Johanne, a healer and stockowner from QwaQwa, argued that although it was technically possible to grow medic-

[37] Mbotyi, 5 December 2012.
[38] Mbotyi, 26 March 2008.
[39] Lungelwa Mhlwazi, Kunzimbini, 19 February 2009.
[40] L. J. McGaw and J. N. Eloff, 'Ethnoveterinary use of Southern African Plants and Scientific Evaluation of their Medicinal Properties', *Journal of Ethnopharmacology*, 119, 3 (2008), 559–74.
[41] William Beinart and Luvuyo Wotshela, *Prickly Pear: the Social History of a Plant in the Eastern Cape* (Wits University Press, Johannesburg, 2011), 157.
[42] Michael Ncobuqa, Lusaka, 3 March 2011.

inal herbs in the back yard, 'it's not good to grow them outside their natural environment as they will not thrive', because 'cultivating medicinal plants is against nature and results in a decline in potency'.[43] However, Anna Pooe from Shakung did grind up herbs to store them over the winter when they were not so easily available.[44] Some stockowners planted aloes in their gardens because they adapted to different environments and were so versatile. Norah Mabena transplanted aloes from the Magliesberg Mountains to her home in Garankuwa so that they were always on hand.[45] Viola Maphosa also reported the planting of aloes in gardens in Alice (Eastern Cape) so that people did not have to hike into the mountains to collect them. Some also conserved medicinal herbs for the winter.[46]

Historically in the Eastern Cape, indigenous aloes were used for hedging so they would have been available for medicine. Similarly, introduced plants like prickly pear, which had a range of medicinal uses, frequently occur in gardens.[47] An Eastern Cape study investigated attitudes to the cultivation of traditional medicines in order to assess whether this could take the pressure off threatened wild species.[48] Researchers found that 40 per cent of their informants cultivated medicinal plants and 89 per cent felt that the healing powers of some species would not be affected by cultivation. But this did not apply to plants that were strongly associated with rituals and ancestors. Traditional healers in KwaZulu-Natal grew a range of species in their gardens in the 1990s.[49] In former Transkei, four out of five traditional healers interviewed in one study planted medicinal plants in their gardens, although some of the most valued species were trees which grew only in nearby forests.[50] Cultivars included *umkhwenkhwe* (*Pittisporum viridiflorum*) used both for gallsickness in animals and colds in people. In the Kat River area, Jenjenzwa found farmers collected herbs from the wild whenever they needed them.[51] In Hewu, in the former Ciskei, *Artemisia afra* was

[43] Rosalina Johanne, Lusaka, 1 March 2011.
[44] Anna Pooe, Shakung, 22 January 2010.
[45] Norah Mabena, Garankuwa, 21 October 2009.
[46] Discussion with Viola Maphosa, Fort Hare University, 23 February 2011.
[47] Beinart & Wotshela, *Prickly Pear*.
[48] K. F. Wiersum, A. P. Dold, M. Husselman and M. L. Cocks, 'Cultivation of Medicinal Plants as a Tool for Biodiversity Conservation and Poverty Alleviation in the Amatola Region, South Africa' in R. J. Bogers, L. E. Craker and D. Lange (eds), *Medicinal and Aromatic Plants* (Dordrecht, Springer, 2006), 43–57.
[49] N. R. Crouch and A. Hutchings, 'Zulu Healer *Muthi* Gardens: Inspiration for Botanic Garden Displays and Community Outreach Projects' (Durban, Department of Botany, University of Zululand, 1999).
[50] J. Keirungi and C. Fabricius, 'Selecting Medicinal Plants for Cultivation at Nqabara on the Eastern Cape Wild Coast, South Africa', *South African Journal of Science* 101, (2005), 497–501.
[51] Vimbai Jenjezwa, 'Stock Farmers and the State: A Case Study of Animal Healthcare Practices in Hertzog, Eastern Cape Province, South Africa', unpublished MA, University of Fort Hare (2010), 142.

grown in gardens and cooked in boiling water as an inhalant. Eucalyptus bark, found near homesteads in Mbotyi, was used for a similar purpose, but a group of men interviewed there said they did not generally grow medicinal plants, because they could get them in the forests; 'gardens are where we grow mealies'.[52] Peaches, as mentioned, doubled as fruit and ingredients for *muthi* and we found *uqangazana* (probably *Clerodendrum glabrum*), used medicinally, in one garden. Stockowners were clearly individualistic in their approach to cultivating medicinal herbs, depending on practicalities and their views on potency. Access to fresh plants was ideal but compromises were made to overcome problems of accessibility and seasonal shortages. Introduced plants such as prickly pear and peach trees, which had some medicinal value, seemed to be especially popular in gardens in some areas.

Although there were similarities in preparation methods amongst different communities, the choice of herbs was personal and also ecologically constrained. Just as farmers showed there was no homogenous understanding of disease causation even within a single village, so too did they reveal a diversity of approaches towards animal health care. As mentioned in chapter 5 many informants classified their knowledge as family knowledge and were secretive about their medicines. This did not mean that every family used entirely different plants to control diseases, but it did result in a diversity of medical treatments even within villages. As with the naming of diseases, it was not always entirely clear which plants people were talking about and we generally had to rely on an individual's description of the herbs and the way they used them. One person's *sekaname* could be another person's *sebete*. There were also regional differences in names. For example, *lepate,* a commonplace herb in North West Province, popular for birthing problems, also went by the name of *tshetlho* and, in Mabeskraal, *makanangwane*. Chris Matone from Sutelong said *lepate* denoted a creeper which bore fruits with four thorns, whereas *tshetlho* only had two thorns. Fallen *lepate* fruits, when dry, could severely injure the hooves of an animal that trod on it.[53] Sometimes stockowners did not know the name of the plants they used, but might be able to point to them in the yard or veld. One informant from QwaQwa referred to a low-lying leafy herb in his garden as the diarrhoea plant, for example.[54]

Environmental differences inevitably affected selection and availability. Some plants that survived in the warm, humid coastal areas of Mpondoland were not available in the colder wind-swept mountains of QwaQwa. Although historically there has been a great deal of incorporation of introduced plants, some of which are used medicinally, we did not hear of deliberate translocation of medicinal herbs from one distinct geographical area to another. Any type of

[52] Mbotyi, 5 December 2012.
[53] Charles Matome, Sutelong, 28 January 2010.
[54] Sidwell Ndlebe, Bethlehem, 4 March 2011.

cultivation, such as raising aloes in the yard, involved plants that had been brought in over short distances. It was striking that farmers in Mbotyi did not use aloes for ticks; most species do not grow there naturally and, in a group report-back to 10 men in the village, none had heard of aloe treatments.[55] Consequently there were regional variations in the medical treatments that people provided for their animals. In the following section, we discuss some of the plants that stockowners administered to their livestock in each region. We also describe methods of treating animals that did not involve herbs. The number of plants we refer to is not exhaustive – we are highlighting the most important varieties mentioned in our interviews. Appendix 4 gives a more comprehensive listing of the flora mentioned by our informants.

Plant medicines in North West Province

Informants used plants as single species, occasionally in combination, and some-times with biomedicines or other substances such as salt and tar. The most popular varieties were aloes (*kgopane* or *mokgopa*), *sekaname, lepate* and *sebete bete* (see below). Information garnered from our field sites bore some similarities with the work carried out by Deon van der Merwe in the Madikwe district located in the north-western part of North West Province. In 2001 van der Merwe published a valuable list of some of the most popular medicinal plants with details about the ways they are prepared.[56] Our testimonies revealed a significant overlap.

Tswana stockowners believed that medicines served a dual purpose. They give their animals plant extracts to build up their strength so that they are more resistant to diseases, and they also administer drugs that expel unwanted substances such as dung and gall from the body. Informants found that aloes were particularly useful for maintaining and restoring health and for cleansing the digestive system. The ready availability of aloes and the fact that stockowners believed they had curative and tonic properties against a variety of diseases, ensured their frequent usage. Respondents did not differentiate between different species of aloes and naming was rather haphazard. For some the term *mokgopa* referred to low-lying aloes and the word *kgopane* meant an aloe tree. However, others reversed this description. Stock owners identified aloes by their height, rather than by the colour or shape of their leaves, or the structure of their flowers.

[55] Mbotyi, 5 December 2012.
[56] D. van der Merwe, G. E. Swan and C. J. Botha, 'Use of Ethnoveterinary Medicinal Plants by Setswana-speaking People in the Madikwe area of the North West Province of South Africa', *Journal of the South African Veterinary Medical Association*, 72, 4 (2001), 189–96; D. van der Merwe, 'Use of Ethnoveterinary Medicinal Plants in Cattle by Setswana-speaking people in the Madikwe area of the North West Province'. MSc thesis, University of Pretoria 2000'.

Photo 6.1 Aloes. Used for digestive ailments, blood purification, tick repellent, gall-sickness in NWP (KB, January 2010)

According to Ida Tsagane, a small-scale commercial chicken farmer from Kgabalatsane (north of Garankuwa): 'people love aloes because they are cheap, versatile and so easy to prepare. All you have to do is chop up the leaves and place them in the water trough. There is no need to measure the dose; chickens just drink when they please.' She thought that aloes were ideal for chickens because 'they never fall sick on a regular diet of aloes'.[57] Norah Mabena from Garankuwa described aloes as 'a traditional African medicine'. 'Every day I give my goats a drink made from aloe leaves and roots. I also place chopped leaves in the water trough to cleanse the goats' insides, to clean the blood and remove the hard faeces so they enjoy good health and fertility.'[58] Johannes Nkosi, also from Garankuwa, felt that aloes had become increasingly important because his goats 'are always feeding on plastic wrappers' and aloes 'are very effective at ejecting inedible objects and poisons'.[59] Itumeleng Sedupane, a school principal and livestock owner from Mareetsane, south of Mafikeng, extolled the purgative properties of aloes, which expelled the impurities and toxins picked up in the veld.[60]

[57] Ida Tsagane, Kgabalatsane, 13 November 2009.
[58] Norah Mabena, Garankuwa, 21 October 2009.
[59] Johannes Nkosi, Garankuwa, 21 October 2009.
[60] Itumeleng Sedupane, Mareetsane, 4 Nov 2009.

Aloes were also a common medicine for gallsickness. Zachariah Maubane explained that farmers gave their cattle aloes to prevent *gala* as they believed the plant stopped animals secreting too much gall. A superfluity of gall impaired digestion.[61] Although few farmers associated *gala* with ticks, many stockowners exposed their livestock to aloes because they believed they made the blood bitter, so as to repel biting acarids and flies.[62] Riana Kleynhans from the plant research institute at Roodeplaat mentioned that many medicinal plants have a bitter taste which gives the impression that their compounds make the blood bitter.[63] In Fafung aloes not only dealt with constipation and sorted out gall-sickness but were also considered to be excellent dewormers.[64] In Mabeskraal stockowners believed aloes cleaned out the uterine as well as the digestive tract, and they administered infusions to animals after they had aborted, or if the placenta had become stuck inside.[65] Many stock owners from Mabeskraal and North West Province in general complained that their cattle did not expel the afterbirth soon after calving. Interviewees had no explanation for this condition, but scientists have suggested a number of causes including miscarriages and difficult births, mineral and vitamin deficiencies, hereditary predisposition, diseases such as brucellosis and unhygienic birthing environments giving rise to infection.[66]

Anna Lebelwane from Mmakau (near Brits) employed aloes to treat skin lesions (orf?) on the lips of her goats. She assumed that browsing on thorny plants and feeding on the rubbish that littered the township caused blisters that hindered feeding. To treat these sores, and other abscesses caused by tick bites or other mites, 'you slice the aloe leaves down the middle and place them on a grill to heat them up. Then you press the warm leaves onto the lips and wait a week for the aloe juice to heal the sores. If necessary you can tie cloths around the leaves to keep them in place.'[67] Prickly pear cladodes were used in a very similar way to treat human boils in the Eastern Cape.[68]

Another broad-spectrum plant with multiple uses was *sekaname* (*slangkop*). This is potentially a poisonous plant but informants discussed its therapeutic

[61] Zachariah Maubane, Mafikeng, 2 November 2009.

[62] Zachariah Maubane, Mafikeng, 2 November 2009; Joseph Gumede, Garankuwa, 21 October 2009; Norah Mabena, Garankuwa, 21 October 2009; Elias Pooe, Shakung, 19 January 2010; Anna Pooe, Shakung, 22 January 2010; Martha Rakgomo, Mabeskraal, 1 February 2010 ; George Malatsi, Mabeskraal, 5 February 2010; Reuben Ramutloa, Disaneng, 11 February 2010.

[63] Discussion with Riana Kleynhans, Agricultural Research Council Institute Roodeplaat, 27 November 2012.

[64] Fafung group meeting, 25 January 2010.

[65] Emily Diole, Mabeskraal, 2 February 2010.

[66] J. C. Moreki, K. Tshireletso and I. C. Okoli, 'Potential Use of Ethnoveterinary Medicine for Retained Placenta in Cattle in Mogonono, Botswana', *Journal of Animal Production Advances*, 2, 6 (2012), 303–9.

[67] Anna Lebelwane, Thethele (Mmakau), 13 November 2009.

[68] Beinart & Wotshela, *Prickly Pear*.

rather than its toxic properties. Timothy Johns, writing about Latin America, noted that 'often the only difference between medicine and toxin is simply a matter of concentration of the ingested chemicals or the circumstances under which they are ingested'.[69] The boundary between medicine and poison was at times blurred and livestock owners believed that some toxic plants could prevent certain diseases as well as cure others. There are analogies here with the use of antelope dung, noted above. Drenching livestock with *sekaname* in advance of spring, when the plant began to flourish, was said to protect animals from its toxins. Aletta Nare in Mareetsane explained that *sekaname* 'can have red or white bulbs and the red ones are less dangerous. We use the safer red bulbs to make a prophylactic infusion. We crush the bulbs and blend them with *phate ya ngaka* (*Hermannia depressa?*) and add boiling water. Once the medicine has cooled down, we give it to the cattle and goats in a plastic bottle. Then if our animals find *sekaname* in the veld they will never get sick.'[70] The Sotho had a similar strategy for dealing with tulp poisoning in QwaQwa (see below).

Therapeutically, farmers found *sekaname* to be very effective for treating wounds. Frederick Matlhatsi from Mabopane used the fine *slangkop* leaves, mixed with water, rather than aloes, to dose his goats when they had sores on their mouth and lips.[71] A former Animal Health Technician had found that it remained common practice to use *slangkop* bulbs in the Pilanesberg and Brits area to cure abscesses as they were so much cheaper than biomedicines. Innocent Setshogoe explained: 'to heal an abscess, stockowners use a razor blade to scour a cross in the abscess and force out the pus. Then they clean the wound with salty water and place ground-up *sekaname* bulb into the cut. Within two weeks the wound will be completely healed.'[72] Lott Motaung from Bollantlokwe used *sekaname* in a similar way to treat venereal warts on his bulls. Mixed with *mathubadifala* (*Boophane distichia?*), Motaung also believed *sekaname* to be an excellent fertility drug for bulls, 'as it enhances their sex drive just like it does for men'![73]

In Mabeskraal local livestock owners valued *sekaname* not only because it was effective at clearing up skin complaints, but also because it could clean the blood and treat animals suffering from a range of diseases. Stockowners ground up the bulbs and placed them in the water troughs to prevent *lebete* (anthrax). They also treated anthrax by dosing animals with a mixture of *sekaname* and *sebete*.[74] Redwater rarely came up in testimonies from North West Province, but Pagiel Sehunoe saw it as a problem and associated it with contaminated

---

[69] Timothy Johns, *The Origins of Human Diet and Medicine: Chemical Ecology* (Tucson, University of Arizona Press, 1996), 10.

[70] Aletta Nare, Mareetsane, 4 November 2009.

[71] Frederick Matlhatsi, Mabopane, 29 October 2009.

[72] Innocent Setshogoe, Bethanie, 26 October 2009.

[73] Lott Motuang, Bollantlokwe, 10 November 2009.

[74] Pagiel Sehunoe, Mabeskraal, 4 February 2010.

blood, because 'the disease turns the urine red'.[75] In this understanding of disease, dirty blood had the ability to affect other organs and turn the urine red. Blood could be a source and disseminator of infection conceptualised as a form of corporal impurity. To treat redwater, Sehunoe administered an infusion concocted from red *sekaname* bulbs 'to cleanse the blood and make animals fit and strong'. Ernest Medupe from the Mafikeng District explained how his family always boiled the *sekaname* because they knew it was a toxic plant. By making a decoction, rather than an infusion, he believed that 'the concentration of poisons is much less', ensuring that the plant had a therapeutic function. He gave it to his animals when they suffered from constipation or had contracted *lamsiekte* (*magetla*), which he identified by lameness and sluggish bowels.[76] Outside Mabeskraal, there was also the belief that the plant had a great medicinal value as a blood purifier. As Anna Pooe, the healer from Shakung, put it, a 'dirty goat is an unfit goat … It needs regular doses of *sekaname* to ensure it feeds and breeds well'.[77]

Another popular plant that farmers often referred to was *sebete bete,* as opposed to *sebete* which is senna. *Sebete* was widely used in Mabeskraal as a laxative and to relieve constipation associated with gallsickness, anthrax and blackquarter.[78] Informants described *sebete bete* as a creeper with small yellow flowers and tiny leaves. In many parts of North West Province stockowners administered it to cattle to clean out the uterine tract after a miscarriage. They uprooted the whole plant and added it to water. After five days the liquid became slimy and was ready to use. Farmers dosed their cows through a funnel in the mouth and after a few hours a gel-like substance would drop from the vulva indicating that the sickness that had caused the abortion had been removed from the animal's body.[79] Moses Tobosi from Mafikeng said that *sebete bete* was commonly used in that area too for treating the cause of miscarriages. In addition, 'it stops gallsickness and its laxative qualities make it an excellent dewormer'.[80]

Farmers administered *lepate, tshetlho* or *makanangwane*, the creeper discussed above, for reproductive problems relating to abortions, dystocia and retained placentas. As with *sebete bete* stockowners used the whole plant, chopped and added to water. *Lepate* was much quicker to prepare than *sebete bete* as the

---

[75] Pagiel Sehunoe, Mabeskraal, 4 February 2010.

[76] Ernest Medupe, Mafikeng, 10 February 2010.

[77] Anna Pooe, Shakung, 22 January 2010.

[78] Mokoboro Rakgomo, Mabeskraal, 1 February 2010; Henry Ramafoko and Elsie Ramafoko, Mabeskraal, 2 February 2010; Emily Diole, Mabeskraal, 2 February 2010; Philemon Ramafoko, Mabeskraal, 3 February 2010; Majeng Motsisi, Mabeskraal, 3 February 2010; Pagiel Sehunoe, Mabeskraal, 4 February 2010; George Malatsi, Mabeskraal, 5 February 2010; Jeremiah Ramakopeloa, Mabeskraal, 5 February 2010.

[79] Innocent Setshogoe, Bethanie, 26 Oct 2009.

[80] Moses Tobosi, Mafikeng, 6 November 2009.

Photo 6.2 *Sekaname.* Blood purifier, tick repellent, used to treat wide range of diseases in NWP (KB, January 2010)

Photo 6.3 *Lepate.* Used for retained placenta and birthing problems in NWP (KB, January 2010)

Photo 6.4 Grinding leaves to make *muthi* in Mbotyi (WB, 2009)

mixture reached the desired slimy consistency after an overnight soaking. The dose was a litre of this viscous infusion which would drive out the placenta and any noxious vaginal fluids within a day. Many farmers, who otherwise used vaccines and antibiotics to prevent and cure infections, swore by *lepate* as not only being readily available in the veld, but also far more effective than the pessaries recommended by scientists and drug companies to deal with these kinds of complaints.[81] In Mabeskraal, *makanangwane* had a wider range of uses. As an infusion, informants said the plant cured gallsickness, blackquarter and anthrax.[82] If an animal died of anthrax, farmers could protect themselves from the disease by disinfecting their hands with *makanangwane*. When soaked in water the plant became a soapy detergent. People also used *makanangwane* as a shampoo especially if they had dandruff.[83]

Informants throughout North West Province frequently mentioned dystocia and retained placentas. Apart from the plants used above, Moses Tobosi explained that an effective cure in the past had been to scratch ash from the roofs of huts and use this to make a black drink to oust the foetus and the afterbirth. However, Tobosi believed that this practice was dying out in North West Province as people 'have swapped thatched for corrugated metal roofs'.[84] We noted the medicinal use of both sorghum porridge and beer above. Peggy Setshedi from Fafung described how she always fed the residual by-products of brewing to her cows and chickens 'as it makes them strong'.[85] In *mohau*-ridden Fafung and neighbouring districts, some farmers gave their animals beer to try to flush out the toxins.[86] As a laxative it was also ideal for treating gallsickness.[87] Its stimulant and eruptive properties encouraged some farmers to give it to drowsy cattle 'to burp them back to life', and to revive animals that had miscarried so that they could conceive again.[88] In the Eastern Cape, pigs were fed the peelings of prickly pear and unwanted dregs of *iqhilika* (prickly pear beer).[89]

Apart from plant extracts there were various other substances farmers used to prevent or cure diseases. African informants thought that salt was a tradi-

[81] Thomas Seemela, Madidi, 23 October 2009; Lott Motuang, Bollantlokwe, 10 November 2009; Shakung Group Interview, 19 January 2010; Elias Pooe, Shakung, 19 January 2010; Fafung group meeting, 25 January 2010; Group meeting, Sutelong, 27 January 2010; Charles Matome, Sutelong, 28 January 2010; Phillimon Mothaba, Kgomo Kgomo, 29 January 2010; Bassie Monngae, Mabeskraal, 1 February 2010; Lucas Moatlhodi, Mabeskraal, 3 February 2010; Philemon Ramafoko, Mabeskraal, 3 February 2010; Jacob Diole, Mabeskraal, 4 February 2010; Pagiel Sehunoe, Mabeskraal, 4 February 2010.

[82] Jeremiah Ramakopeloa, Mabeskraal, 5 February 2010.

[83] Soloman Matome, Mabeskraal, 3 February 2010.

[84] Moses Tobosi, Mafikeng, 6 November 2009.

[85] Peggy Setshedi, Fafung, 27 October 2009.

[86] Fafung group meeting, 25 January 2010.

[87] Abraham Meno, Mareetsane, 4 Nov 2009.

[88] Lott Motaung, Bollantlokwe, 10 November 2009; Nick Lebethe, Bethanie, 26 October 2009.

[89] Beinart & Wotshela, *Prickly Pear*.

tional African medicine – used before white colonisation and western veterinary ideas. Salt could be given on its own as a block that was left in the kraal, near the water trough, so livestock could lick the minerals as and when they pleased. Alternatively, stock owners mixed salt with other vitamins and minerals to boost nutrition. Most informants explained that salt was good because of its laxative effects. July Chiloane a commercial farmer from Garankuwa, provided blocks of salt because it made his cows drink lots of water, 'which keeps the stomach flowing and speeds up defecation'. He believed that salt 'has a cleansing role that is essential for good health' and that sickness occurred when there was an 'accumulation of dirt inside'.[90] David Modiga also from Garankuwa thought that salt prevented livestock becoming sick on green grasses and 'it gives them strength so they do not lose weight'.[91] Ernest Phage, another Garankuwa resident, explained how salt was good for the gallbladder and stopped gallsickness. 'Salt has the power to dilute gall and stop the gallbladder from swelling. Too much gall prevents the absorption of nutrients and leads to death.'[92] Other informants described the merits of salt as an aid in the expulsion of plastics and placentas. Farmers drenched livestock with salty water or else they added salt to fish oil or coke to speed up expulsion.[93]

We asked about salt and licks in Mbotyi. While some had heard of licks, which they called *ilitye* (plural *amatye* or stones), we did not find anyone who used them or gave salt to their livestock. Nonjulumbha Javu, one of the most careful and assiduous cattle owners, said

> I don't buy licks. I know them, I've seen them on the farms, I used to work on a white man's farm near Umzumbe [KwaZulu-Natal]. They buy many of them ... and cattle come from far away to lick these stones [*amatye*]. I have never seen them here. We don't have them here. You have to buy them in shops and the animals would not know them if you bought them and put them out.[94]

It is possible that the veld around Mbotyi is naturally salty from the sea winds and this provides animals with their requirements. Cattle and goats often go down to the beach, where there are fewer ticks and flies, and perhaps they imbibe salt there too.

Aside from salt, North West Province stockowners saw powdered potassium permanganate as good for health, especially in the eastern part of North West Province. Respondents claimed that it cleansed the blood, which was just as essential as a clean gut for good health and reproduction. Users added a pinch of potas-

---

[90]   July Chiloane, Garankuwa, 21 October 2009.

[91]   David Modiga, Garankuwa, 24 October 2009.

[92]   Ernest Phage, Garankuwa, 21 October 2009.

[93]   Lott Motaung, Bollantlokwe, 10 November 2009; Abraham Meno, Mareetsane, 4 November 2009; Andrew Mabaso, Mafikeng, 5 November 2009.

[94]   Nonjulumbha Javu, Mbotyi, 19 December 2011.

sium permanganate to the water trough or else mixed it with salt and ground-up plants to make a lick.[95] For tick bites and fly strike, stock owners might smear the sores with motor oil or petrol, instead of aloes. This seemed to be more common around Mafikeng where the climate is much drier and aloes may be less prolific than in the wetter eastern part of the Province.[96] Blindness caused by pollen grains came up in a number of interviews and there were a range of treatments. Some farmers placed sugar or snuff into the eyes, or else powdered detergents.[97] By far the most common remedy for blindness was to use the ground-up bodies of dead millipedes that had turned white in the sun. The whiteness of the corpse mirrored the white film on the cornea, suggesting that an animal form of the 'doctrine of signatures' was at work here.[98] We heard of crushed millipedes mixed with squid bones, found on the beach, as an eye treatment in Mbotyi.[99]

Another parallel with the doctrine of signatures was the use of tar to prevent or cure blackquarter, a disease that farmers identified by the darkened spongy legs. Since the 1960s this practice has been gradually dying out in North West Province due to the availability of vaccines. Nevertheless, informants from Kgomo Kgomo and Mabeskraal said that local farmers still used tar to ward off disease.[100] To safeguard their cattle from blackquarter, David Modiga from Garankuwa explained:

> We used to make an incision in the dewlap with a stick or sharp instrument and then place tar in the hole. The tar was held in place by a cloth or copper wire. Some farmers also smeared tar on the teeth for added protection against the grasses that cause blackquarter. We used to buy the tar in 5-litre buckets from the coop. But nobody does this nowadays.[101]

Bosie Morapadi from Mabeskraal noted that stockowners 'used to chip tar from the side of the road so it was free like plant medicines'.[102] Informants believed

[95] Joseph Gumede, Garankuwa, 21 October 2009; Johannes Nkosi, Garankuwa, 21 October 2009; Norah Mabena, Garankuwa, 21 October 2009; Frederick Matlhatsi, Mabopane, 29 October 2009.

[96] Otsile Frederick Morwe, Dithakong, 3 November 2009; Moses Tobosi, Mafikeng, 6 November 2009; Patrick Sebeelo, Lokaleng, 3 November 2009.

[97] Aletta Nare, Mareetsane, 4 November 2009; Fafung group meeting, 25 January 2010; Shakung group meeting, 19 January 2010.

[98] Frederick Matlhatsi, Mabopane, 29 October 2009; Shakung Group Meeting 19 January 2010; Bosie Morapadi, Mabeskraal, 2 February 2010; Soloman Mokebe, Kgomo Kgomo, 29 January 2010; Zachariah Tlatsane, Lokaleng, 9 February 2010; Laurence Bodibe, Ramatlabama, 10 February 2010.

[99] Lenox Philani Mkhanywa, 4 December 2012; Dold & Cocks, *Voices from the Forest*, 29.

[100] Moiletswane group meeting 20 January 2010; Simon Moeng, Sutleong, 26 January 2010; Simon Mokonyana, Kgomo Kgomo, 29 January 2010; Bosie Morapadi, Mabeskraal, 2 February 2010.

[101] David Modiga, Garankuwa, 24 October 2009.

[102] Bosie Morapadi, Mabeskraal, 2 February 2010.

calves were particularly susceptible to blackquarter and carried out this procedure when their animals were young.[103] Some also inserted garlic in the incisions to prevent diseases like gallsickness and heartwater.[104] Interviewees had no
explanations as to how the tar and wire stopped blackquarter, but most were
adamant that it had been effective in the past and some still perceived that this
method worked. A possible scientific explanation is that the dewlap is rich in
blood, and the wire might attract spores that enable cattle to produce some
antibodies that enhance their resistance to infection. However, many calves are
naturally resistant to the *Clostridium chauvoei* that causes blackquarter and the
insertions of tar and wire could give the impression that these substances were
valuable prophylaxes.[105]

Blackquarter was a disease that seemed to generate a number of non-herbal
preventatives and cures. Informants said you could prevent or cure it by cutting
off the ear tips of cattle. Samuel Mothapo described this as an 'old remedy' for
blackquarter in Slagboom. People not only excised the tips of the ears but also
the tail and then dosed the animals with a mixture of *sekaname* and *mathubadifala*. Mothapo added that some people have replaced the plants with Terramycin
(oxytetracycline) which he thought was easier to use, but not necessarily more
effective.[106] Some farmers bled their cattle for blackquarter.

Incisions were made for other ailments. Abraham Meno used a razor to cut
open the tail of an animal that was sick from *sekaname* poisoning, believing 'that
spurting blood releases the toxins'.[107] In nearby Mafikeng, people used to cut
the dewlap to let the blood flow in order to expel all kinds of infections.[108] A
way of preventing worms and also curing biliary fever in dogs was to cut the
ligament under the tongue so they did not fall sick.[109] Respondents suggested
that such techniques of bleeding were increasingly rare, though they had not
disappeared, and farmers used antibiotics like Terramycin, rather than herbs, to
complete the treatment. We did not hear of cutting and bleeding in Mbotyi, but
as noted above, stockowners had experimented with excising *umhlaza* with a
knife and treating the wound as well as giving a euphorbia drench. We cannot

---

[103] Thomas Seemela, Madidi, 23 October 2009; David Modiga, Garankuwa, 24 October 2009;
Elias Poo and Richard Modiga, Shakung, 19 January 2010; Group interview, Moilestwane
community centre, 21 January 2010; Fafung group meeting 25 January 2010; Simon Moeng,
Sutelong, 26 January 2010 ; Phillimon Mathaba, Kgomo Kgomo, 29 January 2010; Simon
(cont) Mokonyana, Kgomo Kgomo, 29 January 2010; Bosie Morapadi, Mabeskraal, 2 February
2010; Majeng Motsisi, Mabeskraal, 3 February 2010 ; George Malatsi, Mabeskraal, 5 February
2010; Jeremiah Ramakopeloa, Mabeskraal, Benson Letlhaku, Mantsa, 12 February 2010.

[104] Thomas Seemela, Madidi, 23 October 2009.

[105] Discussions with Christo Botha, Toxicologist, University of Pretoria, 18 November 2009 and
Johan Naude, state veterinarian, Bethlehem (FS), 4 March 2011.

[106] Samuel Mothapo, Slagboom, 10 November 2009.

[107] Abraham Meno, Mareetsane, 4 November 2009.

[108] Itumeleng Sedupane, Mareetsane, 4 November 2009.

[109] Frederick Matlhatsi, Mabopane, 29 October 2009.

be sure which disease this may be but saw one case where a cow had extensive warts or pox on its skin covering most of one front flank. As its owners commented, it was still 'strong' (they used the English word) and it was relatively free of ticks, as it had been dipped, but they believed it was barren.[110] In this case, the Satsha family was reluctant to intervene physically, because they had heard of animals dying from attempts to cut away the warts. The veterinarian treated the *umhlaza* with a surface spray and injection.

Plant medicines in QwaQwa

Veterinary practices in QwaQwa overlapped with those in North West especially in the prevention of blackquarter and also methods to mitigate the effects of toxicoses. Sotho farmers put wires through the dewlap to stop *serotswana* (blackquarter), especially when the calves were young and before they were exposed to the grasses that were said to cause this disease.[111] Some farmers smeared the wire with tar or with herbal medicines.[112] James Motloung, from Lesotho, who worked as a herder in South Africa, covered the wire with ash from the burnt *setimamollo* shrub to make the treatment more effective.[113] Informants here also had no explanation as to how the wire worked, but they too were convinced that it did. This practice did not seem to be dying out in QwaQwa as quickly as it was in North West Province, suggesting that there was possibly a stronger adherence to traditional medicines in this part of the Free State than amongst Setswana-speaking farmers further north.

QwaQwa informants did not talk about *sekaname* which did not flourish in this mountainous area. However, there was a problem with tulp (*Homeria pallida*). Unlike *slangkop* it had no redeeming medicinal qualities and many stockowners viewed it as an especially dangerous weed. They called tulp *tele* and they protected animals from this toxic plant in a similar way to farmers in Mareetsane who dealt with *sekaname*. David Mphuthi said he had had some success in preventing *tele* (tulp) poisoning: 'we burn the *tele* and mix the ash with water to make a drink. Once the cattle have tasted this mixture they never touch *tele* in the veld.'[114] It worked like an inoculation. There were similarities here with thinking about *mohau* in North West Province, except that this treatment was believed to prevent consumption of the poisonous plant rather than to counteract its effects.

[110] Masamekile Satsha, Mbotyi, 20 December 2011 and 20 March 2012.
[111] Ntsoeu Livestock Company, Phuthaditjhaba, 11 October 2010; David Mphuthi, Phuthaditjhaba, 14 October 2010; Chadiwick Mbongo, Lejwaneng, 12 October 2010; Seketsa Mokoena, Lusaka, 1 March 2011; Zulu Mokoena, Lusaka, 2 March 2011.
[112] Johannes Seekane, Bethlehem, 5 March 2011.
[113] James Motloung, Bethlehem, 4 March 2011.
[114] David Mphuthi, Phuthaditjhaba, 14 October 2010.

Mphuthi admitted that this tactic did not always work. If it failed he resorted to modern medicines, in this case the use of activated charcoal to bind and expel the toxins. Treatment has to be delivered swiftly, so herders had to be particularly alert to tulp poisoning and possess a sound knowledge of the composition of the veld in order to detect *tele* amongst the grasses. [115] Like Mphuthi, Ephraim Mofokeng used *tele* ash but administered it as a cure rather than as a prophylaxis. He assumed his cows had ingested the plant if their dung turned hard and round like the droppings of goats and sheep. [116] Rather than ash, James Motloung, gave his cattle crushed fragments of the *tele* bulb in its raw state so that they would become sensitised to the plant and thus avoid it. [117] The idea that livestock develop an aversion to this plant has now been scientifically established. Work at Onderstepoort in the 1990s demonstrated that if cattle were exposed to a sub-lethal dose of tulp (and other toxic flora) they experienced some discomfort, ensuring that they were unlikely to touch the plant in future. [118] This method of tackling the threat of toxicoses appears to have been a long-standing African tradition, especially with regard to tulp. There are accounts in the *Farmers Weekly* dating back to the 1930s which showed that Transkeian stockowners also exposed their cattle to this plant before they turned them out onto the spring grazing lands. White farmers copied this method as H. H. Klette from Umtata in Transkei, recounted:

> Our Natives take from 10 to 12 'tulp' bulbs, smash them up, and then boil them in about two bottles of water, allow to cool off and then give one bottle with the pulp in, followed about five hours later with the second bottle. They never lose a beast. [119]

Whereas Setswana speakers in North West Province often administered plants individually, there was a far greater use of plant mixtures in QwaQwa to tackle a variety of symptoms or to enhance potency. Gallsickness (*nyoko*), like blackquarter, was common in the region and many informants focused their discussion on this disease. As constipation was a symptom of gallsickness, their curative plants had the predictable purgative qualities that informants not only believed cured the disease, but also enhanced health and appetite by cleansing the system. Interviews suggested that there was a far greater reliance on herbs to treat gallsickness than in North West Province and none of our informants said they used Terramycin for this disease.

Aloes (*mohalakane* or *lekgala*) again featured as a major element in the local pharmacopoeia – this time to treat *nyoko*. They placed the aloes in a feeding

[115] David Mphuthi, Phuthaditjhaba, 14 October 2010.

[116] Ephraim Mofokeng, Phuthaditjhaba, 15 October 2010.

[117] James Motlaung, Bethelehem, 4 March 2011.

[118] Correspondence with the late Theuns Naude, retired toxicologist from Onderstpoort, 26 April 2007.

[119] H. H. Klette, "Farmyard Problems: Tulip Poisoning," (South African) *Farmers Weekly* 50 (25 December 1935), 1159.

trough in the kraal or made an infusion by boiling the leaves in water for five to 10 minutes, some added sugar or salt.[120] Other plants in use for gallsickness were *leshokgwa* (*Nasturtium officinale/Xysmalobium undulatum)*, which farmers boiled in water and then dosed in quantities of one litre per adult cow and 500 ml for calves.[121] *Lebejana* (*Asclepius spp.*) grew along the wayside and has milky pods that can be crushed and mixed with salt to make a lick that is said to treat *nyoko* effectively.[122] A litre of infused leaves from the *mofifi* (*Rhamnus prinoides*) tree was also popular.[123] Chadiwick Mbongo grew *kgamane* (*Rumex spp.*) in his yard to treat his animals, although he sometimes went into the mountains to collect fresh *msilwenge* bulbs.[124] The healer Zulu Mokoena drenched his cattle with a litre of *kgware* water. *Kgware* (*Pelagonium caffrum*) grew in the mountains and has small bulbs that turn the water red.[125] April Nhlapo dosed his animals with sugar and vinegar, or gave them salt licks, or a preparation made from *hloenya* (*Dicoma anomala*) and *mositsane* (*Elephantorrhiza elephantina).*[126] Farmers did not necessarily resort to one type of medicine for *nyoko;* it seemed to depend on which plants and products were conveniently available. There was a great variation in the types of medicinal flora people used. Overwhelmingly, stockowners claimed they all worked.

Farmers had a number of plants which they administered for digestive disorders (see appendix 4). *Mositsane* was popular for diarrhoea, constipation and worms.[127] Informants said they crushed up the bulbs and added them to water to make a red infusion. For some *mositsane* was an ideal plant, like the aloe, because of its versatility. Crushed and mixed with salt it made a lick to prevent and treat blackquarter and gallsickness. Like *thobeha*, discussed earlier, the ground-up bulbs were said to speed up the healing of bones.[128] *Mofifi* leaves were not only used for gallsickness, but also applied as a tick repellent: steeped in boiling water for five to 10 minutes, they washed the acarids away.[129] An alternative spray was made from *lebejana* with potassium permanganate. The mixture could also be given as a drink to make the blood bitter and the animal less attractive to biting ticks and insects.[130]

[120] Ntsoeu Livestock Company, Phuthaditjhaba, 11 October 2010; Konese Mofokeng, University of FS Phuthaditjhaba, 11 October 2010; Maria Khoboko, Lejwaneng, 12 October 2010; Abel Lebitsa, Mountain View, 14 October 2010.
[121] Daniel Khonkhe, Ha-Sethunya, 13 October 2010.
[122] Ntsoeu Livestock Company, Phuthaditjhaba, 11 October 2010.
[123] Sethunya Mathu, Ha-Sethunya, 13 October 2010.
[124] Chadiwick Mbongo, Lejwaneng, 12 October 2010.
[125] Zulu Mokoena, Lusaka, 2 March 2011.
[126] April Nhlapo, Lusaka, 28 February 2011.
[127] David Mphuthi, Phuthaditjhaba, 14 October 2010; Alan Malinga, University of the Free State, 13 October 2010.
[128] Gladice Mokoena, Ha-Sethunya, 13 October 2010.
[129] Daniel Khonkhe, Ha-Sethunya, 13 October 2010.
[130] Ntsoeu Livestock Company, Phutadijhaba, 11 Oct 2010.

Plant medicines in the Eastern Cape

Informants in Mbotyi used plants both for curing livestock diseases, and doctoring homesteads and kraals against supernatural forces.[131] We found that different plants were used for these two purposes; protective *muthi* is discussed in the next chapter. Remedies for gallsickness (*inyongo*) were the most widely known because many learnt them as herdboys – the quicker they are administered out in the grazing lands, the more chance of their success. As noted in chapter 5, even this knowledge was by no means universal: a few of the men, and all of the women, interviewed said they did not know the remedy. Most agreed that it consists of four plant ingredients, pounded together on a grinding stone and soaked to make a thick liquid. This could be administered immediately or stored in bottles. They fed the drench to the sick cattle through the mouth in doses of 1 to 3 litres depending on the size of the animal. They made the infusion out of the bark of the *umkhwenkhwe* tree, the stem of a creeper called *injalamba*, and the leaves of two plants, *umlung'mabele* and *umzane*. Some omitted the last plant or replaced it with another.

Umkhwenkhwe (*Pittisporum viridiflorum* or cheesewood) is a common tree along the east coast of southern Africa. Its bark has a bitter taste and strong resinous smell. In addition to its old-established role in combatting gallsickness, it is used for stomach complaints and colds in people, for glanders in horses, and as an addition to beer.[132] We saw a few trees in Mbotyi that had been heavily cut, revealing the bright yellow inner wood. *Injalamba* is probably a species of *ipomoea*, a convolvulus creeper introduced from the Americas, known for its purgative effects.[133] When taken by people, 'you can go to the toilet the whole day' and it also takes out the bile.[134] *Umlung'mabele* is *Zanthoxylum capense* or knobwood, a common tree of the east coast with a large lanceolate leaf that exudes a pungent, aromatic smell with a similar taste. In the nineteenth century it was known in isiXhosa as *umnungumabele* – probably meaning a teat with a frightening appearance, in reference to the spiky light grey knobs that grow on the bark. The humorous usage in Mbotyi now means 'white women's teats'. Andrew Smith recorded the leaves' disinfectant properties in the late-nineteenth century; they were especially effective for meat from cattle with *milt-ziekte*, or anthrax.[135] Kropf, in his 1915 dictionary, lists the root as a remedy for

---

[131] Dold & Cocks, *Voices from the Forest*.

[132] Andrew Smith, *A Contribution to South African Materia Medica*, third edn (Grahamstown, Rhodes University Cory Library, 2011, first published Lovedale Press, 1895); Keirungi & Fabricius, 'Selecting Medicinal Plants'.

[133] Watt & Breyer-Brandwijk, *Medicinal and Poisonous Plants* (1932), 152.

[134] Lenox Philani Mkhanywa, 4 December 2012.

[135] Smith, *Contribution to South African Materia Medica* – in this source and in Kropf, *Kafir-English*

snake and tsetse bites and the tree also features in treatment for flatulence and fevers. *Umzane* is *Vepris lanceolata*, the white ironwood tree, which Kropf records as a medicinal plant for gallsickness; it has other roles in human medicine. *Umzane*, with leaves that smell a little like lemon, is also widespread in coastal sites on the eastern seaboard of South Africa.

Thus the local recipe for *inyongo*, while possibly unique in its combination, draws on some long-established medicinal plants including those thought to act as a laxatives and disinfectants. Informants did not specify to us the function of the various constituents, nor did they say that they used these different plants individually. As mentioned, these plants were not difficult to identify and collect in Mbotyi. There are pockets of forest and woodland in some of the steep valleys within the area of the settlement and also larger patches on the peripheries of the village. One day we set out at 12.10, after an interview, to collect the four plants for *inyongo* and were back at 12.45.[136] We found the creeper, the knobthorn and the white ironwood in a forest patch within 10 minutes from Sonwabile Mkhanywa's home and then walked 10 minutes further to get the bark from an *umkhwenkhwe* tree. There was one closer to his house, but ease of access from the village had led to heavy exploitation of its bark and the tree was dying. Other trees along this route were also heavily cut, such as the wild plum (*Harpephyllum caffrum*); its bark is used for skin treatments and it carries an edible fruit. We then smashed and ground the leaves, creeper stem and bark together on a stone kept for the purpose outside the kitchen hut. When they were partly pulped, we soaked them in cold water for about half an hour, squeezing them further by hand. The greenish liquid was drawn off, but not thoroughly strained; it turned light brown in storage.

Zipoyile Mangqukela, around 90 when we interviewed him in 2008, claimed that many years before he had developed an effective remedy for *umbendeni* (redwater) and he had been able to cure a number of animals.[137] He considered the mixture, consisting of eight different plants, to be particularly bitter and 'strong' – too strong to give to people. Some constituents were the same as for gallsickness: *umlung'mabele*, *umzane*, *injalamba* and *umkhwenkhwe*. He experimented by adding four more to this list. The fifth plant, *uqangazana*, is usually given as the isiXhosa for *Clerodendrum glabrum* (Cat's Whiskers or Verbena tree or Tinderwood). In other parts of South Africa, its leaves are well known as an insect repellent as well as a drench for ticks, heartwater and internal parasites.[138]

[(cont)] *Dictionary*, it is spelled X*anthoxylon capense*.

[136] 26 March 2009.

[137] Zipoyile Mangqukela, Mbotyi, 29 March 2008.

[138] Kedibone Gloria Mawela, 'The Toxicity and Repellant Properties of Plant Extracts Used in Ethnoveterinary Medicine to Control Ticks', unpublished MSc, University of Pretoria (2008); M. A. Bisi-Johnson, C. L. Obi, L. Kambizi and M. Nkomo, 'A Survey of Indigenous Herbal Diarrhoeal Remedies of O. R. Tambo District, Eastern Cape Province, South Africa', *African Journal of Biotechnology*, 9, 8 (2010), 1245–54.

In Limpopo, the sap is 'smeared onto animals for tick removal'.[139] The plant's potential, in particular solutions, as a tick repellent was confirmed in a recent scientific study.[140] No-one interviewed in Mbotyi mentioned its traditional use for this purpose, and it came up in only two interviews. It is of course possible that this was one of the plants that people were reluctant to talk about, but Mangqukela himself, like others, was quite clear that there were no local remedies for ticks: 'we never had a traditional *muthi* for ticks; I never heard about it.'[141] Perhaps this plant featured more regularly in remedies for other diseases in earlier times and indirectly had the effect of diminishing susceptibility to ticks.

The sixth plant in the mixture for redwater, *ulwimi lenkomo* (cow's tongue), is probably a species of gasteria, a succulent that is widely used medicinally and has been found to have some effect on opportunistic fungal infections associated with HIV/AIDS. [142] One gasteria species, *Gasteria bicolor*, is also associated with supernatural protection both in KwaZulu-Natal and the Eastern Cape and is grown in roof pots as a protection against lightning. It is also called *intelezi* – a more generic name for plants with such properties. *Gasteria croucheri* seems to be the most common species north of the Umzimvubu and it is likely he was referring to this plant. It is possible that a plant which helps to control fungal infections, while it may not have any direct impact on tick-borne diseases, enables an animal's immune system to cope better. In addition he mentioned, seventh, *intongana*, probably *Drimia robusta*, a recorded human medicinal bulb, and *ukofukofu*. We cannot find a scientific or English name for this latter plant, which Mangqukela said grew on hills near the sea. The word is translated by Kropf as hard-breathing, and elsewhere as asthmatic breathing; it may be derived from English 'cough', and associated with chest complaints.

Mangqukela said he was able to develop this medicine because he had a good knowledge of plants and experience in healing gallsickness. He did not cook the plants, chopped and crushed them, before placing them in a bucket of water where they infused. He sieved the mix before use. When it worked on one cow, he administered it to others and claims to have cured animals for many (*kakhulu*) people. But it was important that cattle were kept away from drinking for a full day after taking the potion to ensure it did not

---

[139] M. M. Matlebyane, J. W. W. Ng'ambi and E. M. Aregheore, 'Indigenous Knowledge (IK) Ranking of Available Browse and Grass Species and Some Shrubs Used in Medicinal and Ethnoveterinary Practices in Ruminant Livestock Production in Limpopo Province, South Africa', *Livestock Research for Rural Development*, 22, 3 (2010).

[140] Mawela, 'Toxicity and Repellant Properties'.

[141] Zipoyile Mangqukela, Mbotyi, 29 March 2008.

[142] Wilfred M. Otang, Donald S. Grierson and Roland N. Ndip, 'Antifungal activity of *Arctotis arctotoides* (L.f.) O. Hoffm. and *Gasteria Bicolor* Haw: Against Opportunistic Fungi Associated with Human Immunodeficiency Virus/Acquired Immunodeficiency Syndrome', *Pharmacognosy Magazine*, 8, 30 (2012), 135–40; M. L. Cocks and V. Moller, 'Use of Indigenous and Indigenised Medicines to Enhance Personal Well-Being: a South African Case Study', *Social Science & Medicine*, 54, 3 (2002) 387–97.

become diluted: 'water', he said, 'encourages the disease to kill the cattle'. *Umbendeni* had to be caught early if the *muthi* was to work. This multipart medicine is unusual even in Mpondoland, where informants mentioned combining of species more often than in North West Province. But Mangqukela was clearly drawing on old knowledge of plants and blitzing the disease. Indeed, the recorded information on medicinal species suggests that each plant in his mix had slightly different properties. Informants did not consider that treating gallsickness had any impact on ticks themselves, although this may be a possible side effect. We did not hear claims to this effect, as were made in North West Province concerning aloes. Mangqukela was clearly interested in experimenting, at least in earlier days. Matwana said he did not experiment with plants, and he distinguished himself from those who did, such as *sangomas*, who learnt new medicines in dreams. But Sonwabile Mkhanywa, our research assistant in Mbotyi, and the son of a *sangoma*, also talked about experimenting.[143] For example, when he was young, he had suffered from a rash which his family thought was caused by an allergy to meat. He was sent to a herbalist who made a *muthi* from the roots of the common blackberry – an exotic, sometimes invasive plant on the forest margins. Years later, when his sister came out in a rash, he tried the same medicine. We dug some up to see if it would work.

The mix for gallsickness could be deployed against other diseases. Nonjulumbha Javu, who had a sizeable herd of about 45 when we first interviewed him in 2009, used both injections and herbs for *inyongo*. He believed that the gall produced by the disease spread everywhere through the veins. When this happened, the injection was inadequate because the cow could no longer excrete or urinate so that herbs were necessary because they had the curative effect of loosening the stomach – he used the English word 'toilet', common in Mbotyi. He cured worms in calves with a similar mix of three plants, omitting the *umzane* bark.[144] In this case, also, the laxative effect helped to cure the animal. Javu thought, however, that the local plant remedies were no longer so effective, and cattle sometimes remained lean after they were treated. One person in Mbotyi mentioned *intolwane (Elephantorrhiza elephantina)* – widely used in QwaQwa – as an effective dewormer.[145] It does not grow locally but is used for diarrhoea and miscarriages in other parts of the Eastern Cape.[146]

---

[143] Sonwabile Mkhanywa, Mbotyi, 25 March 2008.

[144] Nonjulumbha Javu, Mbotyi, 11 February 2009.

[145] Maphosa, 'Determination and Validation', 37–52.

[146] Victoria Madikane, Mbotyi, 12 December, 2011. The plant does not grow around Mbotyi and was not mentioned there except in this case. See A. P. Dold and M. L. Cocks, 'Traditional Veterinary Medicine in the Alice District of the Eastern Cape Province, South Africa', *South African Journal of Science*, 97, 9–10 (2001), 375–9; Bisi-Johnson et al., 'Indigenous Herbal Diarrhoeal Remedies'; Chris Simon and Masilo Lamia, 'Merging Pharmacopoeia: Understanding the Historical Origins of Incorporative Pharmacopoeial Processes amongst Xhosa Healers in Southern Africa', *Journal of Ethnopharmacology*, 33, 3 (1991), 237–42.

In Mbotyi, remedies for gallsickness were similar throughout the village. A striking feature of interviews elsewhere in the Eastern Cape was the diversity of plants and patent medicines deployed even within particular villages. As part of our project, Vimbai Jenjezwa interviewed 30 livestock owners in Hertzog and Tamboekiesvlei in the Kat River valley north of Fort Beaufort.[147] This area was long settled by coloured families, descendants of members of the Cape's 'Hottentot' regiment, who were largely Afrikaans and English-speaking. Some retained their land when it was incorporated into the Ciskei homeland, but many African families moved in. These villages are more diverse than most in the former homelands. They drew on different knowledge bases, mixing Khoisan, Afrikaner and Xhosa traditions.

In most cases, as elsewhere, the medicine for gallsickness in the Kat River valley was designed to make the animal purge. An 80-year-old coloured farmer, Eric du Preez, knowledgeable in local veterinary medicine, attributed the disease both to ticks and to green grass. He dosed cattle with four plants that he called *nkwenkwebos, galbossie, nieshout,* and *perepram*.[148] Aloe could be added to the mixture. These plants were boiled in about three litres of water then administered in a litre bottle to cattle three times a day. The remedy was green in colour and bitter; he also treated three-day sickness with this mix. (This often passes of its own accord although symptoms can be relieved.) *Nkwenkwebos* is *Pittisporum viridiflorum* or cheesewood, mentioned above with reference to Mbotyi. *Perepram* is correctly *perdepram* (horse teats), a common Afrikaans name for knobwood (*Zanthoxylon capense*), again an element in the Mbotyi mix. *Nieshout* or sneezewood (*Ptaeroxylon obliquum*) has a number of human medicinal uses and it has been recorded as a tick repellent in the Eastern Cape.[149] In the Xhosa pharmacopoeia, *umthathi* (sneezewood) features largely as snuff.[150] Galbossie may be one of a few species. Thus two of his ingredients were the same as those in Mbotyi, but two differed.

In contrast to du Preez's, most local remedies for gallsickness had aloe as a primary ingredient. One 60-year-old coloured farmer said he used a mixture of *slangbossie,* aloe, *balsam koppifer* root and *nkwenkwebos*.[151] The latter, cheesewood, is the only common factor with his neighbour's mix and those in Mbotyi. Aloe, as we have noted, is common elsewhere. *Balsem-Kopiva* was a well-known patent medicine marketed amongst Afrikaners. The name probably

[147] Jenjezwa, 'Stock Farmers and the State'.

[148] Jenjezwa, 'Stock Farmers and the State', 122; Eric du Preez, 18 November 2008, interview by William Beinart and Vimbai Jenjezwa.

[149] B. Moyo and P. J. Masika, 'Tick Control Methods used by Resource-limited Farmers and the Effect of Ticks on Cattle in Rural Areas of the Eastern Cape Province, South Africa', *Tropical Animal Health and Production*, 41 (2009), 517–23.

[150] *Galbossie* is difficult to identify – it may be *Conyza ivaefolia* (mentioned by Smith as the Albany gall-sick bush).

[151] Jenjezwa, 'Stock farmers and the State'; L. Pringle, 2 September 2008.

derives from a globally-used medicinal resin from the South American copaiba tree, but it seems that the word was localised and it is recorded by van Wyk as a popular name for *Bulbine frutescens*. Africans administered an infusion from the bulb and leaves, dried and boiled, to treat diarrhoea. The fresh fleshy leaf yields a gel that sooths burns and wounds.[152] This is also another of the plants that is named in isiXhosa *intelezi*. Scientifically extracted, *Bulbine* species have antibacterial properties, although this is much less marked in water, the only solvent available to traditional users.[153] A related species, *Bulbine alooides* – probably also covered by the Afrikaans common name of *balsemkopiva* – has a red bulb. It is seen as a valuable blood purifier and cure for rheumatism in people and for redwater in cattle.[154] Whatever its properties, this red colouring, like *sekaname* – noted above – seems to be associated by African healers with blood. This *balsemkopiva* was probably a local medicinal plant, though not one which has been recorded previously for gallsickness. The term *slangbossie* is used for a variety of plants, including *Elytropappus glandulosus* and the *Stoebe* genus both of which have medicinal uses.

A coloured farmer who owned 30 cattle mixed dry aloe and a packet of Epsom salts, boiled in water, for gallsickness. Sometimes he replaced aloe with *slangbossie* or *gallbossie*, and sometimes added *balsemkopiva*.[155] An elderly Xhosa stock owner mixed *umxhube* (Kropf gives this as a mix of sorghum and maize), aloe, *ubuhlungu* (see below for snake bites) and *umkhwenkhwe*. She gave a half-litre bottle to the affected livestock in the morning and again in the evening.[156] An elderly Xhosa man used aloe alone, focusing on the laxative effect.[157] Thus even in one settlement, most of those interviewed used a different combination of plants for gallsickness – perhaps an indicator of social diversity and immigration.

Dold and Cocks record that in Masakhane, between Peddie and Alice, less than 100 km from the Kat River valley, the leaves of knobwood were mixed with those of *Grewia occidentalis* and *Olea europaea*, as well as *Aloe ferox* sap, soaked in cold water.[158] Although aloe and knobwood again feature in this recipe for gallsickness, two additional plants, neither recorded for this purpose elsewhere, came into play here. Mike Kenyon, returning to this site, found that *uvendle* (*Pelargonium reniforme*) was mixed with aloe for *inyongo*. *Umkhondo* (*Agapanthus*

[152] Van Wyk and Gericke, *People's Plants*; R. M. Coopoosamy, 'Traditional Information and Antibacterial Activity of Four *Bulbine* Species (Wolf)', *African Journal of Biotechnology*, 10, 2 (2011), 220–24.
[153] Coopoosamy, 'Traditional Information'.
[154] Smith, *Contribution to South African Materia Medica*; Dold & Cocks, 'Traditional Veterinary Medicine'.
[155] Jenjezwa, 'Stock Farmers and the State'; J. Loots, 14 May 2009.
[156] Jenjezwa, 'Stock farmers and the State'; N. Makhapela, 29 April 2009.
[157] Jenjezwa, 'Stock farmers and the State'; K. Rangana, 24 March 2009.
[158] Dold & Cocks, 'Traditional Veterinary Medicine'.

praecox) also seemed to have a role, although it was more significant for a range of other ailments in cattle. In coastal Peddie, again close to Masakhane, gall-sickness was treated with the leaves of *isifithi* (*Baphia racemosa* or violet pea), sometimes with sea water added. One owner

> … administers up to seven 2-litre bottles of sea water per night to a sick animal. When the animal is released from the kraal the next morning, it goes straight to the trough to drink water and you can see it feels better. In the kraal, you will find a pile of little red worm-like things in its shit, but they are not worms.[159] (Sea water is used in QwaQwa to counteract witchcraft – see chapter 8.)

Some used *isifithi* with Terramycin and added other herbs, notably *mathunga*. This is probably *Haemanthus albiflos*, a large bulb with a striking flower, which is recorded as healing broken limbs in humans, but not for *inyongo,* elsewhere.[160]

It is intriguing that gallsickness is treated with a wide range of plants in different parts of Eastern Cape. Two or three key species recur in different sites, but are not a constant and, with the partial exception of Mbotyi, informants told us of diversity within villages as well as between them. This diversity suggests that the local medicines are non-specific, and perhaps work at a general level as laxatives, analgesics, anti-inflammatories, anti-insecticides and anti-micro-bials. As noted above, scientific researchers have made attempts to isolate effec-tive chemical agents in some plants, and positive indications have been recorded, but most find that solutions in water – as opposed to other solvents – release fewest of the active ingredients. Kobus Eloff, a phyto-chemist at the University of Pretoria, commented that the use of water as the only solvent was a major limitation in the preparation of local plant-based medicines.[161]

In Mbotyi, Myalezwa Matwana specialised in birthing problems and retained placentas, which, as we noted above for North West Province and QwaQwa, were considered especially problematic. He mixed and ground four plants for this, all different from those for gallsickness and from those used in other sites. *Umkhomakhoma*, mentioned above as a wet plant, is readily available in the local forests; a plant with a similar name is used in KwaZulu-Natal against tape-worms. To this is added *umsintsi* bark (*Erythryna* species, possibly *E. caffra*, the coastal coral tree) from another common tree on the east coast that is also a treatment for wounds with possible anti-inflammatory effects. *Mhlolo* (*Grewia lasiocarpa*, a common small-forest-margin tree) and *umkhazi* complete the combination.[162] *Umkhazi* is *Typha capensis,* the common bulrush, very similar to

[159] Andrew Ainslie, 'Hybrid Veterinary Knowledges and Practices among Livestock Farmers in the Peddie Coastal Area of the Eastern Cape, South Africa' (unpublished report, 2009).

[160] Dold & Cocks, 'Traditional Veterinary Medicine'.

[161] Discussions with Kobus Eloff, University of Pretoria, 25 February 2010.

[162] Dold & Cocks, 'Traditional Veterinary Medicine', recorded that a similar Grewia species featured in one of the gallsickness remedies in the Eastern Cape.

the northern hemisphere species *Typha latifolia*. Its rhizome is edible, and was once a valuable indigenous starch for human consumption. This very widespread wetland reed is also believed to have a number of medical properties, notably as an aid to human as well as animal birthing. Scientific experiments have indicated antibacterial properties though these were not very conclusive.[163] Again, as in the case of the *muthi* for *inyongo*, these plants are relatively easily found in the vicinity of Mbotyi village. This combination, which Matwana said he learnt from his father, seems to include plants with different and complementary properties.

When snakes bit livestock, people in Mbotyi used *isihlungu* in the wound. This word can apply both to poison and to medicinal plants used to counter it but the specific plant is uncertain because the word refers to a number of species.[164] The term includes plants of the genus *Acokanthera* which have a poisonous sap, used in earlier years by the San for arrows.[165] In the nineteenth century, *Acokanthera spectabilis* was sometimes administered to animals during transhumance when they were moved from inland sweetveld pastures to coastal sourveld.[166] The isiXhosa word could also refer to *Teucrium africanum* (*ubuhlugu benyushu*) used throughout the Transkei as a snake bite antidote, especially for puff adders.

For birthing problems, Myalezwa Matwana would visit the animal in difficulty and he specialised in assisting where the foetus was presenting irregularly. In these cases he would act as midwife, manipulating the calf with a soapy hand. He and others also performed a simple operation on what they called *impena*, a growth or thickening inside the vagina of the cow that they believed hampered fertility and delivery, killing calves in the womb. It could affect cows in that bulls troubled them continuously. He cut this out with a knife then dosed the animal with a mixture of four plants; the calf would then grow properly. He and others also had plant medicines for wounds from horns. These are not uncommon as, in contrast with North West Province, cattle are seldom dehorned and are often cooped up together in kraals. After a slaughter, Sonwabile Mkhanywa said, the stomach contents are usually left spread on the ground next to the animal; this discouraged cattle from fighting.[167]

---

[163] P. Masoko, M. P. Mokgotho, V. G. Mbazima and L. J. Mampuru, 'Biological Activities of *Typha Capensis* (Typhaceae) from Limpopo Province (South Africa)', *African Journal of Biotechnology*, 7, 20 (2008), 3743–8.

[164] Kropf, *Kafir-English Dictionary* 162–3; R. B. Bhat and T. V. Jacobs, 'Traditional Herbal Medicine in Transkei', *Journal of Ethnopharmacology*, 48, 1 (1995), 7–12, identify it as *Dias cotinifolia* for human use but don't mention its use in snake bites.

[165] Simon & Lamla, 'Merging Pharmacopoeia'.

[166] Smith, *Contribution to South African Materia Medica*.

[167] Sonwabile Mkhanywa, Mbotyi, 24 March 2012.

Conclusion

African livestock owners use a wide range of plants in different parts of South Africa. A few stand out in some areas, such as aloes and *sekaname* in North West Province. With regard to gallsickness two or three key species – particularly aloes and pittisporum – recur in different sites, but are not a constant, and we found considerable diversity, especially in the Eastern Cape, within villages as well as between them. This diversity suggests that the local medicines are often non-specific, and a range of species may have the desired effect. They seemed to work at a general level as laxatives and some may also impact on pain and fungal infection. Researchers into human diarrhoeal remedies found 32 different species used in one Transkeian district, O. R. Tambo (in which Mbotyi is located) alone.[168] Overall, as we discussed in chapter 3, ensuring that the animal eats and excretes well, that its external symptoms are relieved and its condition improves are the key desired effects of medicinal plants.

Most plants are ground and infused or boiled in water – either alone or in combination – and administered through the animal's mouth as a drench. Some are smeared externally or inserted into cuts in the skin. A few are burnt, though this seems to apply more to those used for supernatural protection. Our evidence suggests that on the whole, different plants are used for healing and for protection against supernatural forces. There is some overlap, but particular combinations of plants are associated with particular diseases in different areas and it appears that these have been well-established for some time – our older informants say they learnt them from their parents. People may see some of this knowledge as deriving from ancestors or dreams, or associated with other forms of spirituality. But most of those we interviewed see the key plant reme-dies playing a natural rather than a supernatural role, and think about them in relation to their efficacy towards particular medicinal ends. Expanding access to biomedicines reinforces these ideas.

Scientific researchers have made attempts to isolate effective chemical agents in a limited number of African medicinal plants. They record some positive indi-cations but most experiments find that solutions in water release fewest of the active ingredients. All of our informants used only water in which to mix their *muthi*s. Such research has, however, generally been done with one plant under laboratory conditions and not in the field at particular seasons, with a mix of plants in the same combination and quantity that are deployed in local veterinary treatment; it may also be necessary to treat animals – which may be used to these drenches – more than once to see the impact and possible side effects. Their value is likely to be closely related to partial immunities and nutrition as well.

[168] Bisi-Johnson, et al., 'Indigenous Herbal Diarrhoeal Remedies'.

Plant remedies might relieve symptoms while animals' natural immunities and resistance enable them to recover. (Biomedicines can also act in this way.) Some popular local treatments can be highly toxic depending on the concentration or dose.[169] *Sekaname* or *slangkop* can kill humans and livestock if taken in large quantities. Local remedies are less subject to standardised dosing and the concentration of plant extracts in any particular mixture can vary greatly. Stock-owners used rough guides for dosage, not only with respect to the amount of plant material in the solution but in the amount administered – usually between 500 to 2000 ml, depending on the size of the animal. Overuse of laxatives can cause lazy bowel syndrome in humans and the complaint by North West Province informants that their animals were often constipated could be related to frequent use of aloe purgatives. We noted the possible misuse of motor oil in Mbotyi.

It is thus very difficult for us to judge the effectiveness of local medicines, especially because they are sometimes administered along with biomedicines. Some informants were themselves uncertain. A properly informed analysis would require a different kind of research. In the final two chapters we turn to the supernatural sphere and the treatments applied to contain and control its threats. They also include plants, and thus make the issue of efficacy even more complex.

---

[169] Johns, *Origins of Human Diet and Medicine,* 10.

7
Animal Health
& Ideas of the Supernatural

The cause of disease in animals and humans is the same: bad food, worms and witch-craft.'[1] (Anna Pooe, traditional healer and goat breeder, Shakung, North West Province)

Introduction: Witchcraft and the ambient supernatural

Throughout our field sites there were those who attributed some animal diseases to types of witchcraft, or to supernatural causes, but occult explanations were far rarer than environmental and nutritional aetiologies. As discussed in chapter 3, many of our respondents ascribed familiar diseases like gallsickness to the state of the veld and seasonal changes in the texture of the grasses. We cannot definitively assess how commonplace beliefs in the supernatural and ritual pollution are amongst African rural communities. In part this is because we interviewed across a number of sites, rather than observed practices in one village over a long period. We are thus largely dependent on what people said about their ideas in connection with animal diseases. It is possible that some stockowners regarded witchcraft as a private matter and were reluctant to discuss it. However, many informants spoke openly about supernatural ideas and ritual pollution. Few explicitly refused to talk about these issues and a good range of our interviews at least touched on them.

In general, few of our respondents attributed animal deaths to witchcraft – at least if we mean by witchcraft malevolent action by an individual against another individual via their animals. In North West Province only three out of over one hundred informants described animals dying of diseases that they

[1] Anna Pooe, Shakung, 22 January 2010.

197

directly ascribed to witchcraft. In QwaQwa informants claimed that witches could make animals infertile, but nobody linked witches directly with livestock diseases and deaths. In Mbotyi, stockowners attributed one particular disease, with specific symptoms, to witchcraft – but not others.

By contrast, informants expressed more general beliefs about a diffuse or ambient sense of supernatural forces. In Setswana-speaking areas this was generally called *mohato* and associated most strongly with the polluting potential of women. We will discuss *mohato* in detail in chapter 8. In Mbotyi, many informants mentioned a disease or condition called *umkhondo*, which included multiple meanings. As Myalezwa Matwana said: 'In Xhosa, we use the same word for a lot of things.'[2] Livestock owners associated it most strongly with traces left on the ground by supernatural forces which could harm animals. Edward Green classified similar ideas in other parts of Africa, including 'pollution' and less identifiable 'environmental dangers' as part of a more naturalistic 'indigenous contagion theory'.[3] We are not seeking to establish tight divisions between natural and supernatural and we argue that *umkhondo* and *mohato*, which may be relatively new in their current meanings, could include environmental and biomedical elements. Yet they seemed to us, and our informants, closer to supernatural understandings.

The social background of people who said they followed the older 'forms of knowing' was not necessarily predictable. By and large people with higher levels of education tended to have a more biomedical approach to livestock diseases. Those who had attended courses organised by the Department of Agriculture might allow women into the kraal and were less likely to accept the idea that they could cause miscarriages or annul the potency of the *muthi*. Formal education taught farmers that animals aborted because of disease or malnutrition, not because they had encountered the pernicious spoor of women. Nevertheless there were educated farmers who used modern drugs, let women into the kraal but accepted some elements of the supernatural. Belief in witchcraft was not something that automatically became eroded through education or exposure to scientific ideas. Reuben Ramutloa, for example, the well-educated principal of a school at Disaneng near Mafikeng, had a traditional healer visit his kraal every year to protect his animals against witchcraft. He believed that his livestock had escaped harm because of this treatment.[4] Age was not necessarily a determining factor either; some younger informants said they held by these concepts. Many women also accepted the idea that menstruation could be dangerous to livestock. Nevertheless, it was clear from the testimonies that some women did challenge this premise, and in a number of families the taboos against women handling cattle had disappeared.

[2]   Interview, Myalezwa Matwana, 7 December 2011.
[3]   Edward Green, *Indigenous Theories of Contagious Disease* (Walnut Creek, CA, Alta Mira Press, 1999), 13–14.
[4]   Reuben Ramutloa, Disaneng, 11 February 2010.

As in all aspects of veterinary knowledge and practice that we encountered, there was no homogenous African approach to the supernatural. One stock-owner might allow women into his cattle kraal, while his neighbours would not. Some laid greater emphasis on witchcraft, pollution or the supernatural than others. We were certainly struck by the pervasiveness of concern about these forces, despite the depth of social change. Local ways of thinking remain dynamic, but fragmented, in this sphere as well as others. We will argue, however, that in contrast to some recent literature on the modernisation of witchcraft ideas, these may be waning in influence, and taking a different form.

In her ethnography of the Mpondo researched in 1931–32, Monica Hunter suggested that witchcraft and supernatural beliefs were very generalised. She did find that some categories of human disease were associated with natural causes and a few, such as leprosy and smallpox, were recognised as passing between people. Her research notes also record examples of scepticism. In a file headed 'extent of criticisms of customs', she indicated that one traditional healer (*igqira/sangoma*) whom she interviewed mused that sacrifices or smelling out of witches were no longer effective against disease.[5] Another mentioned that he no longer believed in the ideas that he was originally taught: 'because of our beliefs … we like to blame someone, we sometimes blame someone when we know [it is] really natural causes'.[6] She also intimated an element of local doubt in the skills of traditional healers evinced by the fact that people went to several *sangoma*s until they found one whom they trusted. However, Hunter's general conclusion in her published text was clear: 'Much space has been devoted to witchcraft and magic but it is commensurate with the part they play in Pondo life. The belief in them permeates the whole of life.'[7]

Modern witchcraft studies tend to emphasise its dynamic and contemporary features and also insist on its pervasive presence in people's lives. Some suggest that supernatural explanations may become more intense in periods of rapid social change. Isak Niehaus has argued that witchcraft has increased in the South African lowveld since the 1960s due to social upheavals.[8] Apartheid poli-cies of betterment and villagisation pushed people together from various back-grounds with different traditions and social alliances. He also connected the Pentecostal Christian revivals in South Africa with a popular perception that witchcraft was rife. Adam Ashforth, who worked in Soweto in the 1990s,

[5] University of Cape Town manuscript collection, BC 880, Monica and Godfrey Wilson papers, Box 10, Doctors, notes on 'Diseases Distinguished by Doctors'.
[6] University of Cape Town manuscript collection, BC 880, Box 10, interview with Majelu.
[7] Monica Hunter, *Reaction to Conquest: Effects of Contact with Europeans on the Pondo of South Africa* (London, International Institute of African Languages and Cultures, 1936), 319.
[8] Isak Niehaus: *Witchcraft, Power and Politics: Exploring the Occult in the South African Lowveld* (Cape Town, David Philip, 2001); 'Witches and Zombies of the South African Lowveld: Discourse, Accusations and Subjective Reality', *Journal of the Royal Anthropological Institute*, 11, 2 (2005), 191–210.

described witchcraft and ritual pollution as expressions of insecurity.[9] More specifically he associated witchcraft with social misery and unemployment that exacerbated tensions and grievances in African communities. Those who were not able to enrich themselves under the ANC alleged that the greedy rich used witchcraft and zombies to prosper, while the beneficiaries of transition saw the envious poor as deploying *muthi* to destroy their wealth. The devastating impact of HIV/AIDS, which some people ascribed to witchcraft, generated new insecurities as individuals searched for explanations as to why they had become victims.[10] Ashforth also suggested that black people saw witchcraft as an 'African disease' which could not be cured by western medicines. He went as far as implying that both witchcraft, and its antithesis, healing, were aspects of African science and reflected a specific interpretation and approach to understanding the world, misfortunes and disease. *Muthi* represented both beneficial healing and malevolent poison and both were born out of African medical traditions, which nonetheless were evolving in response to new technologies.[11]

Ashforth and other anthropologists have also questioned why witchcraft has survived as a set of explanations for ill health and misfortune. Typically the answer seems to reside in questions surrounding 'why me?', as opposed to anybody else, and 'why now'? In such ideas of causation, nothing is left to chance. Things don't just happen.[12] The general consensus in the anthropological literature is that the main motivating factor behind witchcraft is jealousy. Niehaus argued that witchcraft accusations not only reflected various social tensions in the lowveld, but also symbolised the ambiguity of wealth. As in the case of Ashforth's interpretation of Soweto, narratives of witchcraft 'present a complex discourse on social inequality, deprivation and envy'.[13] Poverty fuelled jealousy of the rich, and the wealthy felt vulnerable when faced with the real or imagined hostility of the poor.

Some of these ideas were echoed by our informants. However, our discussion deviates from these analyses in important respects. First, as Hunter's notes suggest, we should be cautious in projecting back an entirely generalised or consistent belief in all of these explanations and practices. Second, as we have argued above, our informants did not ascribe disease and death of animals primarily to witchcraft. It is possible that these forms of explanation are declining in significance. Third, Hunter also suggested that there was no clear line between the natural and supernatural, the therapeutic and the magical – although some of her informants had a stronger sense of

[9]   Adam Ashforth, *Witchcraft, Violence and Democracy in South Africa* (Chicago, Chicago University Press, 2005), 155.

[10] Niehaus, 'Witches and Zombies'; Ashforth, *Witchcraft, Violence and Democracy*, 88–110.

[11] Ashforth, *Witchcraft, Violence and Democracy*, 133–53.

[12] Max Gluckman, 'The Logic of African Science and Witchcraft' in Max Marwick (ed.) *Witchcraft and Sorcery: Selected Readings* (Harmondsworth, Penguin, 1970), 443–51

[13] Niehaus, *Witchcraft, Power and Politics*, 111.

natural causes.[14] Our impression is that natural explanations for disease were distinguished from the supernatural and that the increasing adoption of biomedical treatments has probably contributed to a stronger differentiation between the supernatural, natural and biomedical.

According to Nongede Mkhanywa, a *sangoma* in Mbotyi, local people made a clear distinction between conditions such as *umkhondo* (see below) and those which were a direct result of 'natural causes'.[15] Amongst the latter, he included the great majority of the livestock diseases they experienced: '*isiDiya* [blackquarter], *nonkhwanyane, iskelem* [worms], *inyongo* [gallsickness] and *umbendeni* [redwater] – all of these, when you open the dead cow you can see what went wrong inside the animal. With *umkhondo* it is difficult to see what killed it.' Our analysis is echoed in Edward Green's overview of disease explanations in 'human and animal ethnomedicine'. He contests the dominance of supernatural ideas in anthropological analysis of most illnesses in Africa. In fact, he argues, 'there appears to be a great deal of naturalism in ethnoveterinary medicine'.[16] Ideas are certainly plural in character, but, as noted in the Introduction, the balance between the different explanations and treatments has changed. In this chapter, we outline some of the supernatural ideas that livestock owners mobilised in discussing and understanding disease. But in doing so, we are not arguing that they had an overarching and powerful effect on approaches to local veterinary practices.

Animal deaths and diseases associated with witchcraft

In this section, we will explore those minority of cases in which animal deaths and disease were attributed to witchcraft. Especially in QwaQwa and North West Province, informants expressed generalised concern about the threat of witchcraft. This was reflected firstly in discussions which invoked supernatural causes for unusual deaths that were very difficult to explain and secondly in the widespread treatment of kraals against sorcery. Much of this discourse tended to be generalised, citing cases outside the immediate experience of the respondents.

Our findings mirror those of Niehaus in that some of the farmers who attributed livestock deaths to witchcraft were more prosperous than many of their neighbours. They assumed that an envious person had callously used witchcraft to kill their animals. In QwaQwa, where several respondents associated barrenness in livestock with witchcraft, as opposed to ritual pollution, the reason cited

---

[14] Hunter, *Reaction to Conquest*, 274.

[15] Nongede Mkhanywa, Mbotyi, 7 December 2011. The translator, Sonwabile Mhkanywa, used the term natural causes.

[16] Edward Green, 'Etiology in Human and Animal Ethnomedicine', *Agriculture and Human Values*, 15, 2 (1998), 127–31, quotation from page 129.

was 'jealousy'. Malicious neighbours would do anything to stop their herds multi-plying. Some informants there claimed that growing urbanisation, as well as competition for jobs and access to grazing lands, was fuelling community strife, which could translate into occult explanations if livestock were stolen or cattle failed to reproduce. Similarly, in North West Province, two of the three accounts of livestock deaths through sorcery, emanated from densely populated areas – Winterveld (north of Pretoria) and Mmakau (near Brits) – which had been settled partly by people forcibly removed from elsewhere since the 1960s.

Interviews in North West Province and QwaQwa did not indicate that witchcraft had increased, but some people felt vulnerable. If the application of *muthi* to protect the kraal from witchcraft and stock theft is an indicator of belief, then it appears that fears of maleficence remain important. In 2011, 19 out of 21 informants from QwaQwa said that they used *muthi* to safeguard the kraal. Some made their own medicines while others invited a traditional healer to their homestead each year to sanctify the cattle and the enclosure. In North West Province a group of agricultural students at the University of Mafikeng claimed that Tswana farmers were still more likely to attribute animal diseases to witchcraft than they were to germs or ticks.[17] However, the testimonies did not bear that out. Whilst many farmers thought witchcraft could potentially lead to livestock diseases, few admitted personal experience of that. So there was a conceptual gap between possibility and experience. On the other hand, ideas about ritual pollution have endured in North West Province, and indeed in QwaQwa. In both regions they were frequently offered as the principle reason given for abortions and infertility.

Although witchcraft was not often invoked as a specific explanation of disease, it was mentioned in relation to more general misfortunes, especially in QwaQwa – for example livestock theft or conflagrations. Witchcraft took a number of different forms, such as 'making *muthi*' to attack others. Ethno-graphies tend to differentiate between witchcraft, categorised as caused by a witch with specific, sometimes inherited evil powers, and sorcery which is a more open category involving the use of magical substances by a wider range of individuals. Linguistically our Tswana informants did not differentiate between witchcraft and sorcery and they referred to both as *boloi*. In North West Province and QwaQwa the translators used this term for both men and women. Interpretations of how witchcraft manifested itself could be highly varied and individualistic. Witches might poison or doctor animals and kraals with harmful *muthi*, or launch a lightning strike as a dramatic demonstration of their evil powers. Some believed that witches used familiars or helpers, which took the form of real or mythical creatures, to carry out their work. In Mpondoland metaphorical baboons and snakes or serpents still serve in the popular imagination as carriers of misfortune. In QwaQwa and Mpondo-

---

[17]   Julius Maubane, Lucky Mogapi, Lebang Molefe, North West University (NWU), 6 Nov 2009.

land some informants also referred to the less definable *tokoloshe*.[18] People sought the advice of *sangomas* (diviners) and herbalists if they believed that they, their animals or their kraal had been bewitched. They sacrificed goats to the ancestors as a means of counteracting spells. Some Tswana stockowners deliberately reared white goats because they were particularly popular with traditional healers and with individuals wishing to commune with the departed (*badimo*). The agricultural students from Mafikeng said that the collar bone (*kgetlane*) of a cow, which had been slaughtered for rituals or celebrations, was especially dangerous as it could be mixed with *muthi* to bewitch people. Slaughterers, they explained, always tried to hide the bone, lest it be stolen for witchcraft.[19]

Joseph Mapetla who farmed in Kestell, near QwaQwa, described himself as a devout Christian, who looked to God to protect him from witches and their emissaries, the *tokoloshe*. He said that *tokoloshes* were responsible for transmitting diseases and were implicated in cattle theft: 'if a cow falls sick overnight in the kraal, or is found dead in the morning, that is the work of a *tokoloshe*. The *tokoloshe* use poisonous *muthi* to destroy livestock. That is the only explanation for a sudden death.'[20] Rosalina Johanne, a traditional healer from Lusaka, QwaQwa, also believed that witches administered poisons to kill cattle and they were the reason for any swift or mysterious deaths.[21] Abram Lebesa spoke of *vutha* a type of *muthi* that witches employed to burn down the kraal and kill all the cattle. Witches also used lightning: 'if lightning kills five animals at a time, then it is not normal. It is the work of witches.'[22]

Michael Majoang, a Sotho farm worker in Koppies, exemplified a mixture of beliefs. Having worked for white farmers for many years, he had a relatively strong biomedical perspective. But he also said that witches could kill animals by ordering a *tokoloshe* to send lightning. It was easy to differentiate between natural lightning and witch's lightning, he explained: in the latter case the carcass went rotten within two hours, which was not normal. Furthermore, Majoang continued, witches could 'send diseases that spread from the hooves to the head'. They could make animals aggressive so that they kicked you when you tried to milk them. Witches, he said, were always adult men and women; they could be rich or poor and were normally motivated by jealousy. Witches were always black; witchcraft is an African sphere. He thought that witches drew their powers from the ancestors who told them how to make dangerous medicines. For protection, people saw *sangomas* who would try to make a stronger type of *muthi* to defy the witches. In the Free State, healers particularly valued

[18] Lesley Fordred-Green, '*Tokoloshe* Tales: Reflections on the Cultural Politics of Journalism in South Africa', *Current Anthropology*, 41, 5 (December 2000), 701–12.
[19] Julius Maubane, Lucky Mogapi, Lebang Molefe, NWU, 6 Nov 2009.
[20] Joseph Mapetla, Phuthaditjhaba, 15 October 2010.
[21] Rosalina Johanne, Lusaka, 1 March 2011.
[22] Abram Lebesa, Ha-Sethunya, 2 March 2011.

goat skins because they had magical powers to overcome bad *muthi*.[23]

From time to time veterinarians would be called out to investigate deaths that stockowners had ascribed to witchcraft. According to Johan Naude, the state veterinarian based in Bethlehem who oversaw QwaQwa, 99 per cent of the cases attributed to witchcraft were a result of lightning strike. He explained that lightning causes congestion in the arteries and veins, the heart stops suddenly and the animal drops dead. The blood is dark due to the lack of oxygen and it puddles under the skin. The electricity precipitates rapid decay and the flesh smells rancid. In his experience, people refused to eat animals that had been struck by lightning because they linked these deaths to witchcraft.[24] This analysis explained Majoang's description of the putrefying carcass, although it did not elucidate why he differentiated between natural and supernatural lightning. Solomon Ndlodu, from Lusaka in QwaQwa, lost a large number of goats to lightning strike and attributed this to witchcraft. Whilst carrying out post-mortems he noticed that blood oozed out and rigor mortis took a long time to set in. That is how he knew it was witchcraft; if the body did not become stiff and dry within the expected amount of time, malice was at work. He corroborated Naude's comment that Africans would not eat an animal that had died of lightning strike because it was possessed by evil forces.

Simon Mosenogi, the state veterinarian in Mafikeng, also narrated how farmers often ascribed sudden deaths to witchcraft. His examples included toxic weeds like *gifblaar* and *slangkop* that killed animals by damaging the heart muscles. Death could be very sudden, giving the impression that the livestock had not died of natural causes. On a number of occasions he had been called out to examine cattle that, in the report of their owners, had just dropped dead in the kraal overnight. In addition to toxic weeds, anthrax was a possibility especially if a number of animals died at once. In some cases where he had tested for anthrax and the blood smears came back negative, rat poison seemed a likely cause of death. Sometimes he saw black granules of a type of rat poison he called *ga le phirimi* in the kraals. He suggested that the poisoning might be deliberate and that stockowners saw it as witchcraft, inflicted upon them by a jealous neighbour.[25]

The evidence from Mpondoland suggested that witchcraft was associated with a specific disease. There were some similarities with the testimonies from North West Province and QwaQwa in that dramatic or unusual symptoms surrounding death appeared to be indicators of maleficence. Zipoyile Mangqukela described how he had lost a large number of his cattle in the 1980s and he felt singled out because other stockowners had not experienced the

---

[23] Michael Majoang, Koppies, 6 October 2010.

[24] Johan Naude, State Veterinarian, Bethlehem, 4 March 2011.

[25] Simon Mosenogi, Mafikeng, 5 Nov 2009. Christo Botha, a toxicologist at Onderstepoort, suggested the granules were possibly aldicarb, a poison burglars often use to kill dogs before entering a property (Discussion, Onderstepoort, 18 November 2009).

same catastrophe at that time.[26] Sickness appeared rapidly and there was no time to treat the ailing. Blood dripped from their nostrils and they died on their backs, instead of on their sides. When he carried out the post-mortems there was internal bleeding around the lungs. He believed that a wizard had been responsible and had sent the *impendulu* (lightning bird) to attack his animals, because they died in such a sanguineous state, as if beaten behind the head with a heavy stick or club. Mangqukela was an ambitious farmer who said he had owned 58 cattle and also made money ploughing for other people. Surprised by how many cattle had died in quick succession, which seemed to him abnormal, he sensed that someone had been jealous of his drive and relative prosperity. He responded by burning *intelezi* herbs in his kraal and splashing an infusion made from the same plants around the cattle enclosure. The description of a sudden death and blood oozing from the orifices suggested that the cattle might have died of anthrax. At this time the state provided an annual anthrax-blackquarter inoculation. It is possible that his animals died, and not those of others, because he failed to vaccinate his cattle. Richard Msezwa similarly described an animal that had been struck by witchcraft and died quickly with blood dripping from the nostrils as if it had been violently kicked.[27] This was the only disease that was specifically attributed to witchcraft in the interviews in Mpondoland.

In North West Province some informants talked in general about witchcraft not only in relation to sudden and seemingly unexplainable deaths, but also in relation to infertility and birth defects. Innocent Setshogoe from Bethanie learnt a lot about local beliefs when he worked as a stock inspector in the 1980s. He suggested that many Tswana stockowners in this district associated the failure of cattle to conceive with witchcraft. Farmers would be particularly suspicious if they had invested in a pedigree bull which failed to impregnate their cows. This they attributed to jealous neighbours using *muthi* to suppress the bull's fertility. Alternatively these potions could result in abortions.[28] In QwaQwa a number of informants also associated sterility in cattle with the envy of others.[29]

Six farmers in Moiletsane (North West Province), whom we interviewed jointly, debated the reality of witchcraft and whether it was possible to bewitch cattle and the kraal.[30] Only one, who had had some agricultural training, felt confident that many of the problems associated with witchcraft had natural or scientific explanations. In Moiletsane a number of livestock had recently been born with various deformities: a kid with paralysed hind legs; a three-legged calf; and a cow with a strange skin disease that made it look like an elephant. Interviewees claimed that people attributed these occurrences to witchcraft

[26] Zipoyile Mangqukelan, Mbotyi, 26 March 2008.
[27] Richard Msezwa, Mbotyi, 10 February 2009.
[28] Innocent Setshogoe, Bethanie, 26 October 2009.
[29] April Nhlapo, Lusaka, 28 February 2011; Moses Mahlamba, Mountain View, 2 March 2011.
[30] Group Interview Moiletswane Community Centre, 21 January 2010.

because they deemed them unnatural. They assumed that jealous neighbours had acquired *muthi* from an evil traditional healer and doctored the cattle so they would not produce healthy offspring. Witches did not just kill animals; they also tried to show their powers in other ways that were clearly visible, and for some, alarming. The net result was the same. The deformed animals would not survive, jeopardising the economic and cultural fortunes of particular live-stock owners.

In none of the above discussions, except for that in Mbotyi, did informants directly ascribe the death of one of their animals to witchcraft. In North West Province, only three stockowners described their own losses in this way. In each case, victims believed jealousy was the reason, though not all were rich. None said they knew who had been responsible for bewitching or destroying their animals and all three blamed an anonymous neighbour, rather than members of their own extended families. For these people, witchcraft was a reality and they articulated both fear and helplessness when faced with its vicissitudes.

Frederick Matlatsi, from Mabopane near Garankuwa, had been keeping cattle and goats on communal lands near Winterveld since the 1950s to provide some financial security in old age. However, over the years all his goats had been stolen and in 2000 someone took 13 pregnant cows, about half of his herd. He did not attribute stock theft to witchcraft, but in 2003 when his remaining cows failed to conceive, he sought help from a Christian prophet rather than a traditional healer. The prophet sprinkled a secret concoction around the kraal and buried a mysterious object deep into the ground at the entrance of the enclosure. The cows soon became pregnant. But unfortunately four calves died after birth. Matlatsi carried out post-mortems on all his dead calves, which revealed 'beautiful shiny balls' in the stomach. He could not identify the balls and believed they had entered the calves through witchcraft. He scoured the kraal to try to find the *muthi* that the prophet had buried at the gate, but to no avail. He took the balls to a *sangoma* who confirmed Matlatsi's suspicions that the original *muthi* was from a 'bad prophet' who had bewitched the kraal.[31] On this occasion the presence of strange but tangible material in the innards of a dead animal pointed to witchcraft. Veterinarians with whom we discussed this case were unable to give a scientific explanation, based on this testimony. However, animals are known to get hairballs in their digestive tracts which can harm them.

Matlatsi, a Lutheran, had sought the help of Christian prophet because he was faced with a medical situation that he had not encountered before. Low fertility rates indicated the need to call someone whom he thought had protective *muthi* because something was clearly wrong if otherwise healthy cows were unable to reproduce, despite apparently copulating with bulls in the veld. Sudden myste-rious deaths in his calves later encouraged him to visit a more traditional

[31] Frederick Matlhatsi, Mabopane, 29 Oct 2009 (none of the vets we have spoken to have been able to provide an explanation for the balls).

authority for an explanation. In both cases he vested his trust in local healers, who had gained the respect of the community in fields that were not directly related to livestock health. Matlatsi did not consult veterinarians, he said, because they were too expensive. Prophets and traditional healers were part of that group of people whom he felt he could consult in times of crisis. However, his subsequent conviction that the prophet had deliberately misled him made him feel uncertain about how best to treat his animals in future.[32]

The death of newborns also featured in the testimony of Elizabeth Serema, who also felt that she had been a victim of witchcraft. Serema had been forcibly removed to what is now the impoverished settlement of Mmakau near Brits in the 1960s. Her husband had worked at a factory in nearby Rosslyn and they had invested in livestock for financial security. When her husband died in 2002 Serema was left with about 30 head of cattle, but very little knowledge of animal husbandry, as that had always been 'men's work'. Like Matlatsi she complained about stock theft, as well as a lack of grazing in this densely populated village. She was afraid to take her animals to the grazing lands because she sensed 'thugs lie in wait, planning to attack small and frail widows like me'. She therefore hired a herder whom she could ill-afford. She explained that Mmakau was a tense place to live. People had been flooding into the settlement since the 1980s and many of the newcomers were poor and uneducated, having lost their work on nearby white farms. She believed there were many people who deliberately directed their maleficent activities against animals in order to hurt their owners. When she lost her husband, she found that she could not call on her neighbours to help her with the livestock and they refused to share their knowledge about animal medicines. She regarded this as symptomatic of the lack of good relations and harmony in the village and was disturbed to find that she would have to struggle on her own.

In Serema's case, she had lost seven new-born calves over the last five years. What was odd to her was that they had all died rapidly, foaming at the mouth. The calves had not been weaned and had not been exposed to diseases in the veld or to toxic weeds. In fact, they had spent their whole lives in the kraal and, as far as she was concerned, it was impossible for them to have contracted any diseases, especially as the mothers were healthy. Serema was convinced that someone had bewitched the kraal as never before had so many animals died so quickly and mysteriously. She had never seen a cow die with spumous substances exuding from its muzzle. As a deeply religious woman, she believed that God had sent a witch to challenge her faith. She therefore decided to call on a traditional healer to protect the kraal and her cattle. She explained: 'the *ngaka* (healer) is an intermediary between myself and God. If the *ngaka* could make my herds grow, then that was God's will.' At the time of the interview in November 2009 she was hopeful that her pregnant cows would soon give birth

---

[32] Frederick Matlhatsi, Mabopane, 29 Oct 2009.

to healthy offspring.[33] Serema's notions of witchcraft reflected a mixture of ideas: beliefs in supernatural healing traditions and Christian teachings, as well as an underlying ignorance of how to treat animals, compounded by the feeling that she could not call upon her community for support and assistance. She was also very bitter about the ANC governments which she felt had done nothing for poor and disempowered people like herself. Disease and witchcraft seemed to have become synonymous with her sense of powerlessness and despair.

Finally, George Malatsi who farmed in Mabeskraal gave a very different narrative about witchcraft. He was familiar with biomedical treatments and his testimony showed how stockowners could interpret diseases as having both natural and supernatural causes. Although Malatsi dipped his animals regularly and bought injections for heartwater, he invited a traditional healer to his kraal every year to protect it from witchcraft, disease, lightning and the polluting presence of women. In the past his efforts had meant that heartwater had never been a problem. However, in 2008 nine of his cattle died of heartwater: 'When they were sick they became deranged. They ran around, hitting trees before they dropped down dead suddenly in the veld.' Malatsi was convinced that they had all died of heartwater, but he did not call on a veterinary surgeon to investigate as he was sure that he could identify the symptoms himself. Given he had taken both medical and spiritual precautions to protect his animals from diseases, he surmised that witchcraft could be the only cause. He was consequently looking for a new healer who could provide him with stronger *muthi* to ensure that this could not happen again. Malatsi was a relatively well-off farmer, owning over 70 head of cattle with the income to lease a 500-hectare farm from the government. He was concerned about jealous neighbours and his success as a farmer led him to interpret unexplained outbreaks of disease as malevolence.[34] However, his land was infested with *mohau*. The camp had formerly belonged to a white farmer who had sold up due to *gifblaar* poisoning. Malatsi had tried to clear the plant, but it kept growing back. His description of symptoms sounded like toxicoses (or rabies). He was one of two farmers we met who had leased land abandoned by whites due to *gifblaar* poisoning, raising questions about the quality of the new land that some black farmers were acquiring.[35]

In QwaQwa many stockowners said they believed in witchcraft and doctored the kraal. Johannes Motaung was also a comparatively wealthy farmer who bred animals on a commercial basis. He owned about 100 cattle, 135 goats and 30 sheep in March 2011. He kept his animals at a cattle post in the mountains, manned by a group of four herders and their dogs. Motaung had made his money as a taxi driver in Phuthaditjhaba and 'loved animals' so he wanted

[33] Elizabeth Serema, Thetele (Mmakau), 9 November 2009.
[34] George Malatsi, Mabeskraal, 5 February 2010.
[35] See also Paulus Mmotsa, Fafung, 25 January 2010.

to invest in livestock. He had the resources to buy pharmaceutical drugs and he purchased whatever the AHTs recommended. However, in 2007 he lost 22 head of cattle. They had all died with their intestines hanging outside their anuses – a scene he had never witnessed before and for which, in his opinion, there could be no other explanation apart from witchcraft. He noticed that the floor of the kraal was littered with a substance that smelt of salt and assumed the killer had sprinkled these granules to attract the cattle to the toxic material.[36] This suggested some form of poisoning similar to the cases Simon Mosonogi referred to in Mafikeng. As a relatively wealthy African farmer, Motaung felt vulnerable. Taxi driving had been fraught with rivalries over routes and customers and it was possible that he had made some enemies.

Solomon Ndlodu, the second interviewee in QwaQwa who specifically associated witchcraft with disease and death, was sure that witches made their *muthi* out of poisonous snakes. He explained how in 2002 he had sold all his goats 'because their kids were always born dead'. He could only assume that that was due to witchcraft. He thought that there was no cure for witchcraft – sorcerers by their very nature could always make more powerful *muthi* than a traditional healer. Ndlodu explained that goats were easy to harm and to kill because they were sensitive to snake bites. However, they were not as vulnerable as cows. Cows were exceptionally simple to bewitch and it was pointless investing in cattle in QwaQwa. So he had decided to replace his goats with pigs, as pigs were the most resistant to a snake's venom and it was very difficult to bewitch a porcine.[37] This was an unusual analysis as Ndlodu had created a hierarchy of susceptibility to maleficence based on species. However, pigs were rare in QwaQwa and no-one else suggested that particular animals were more vulnerable to snake bites, or nefarious *muthi*, than others. Nevertheless because of the economic and cultural value stockowners attached to cattle, some may have assumed that they had been victims of witchcraft if their cows died in suspicious circumstances.

Although most farmers in QwaQwa did not associate witchcraft directly with disease, they did see it as prevalent and as manifesting itself through cattle theft (see chapter 4). Some believed that witches, or their familiars, had *muthi* that could make themselves invisible to herders and their guard dogs.[38] Their *muthi* was said to be stronger than the medicines that *inyangas* prescribed or administered to the kraals. For that reason, some stockowners had lost all trust in traditional healers, believing their potions were useless. In this case, a sense of witchcraft as cause may be related to the difficulty of resolving the problem of theft. Farmers saw policing as ineffective and they themselves could not easily challenge those they believed to be thieves. Understanding the position through witchcraft seemed to reflect their powerlessness.

[36] Johannes Motaung, Makwane Tebang, 1 March 2011.
[37] Solomon Ndlodu, Lusaka, 3 March 2011.
[38] Joseph Mapetla, from Kestell, 15 October 2010.

To summarise, our interviews suggest that witchcraft was not often blamed for specific animal diseases, but that some stockowners, whether rich or poor, felt vulnerable and that in QwaQwa, particularly, stock theft was closely associated with individuals alleged to possess occult powers.[39] Informants claimed that witches could damage the fertility of their animals by using *muthi* to destroy their libido or prevent cows from conceiving. Occasionally, specific symptoms were associated with witchcraft. Those who invoked witchcraft generally associated it with jealousy but it could also be an explanation of their perceived powerlessness.

The ambient supernatural: Umkhondo *in Mpondoland*

The term *umkhondo* came up frequently in discussions about the supernatural and animal health in Mbotyi. While a few people described *umkhondo* as sharing certain characteristics with witchcraft, and individual malevolence, it generally included broader meanings which we will call the ambient supernatural. By ambient, we mean that the supernatural force or cause was not directly linked to intentional malevolence on the part of a witch or a familiar – nor attributed to jealousy and powerlessness. It was found in the environment or in some more non-specific agency. The word is used in Mbotyi both to describe such influences, as well as a range of symptoms that animals suffer when they are affected. *Umkhondo* is translated in the 1915 edition of Kropf's isiXhosa dictionary to mean a track, a trace, or the 'footmarks of a man or beast'.[40] It is also now used to mean the spoor of an animal, for example prints left by wild animals being chased by hunting dogs. In this sense, it includes smell, or other traces, as well as physical footprints. The word can have metaphorical associations – such as following in someone's footsteps. However, *umkhondo* is not associated in Kropf with illness in people or animals. Nor does it appear in Hunter's *Reaction to Conquest* (1936) although she discusses many words associated with supernatural causes of disease.

According to Harriet Ngubane, researching in the early 1970s in the Valley of a Thousand Hills near Durban, Zulu informants believed that people and animals left something on the ground as they crossed the earth. These tracks, for which she recorded the word *umkhondo*, could be benevolent or hostile. People and animals picked up the spoor of others as they crossed over them, or else this

[39] Kenneth Mokoena, Lusaka, 28 February 2011; Johannes Motaung, Makwane-Tebang, 1 March 2011; Seketsa Mokoena, Lusaka, 1 March 2011; Zulu Mokoena, Lusaka, 2 March 2011; April Nhlapo, Lusaka, 28 February 2011; Johannes Mofokeng, Lusaka, 3 March 2011; Moses Mahlamba, Mountain View, 2 March 2011; Abram Lebesa, Ha-Sethunya, 2 March 2011; Solomon Mokotla, Ha-Sethunya, 2 March 2011.
[40] A. Kropf, *A Kafir-English Dictionary* second edn, ed. R. Godfrey (Alice, Lovedale Press, 1915), 192.

type of metaphysical contagion could spread through the air, in the wind, or pass from one being to another through touch. Crossroads and highways were particularly dangerous places where people, without knowing it, could pick up and discard diseases.[41] Men in Mbotyi in their sixties and seventies said that they grew up (in the 1940s and 1950s) with this concept. It is possible that Hunter failed to record it, but unlikely given her thoroughness. Perhaps it is an idea that seeped in from neighbouring isiZulu-speaking communities, or was picked up by the many migrant workers from Mpondoland on the Natal sugar fields. It could thus be a new local concept, at least since the 1930s, to some degree fusing supernatural and natural or environmental ideas about the cause of illness. New associations, ideas and terms were flooding into Mpondoland during this period. For example, the 1930s marked the origins of the *indlavini* – a male organisation welding traditional elements with the experience of migrant labour to the gold mines and sugar fields.[42] Hunter recorded many new words adopted from Afrikaans and English, including popular conceptual ideas. For example, she noted in the 1930s that 'with the word isporo, (Af spoek) [Afrikaans for ghost] a new idea was introduced quite distinct from that of itongo (ancestral spirit) or isitunzela (deceased raised by sorcery) but isporo has been interpreted in terms of native culture, and has many attributes which spoek did not have in Africaans [sic]'.[43]

Umkhondo came up first in the context of discussions of transhumance when informants in Mbotyi explained why people did not like to keep livestock in the village during the wet summer months. Sidwell Caine said that the kraals got very muddy and the cattle suffer from *umkhondo*: 'the cattle get skinny and when they died, we found water in the belly and in the knee joints'.[44] Similarly, Thulebona Malindi, when we asked about *umkhondo*, responded immediately that it was caused by wet kraals: 'the manure in the kraals becomes mud and the cattle are up to their knees – then they get *umkhondo*.'[45] The term was also used to explain why some people would not allow goats – and occasionally cattle – out of the kraal before about 10 am in the morning. They gave as their reason that animals would suffer from *umkhondo*.[46] Goats could also become more susceptible to gallsickness if taken out early. People said that dew, or small spiders that are found in the wet grass, could be dangerous to livestock. Thus for some *umkhondo* had a meaning at least partly associated with

---

[41] Harriet Ngubane, *Body and Mind in Zulu Medicine: An Ethnography of Health and Disease in Nyuswa-Zulu Thought and Practice* (London, Academic Press, 1977), 24–29.

[42] William Beinart, 'The Origins of the *Indlavini*: Male Associations and Migrant Labour in the Transkei', *African Studies*, 50, 1 (1991), 103–28.

[43] University of Cape Town, Monica and Godfrey Wilson papers, BC 880, typescript entitled 'Notes on Changes in IsiXhosa resulting from contact with Europeans'.

[44] Sidwell Caine, Mbotyi, 22 March 2008.

[45] Thulebona Malindi, Mbotyi, 12 December 2011.

[46] Sonwabile Mkhanywa, Mbotyi, 8 February 2009.

natural causation of diseases, such as muddy kraals or the dangers of morning dew. It is possible that this practice minimises exposure to midges, most active at dawn and dusk, which can carry gallsickness. Simphiwe Yaphi, a young man with junior certificate education, who worked as an English-speaking guide for the hotel, and owned five head of cattle, said explicitly: 'the mud is the cause, and the cattle get ill and thin within a few weeks even if they had been fat and well.' He adamantly rejected the supernatural elements.[47] In North West Province (see chapter 3), stockowners had spoken of a worm that contaminated the grass with a toxin so it was unsafe to let the animals graze until the dew had disappeared. They associated the worm with naturally occurring poisonous veld.

Others saw *umkhondo* as physically manifest in trails or traces in the dew. If livestock crossed these trails they would fall sick. It was therefore only safe to allow them out of the kraal once the sun had burnt off the dew. Lungelwa Mhlwazi, a specialist healer, explained that the trails were left by a serpent called *ichanti*.[48] The *ichanti* serpent, which is capable of metamorphosing into other forms, is a long-established and well-known agent of illness and misfortune amongst people.[49] Hunter suggested that *ichanti* could act both as a witch's familiar and also by itself. The latter seems to be the case in explanations offered in our interviews. Mhlwazi described the *ichanti* as a distinctive serpent that resembled no snake and whose spoor looked like shards of glass or diamonds. (This description in itself is significant, because there were neither glass nor diamonds in pre-colonial Mpondoland.) The serpent deposited dangerous *muthi* in its wake, which became absorbed in the dew. If livestock crossed this trail they would fall sick. It was therefore only safe to allow the cattle and goats out of the kraal once the sun had dried the veld.[50] *Umkhondo* had yet other associations for Nongede Mkhanywa, a *sangoma*, which bridged the serpent and the potential dangers of traces left by pregnant and menstruating women:

> When a woman is pregnant, before she gives birth, if cattle walk where she has been walking, they will be sick. Here in Mpondoland, a woman who has a period will not go into a kraal. Both cause *umkhondo*. You can see the symptoms of *umkhondo*, the cattle will not have enough blood and will not be healthy in fact. You can see them getting lean, even though they are grazing well. If pregnant cattle are affected by a serpent, they will give premature birth. It also loosens the stomach [*uyahambisa*], and causes a cough [*kohlela*].[51]

---

[47] Simphiwe Yaphi, Mbotyi, 18 December 2011.

[48] Although our interpreter used the term snake, we will call this a serpent, in order to capture the metaphorical connotations.

[49] Hunter, *Reaction to Conquest*, 286.

[50] Sonwabile Mkhanywa, Mbotyi, 8 February 2009.

[51] Nongede Mkhanywa, Mbotyi, 25 February 2009; Sidwell Caine, Mbotyi, 22 March 2008; Myalezwa Matwana, Mbotyi, 7 December 2011.

Similarly Matwana directly associated *umkhondo* with menstruating women: 'if cattle cross where the woman has walked, then they can catch *umkhondo* … and they are not happy [*ukubhukaxa*].'[52]

Descriptions of the symptoms of *umkhondo* varied between informants. In addition to weak blood and inexplicable leanness, some said that animals suffering from *umkhondo* appeared feeble with weakness in their joints.[53] They lay down and found it hard to stand up as if the joints were swollen. Myalezwa Matwana said that a cow's ears drooped.[54] Animals would cough in the middle of the night, unable to sleep, or their hair would stand up at the back of their spine. Delinkosi Soyipha associated it with loose stomachs – translated as 'toilet' in Mbotyi, a word that is sometimes used in isiXhosa.[55] The dung could also have blood. One person associated it with worms, and it included symptoms that are associated with worm infection, but people did not consciously make this connection. For Mkhanywa, evidence of *umkhondo* was revealed by overall weakness, as if the snake had drunk the animals dry. Their blood was 'weak', they failed to graze properly and pregnant stock aborted. The serpent could also drink cow's milk, draining the mother so that the calves suffered too.[56] This account bore some resemblance to the *mamlambo* snake that Isak Niehaus discussed in relation to the Pedi which could enrich its owner and feed on blood in order to survive.[57]

Umkhondo also came up in accounts from other parts of the Eastern Cape, such as Masakhane. Again there were differences in understandings of the word. In that village, Mrs Nkqayi took active measures to insulate her kraal against snakes and used the term *umkhondo* to explain a type of protective *muthi*. The curse of a serpent could manifest itself in many abortions and inexplicable deaths in livestock. To safeguard her animals she cut pieces of *yakayakana,* a large indigenous climber, and laid them across the entrance to the kraal 'so each beast must pass over it and get its power'.[58] She also had a variety of plants that could dispel curses. By contrast, other informants in Masakhane and Peddie, who mentioned *umkhondo* – or responded to questions about it, associated the term with heartwater or the agapanthus plant. As we note, the word is in common use in the western parts of the Eastern Cape as the isiXhosa translation for heartwater and this is incorporated in veterinary texts and posters. Dold

---

[52] Myalezwa Matwana, Mbotyi, 7 December 2011.

[53] Lungelwa Mhlwazi, Mbotyi, 19 February 2009.

[54] Myalezwa Matwana, Mbotyi, 7 December 2011.

[55] Delinkosi Soyipha, Mbotyi, 18 December 2011.

[56] Nongede Mkhanywa, Mbotyi, 25 February 2009.

[57] Niehaus, *Witchcraft, Power and Politics,* 56–59. Hunter, *Reaction to Conquest*, 286–7 reports the *mamlambo* in Mpondoland in a way that suggests it is a new idea from outside the area, but does not associate it with this particular relationship.

[58] Mike Kenyon, 'Approaches to Livestock Health and Sickness in Masakhane' (unpublished report, July 2009); Andrew Ainslie, 'Hybrid Veterinary Knowledges and Practices among Livestock Farmers in the Peddie Coastal Area of the Eastern Cape, South Africa' (unpublished report, 2009).

and Cocks, who did earlier research in Masakhane, south of Alice, found that *Agapanthus praecox* was called *umkhondo* in their area.[59] The root of this plant was boiled for a *muthi* to cure diarrhoea and this may be a link with the meaning in Mbotyi. The absence of more supernatural elements in these localities may have resulted – as Ainslie found – from stockowners reluctance to discuss this realm. By contrast, Sesotho-speaking interviewees from QwaQwa, and some informants in Mbotyi, were happy to discuss witchcraft, pollution, supernatural ideas and the *muthi* used to protect kraals.

In Mbotyi the idea of *umkhondo* was tied in with the old patterns of transhumance, reinforcing the belief that animals should be removed from muddy kraals in the wet summer months and taken to the grazing grounds at Lubala on the Lambasi plain, or southwards to Ngquka. *Umkhondo* is generally connected with the kraal and the homestead, rather than with the pastures. However, Nonjulumbha Javu, who grazed his animals locally near the village, said that previously cattle were free from *umkhondo* at Lambasi, but 'now things are different: the cattle here are OK but those out at Lubala are being attacked by *umkhondo*'.[60] He thought that this was because of neglect, and ticks. So in this case too, the idea reinforced his decisions about grazing his cattle near the village, rather than at Lambasi. The associations with *umkhondo* are, however, much more diverse and in some respects it seems to be a general category for understanding illness in animals where there is no obvious natural cause, and where post-mortem symptoms are not evident, but malevolent witchcraft is not suspected. It may well be a relatively new term for illness and seems to operate as a multifaceted in-between or hybrid category, mixing notions about the ambient supernatural and natural influences.

Milking and the supernatural in Mbotyi

Another manifestation of the ambient supernatural emerged in interviews about milking in Mbotyi. Livestock owners said that they no longer milked their cattle. Only one owner amongst those interviewed still milked regularly and others spoke about the demise of milking as a general phenomenon.[61] Yet in earlier times, soured milk (*amasi*) was a central element in the rural diet and most families with cattle milked their cows up to recent decades. This rural community, who lived in an area with unusually rich grazing resources, was losing out on one of the key benefits of livestock ownership. In this respect, Mbotyi livestock owners differed radically in their practices from those interviewed in North West Province and QwaQwa where milking was still

[59] A. P. Dold and M. L. Cocks, 'Traditional Veterinary Medicine in the Alice District of the Eastern Cape Province, South Africa', *South African Journal of Science*, 97, 9–10 (2001), 375–9.
[60] Nonjulumbha Javu, Mbotyj, 19 December 2011.
[61] Tata (Alfred) Banjela, 20 December 2011.

common. When we asked why, a few responses suggested that there was no longer enough milk: 'if we milk for ourselves, there will not be enough for the calves.'[62] However, a more complex supernatural explanation was also generally offered. It seems to be widely shared in the village, although – as in the case of *umkhondo* and *mohato* – the precise explanation differed from person to person.

We noted above that Michael Majoang in Koppies described how witches could make cows aggressive and difficult to milk. This had become a general belief in Mbotyi although it was associated with the agency of a baboon (*imfene*) and not always with witchcraft. People said: 'baboons came from the forest to the place where calves are drinking from the teat. The baboons then also drank causing burns on the nipple. The cows found this painful and kicked the calves when they tried to suckle. They could not drink properly and the calves could die.'[63] For the same reason, people could not milk cattle either. Some people talked about baboons, but others conceived an imaginary agent: 'No-one can see this baboon from the forest, it is something different. If it is killed you will find *amasi* [soured milk], porridge [*umvubo*] and tinned fish in the stomach.'[64] In this story about the nearby settlement of Hombe, the baboon seemed to be both material and supernatural, consuming products that could only be found in the household and thus crossing boundaries into the homestead. Baboons have long played a role as key witches' familiars in the area and Hunter heard in the early 1930s that they could scratch and damage udders.[65] Udders were clearly a sensitive and vulnerable zone in Mpondo thinking. At that time, however, they did not stop milking.

Caine, a man of about 40, started keeping cattle in 1985 when he was a teenager. He was particularly skilled with horses, and worked with the Mbotyi River Lodge in organising horse-riding for guests.[66] He used biomedicines, including vaccinations against the scourge of African horsesickness, which were made available by the Lodge managers. Although he used biomedicines for some purposes, he said that even younger men like him were still using traditional medicines. Myalezwa Matwana was his key source for these, 'the most trusted man here'. Caine recalled that they milked regularly at his parents' place on the southern edge of the village when he was a boy in the 1970s. The exception was the cow that he received through a loan (*sisa*). This used to be a common method of starting a herd. Wealthier owners would loan cows to poorer families or to youths and in the longer term, they would claim back some of the increase. Owners of loaned cows, Caine said, did not like them to be milked for human consumption because they thought that the calves would lose out, and they were particularly concerned to facilitate rapid increase.

---

[62] Nonjulumbha Javu, Mbotyi, 19 December 2011.

[63] Nondege Mkhanywa, Mbotyi, 7 December 2011.

[64] Nondege Mkhanywa, Mbotyi, 7 December 2011.

[65] Hunter, *Reaction to Conquest*, 66.

[66] Caine, Mbotyi, 8 December 2011. He used, and was described by, his surname.

Caine indicated, however, that in recent years people at Mbotyi were 'scared to milk' all cattle because 'evil spirits might come at night and damage the cows'.[67] He also talked about the cows having sore teats in the morning, so that neither calves nor people could milk them. Caine reckoned that 'this is a result of treating animals with western medicines. Our parents were relying on old medicines and these prevented the evil spirits (*umoyo*) from coming. ...The gates are open for spirits to attack the livestock.'[68] He reported this story as a general explanation, and then mentioned that some people believed in this, and some not. He occasionally milked cattle, but said he found it difficult:

> If you separate the calves from the cows at night to have milk for people, then a baboon will come at night to drink. They are happy to drink at night. The baboons that milk cows at night have sharp teeth and they scratch and burn the teats. When the calves try to suckle, the mothers kick them because the teats are so painful.

Caine hovered between scepticism and belief in this explanation. Patterns of milking that aim to share the cow's milk between people and calves require such temporary separation of the calves and informants who spoke about baboons generally saw this as the moment when they would intrude.

Mayikalisa Jikijela, a substantial owner with 23 cattle and 18 goats, said that she was milking some years ago 'but the baboon was coming to milk the cows'. She never saw them herself, but thought that there were baboons nearby and if you left food out, they would come and eat. Most people stopped milking because of the baboons, she thought, and one reason they no longer kept animals in their homes, but only in the grazing lands, was because of the threat of these baboons. Simphiwe Yaphi had a particularly literal sense of the baboons, wondering whether the nearest troop he knew of, about 10 kilometres away in the forest, could actually get to the village at night. He said that he himself had seen the deep scratches on the cows' teats, as had other people. Mamcingelwa Mtwana believed: 'if they milk cows, and keep the calves separate, then the baboons come at night to milk the cows and when the calf is hungry or you touch the teat, the cow kicks them away.'[69] We asked her if everyone milked their cattle, would there be enough baboons to come and suck; she said more baboons would come. These accounts were not linked to witchcraft or the agency of any particular malevolent person, except in one conversation where a specific person was named as the agent directing the baboons.[70]

Delinkosi Soyipha, who owned a sizeable herd of 37 cattle, including 7 that he kept at home, offered a more material explanation.[71] 'There is something wrong with the cattle, they fight with you' he said. 'People don't like to milk,

[67] Caine, Mbotyi, 8 December 2011.
[68] Caine, Mbotyi, 8 December 2011; Nonjulumbha Javu, Mbotyi, 19 December 2011.
[69] Mamcingelwa Mtwana, Mbotyi, 13 December 2011.
[70] Conversation in Mbotyi, 18 December 2011.
[71] Delinkosi Soyipha, Mbotyi, 18 December 2011.

they prefer buying milk from the shop.' He recalled that his family stopped milking when he was a herd boy, which was in the early 1980s. He was emphatic that they did not stop milking because of the quality of the milk; it was 'alright' (a term often used in isiXhosa as well as English): 'they did not stop for that reason but rather they said that if they milked the cows the calves will die.' Sonwabile Mkhanywa remembered that they milked when he was a child in the 1980s but most cows gave very little, 'just a cup', so that it was hardly worth it. Although no interviewee mentioned a specific date, or event, it seems that milking stopped in the 1980s.

Nonjulumbha Javu, who owned over 40 cattle in 2009, lived in an outlying homestead, about 5 km away at Makwane, until about 1997.[72] He was milking there, and got up to 20 litres in a day, more than enough for home consumption. He then moved into the village. When he came to Mbotyi, he stopped milking even though he continued to own the same cattle and still took some of them to pastures away from the village, nearer to his old settlement. Javu said he was not told to stop, but he had seen that other people were not milking, and so he gave up; he implied a certain amount of social pressure. He did not invoke the nightly visits of baboons, and we did not ask directly about this. Rather he mentioned a range of more prosaic causes. The cattle here had less milk, he said, and 'if they continue milking then the calf will die'. The milk from some cattle was very thin, as if diluted by water, and they also stopped for those reasons. He and a couple of others mentioned *umkhondo* in connection with milking. If he continued to milk, Javu thought, the cattle would attract *umkhondo* and get thin. Masamekile Satsha said that her husband had stopped milking because 'there would be *umkhondo* from the baboon that will come and milk the cows'.[73]

Alfred Banjela, alone amongst those interviewed, milked regularly. He was the chair of the dipping committee, close to the old sub-headman and seen by some in the village as one of the key figures in the conflict over dipping. His herd of 28 was by no means the largest, but he kept them locally and kraaled them every night. At Lambasi he said the animals were dying more frequently and no-one was looking after them; sending them there was 'like throwing away cattle'.[74] He got up to 5 litres from one cow. He claimed people did not milk because they were 'lazy [*bayanqena*]'. Banjela may have been a man with some local power, but he was not a particularly wealthy accumulator, and his house was similar to many others with established homesteads. In most respects, he shared local attitudes and knowledge. For example he believed that *umkhondo* was caused by livestock going out into the dew or by a little spider (*isigcawu*), which infected animals that ate it. He used both biomedicines and *muthi* from the forests and he was adamantly opposed to selling livestock except in an emer-

---

[72]   Nonjulumbha Javu, 19 December 2011.
[73]   Masamekile Satsha, Mbotyi, 20 December 2011.
[74]   Tata (Alfred) Banjela, Mbotyi, 20 December 2011.

gency. Despite this he disbelieved the more general baboon narrative.

It is tempting to link this supernatural explanation to natural processes. There may be a number of different causes for the abandonment of milking. Although cattle numbers in the village have almost certainly not declined, and Mbotyi livestock owners generally have adequate grass, the condition of animals is not always conducive to the production of milk. Brahman type cattle have probably been the most widely introduced in recent years and these are a beef breed which then mingled in a relatively uncontrolled way with the existing mixed animals. The local herd is probably shifting towards a greater predominance of beef types and cross-breed cattle sometimes do suffer from diminished lactation. As Thulebona Malindi said: 'our cattle don't produce much milk any longer and people now keep cattle for meat and for slaughter.'[75] Alongside the ideas about baboons, the most consistent explanation for the demise in milking was to favour the calves. The two ideas could be connected. If cows were milked for human consumption, they had to be separated regularly from their calves. The supernatural narrative seemed to suggest that it was in the immediate after-math of separation, at night, that the baboons would strike. If this was in some senses a functional story, it was to instil anxiety about separating calves from the cows.

Another element in the demise of milking was probably related to the avail-ability of labour. Many, though not all cattle, were kept away from the village for long periods of the year. They used to be brought back more regularly for dipping and milking during the summer months. When they were herding, boys would drink milk out in the pastures. Sometimes they would bring back containers with milk for the homesteads. However, the gradual decline in regular dipping, which became more marked from the 1970s, and in herding, probably resulted in increasingly irregular milking.

Disease patterns may be a central explanation. Mastitis is almost certainly an important cause of sensitive and damaged udders. This contagious infection, spread by microorganisms in dung and dirty water, especially in wetter districts, is probably widespread in the coastal Transkei.[76] Tick bites around the udder also facilitate the spread of mastitis. As noted in chapter 2, tick infestation can close udders up; even when they are not sealed, the bites make teats sensitive. As Masemakile Satsha said: 'ticks can damage udders – a cow can give birth but that calf is dying as it has nothing to drink. The udder is big but it has nothing to drink.'[77] The decline of dipping from the 1970s seems to coincide with the demise of milking and it may be that the growth in tick damage and mastitis – especially when the bont tick returned – are major causes of the problems in milking. A few informants also reported small red ticks on the underside of

---

[75] Thulebona Malindi, Mbotyi, 11 December 2011.
[76] I. M. Petzer et al., 'Trends in Udder Health and Emerging Mastitogenic Pathogens in South African Dairy Herds', *Journal of the South African Veterinary Association*, 80, 1 (2009), 17–22.
[77] Masamekile Satsha, Mbotyi, 20 December 2011.

cattle in recent years. Considered in this context, the assault by metaphorical baboons was a means of explaining mastitis and tick damage by using a familiar threat to social wellbeing. Informants in Mbotyi did not seem to be aware of mastitis or to have a local name for this disease. But in a few versions, the baboon narrative also seemed to act as a form of social pressure, so that even those who could milk were discouraged because conditions made milking increasingly difficult for the majority of cattle owners. Although people did not talk about the old familiars, baboons, being sent by jealous agents in the mode of witchcraft aimed at accumulators, the story may have served in a general sense to discourage differential access to local milk.

We touched on the apparent wildness of cows left out in the pastures at Lambasi and Ngquka for long periods (chapter 4). Mamcingelwa Mtwana noted that 'old people concentrated on milking animals and caring for animals', but neglect had made it difficult to handle livestock. She associated this with a changing consumption patterns: 'the new generation don't eat the old things, they buy from shops.'[78] This was a refrain in a number of interviews. In some communities there is a reluctance to drink unpasteurised milk but we did not hear this from anyone in our interviews.

Conclusion

Dealing with this material on supernatural explanations has been difficult given that our research relies largely on interviews in a number of settlements rather than deep ethnographic immersion in a particular village. It is sometimes difficult to distinguish between discussions of ideas about witchcraft and pollution on the one hand, and actual practices on the other. There is sufficient consistency across many interviews in different places to suggest that ideas and beliefs about the supernatural remain reasonably strong, even if they are to some degree hybrid and individualised. Similarly, as we will illustrate in chapter 8, it is clear that some families do use herbal remedies and preventatives for these conditions and some, especially in North West Province and QwaQwa, try to enforce old taboos.

Older ethnographies, written between the 1930s and 1950s, suggested strongly that African societies in southern Africa did not distinguish sharply between natural and supernatural causes. We can see the continued interweaving of these elements, but our interviews indicated that there is a stronger distinction made now, and some people have rejected supernatural explanations or ideas about pollution. It is difficult to analyse systematically how these ideas have changed through time and how they interact with the more dominant views about the environmental origins of disease. Some communities and indi-

---

[78] Mamcingelwa Mtwana, Mbotyi, 13 December 2011.

viduals have a stronger sense of environmental causation, and some have incorporated elements of biomedical understanding of disease. Environmental explanations tended to be affirmed and tested by experience to a greater degree than those associated with the supernatural and pollution.

In relation to some diseases, perhaps those especially associated with sudden deaths, where informants were hard pressed to find environmental explanations, informants invoked witchcraft more frequently. Our overall impression was that supernatural explanations were fading slowly but unevenly. Witchcraft was infrequently mentioned as a specific cause of disease or death, but a wider range of ailments were attributed to a more ambient set of supernatural causes, expressed in concepts such as *umkhondo*. This included – in the minds of some informants – a strong sense of environmental causation and we see it as a relatively new concept, or at least an idea that has gradually widened in scope to incorporate natural, supernatural and even scientific elements. Supernatural ideas remain dynamic and may be taking new forms – for example in explanations about the demise of milking in Mbotyi. Compared with Hunter's analysis in the 1930s, however, we found less attribution of blame attributed to witchcraft and malevolent individuals and a clearer distinction between the natural and the supernatural. In relation to recent anthropologies of witchcraft, we suggest an overall diminution in the power of these ideas, even though elements of traditional culture are still strongly reasserted in some rural villages. Our arguments may be specific to animal, rather than human, health because this is an area where observation and experience of the natural world, as well as post-mortems, play such a large role.

8

Gender, Space & the Supernatural

Introduction

> You can't bewitch a cow and the idea that women can spoil the fertility of cattle is totally discredited now.[1] (Gideon Morule, Mafikeng)

> If black men say they allow women into the kraal they are probably lying.[2] (Majeng Motsisi, Mabeskraal)

These two quotations, both from commercial farmers from North West Province reflected contrasting viewpoints on the aetiology of diseases. Morule was president of North West Province Farmers Union – an organisation he believed could be an instrument for agricultural education and rural development. He described himself as a progressive farmer and believed that all Tswana stockowners held a modernist view of the world, in which superstitions and supernatural explanations derived from pre-colonial beliefs and traditions had disappeared. These beliefs included witchcraft and what ethnographers have termed 'ritual pollution' – the idea that people, most often women, could undermine human and animal health, and even kill, by being in an 'impure' condition. Motsisi, by contrast, was an elderly man in his 80s, who had worked for the government and accumulated savings to invest in livestock. In 2010 he had a herd of 75 cattle and rented a government farm. He used some biomedical drugs but also prepared medicinal herbs. He thought that many diseases arose from the environment, but women were responsible for a commonplace

---

[1] Gideon Morule, Mafikeng, 11 February 2010.
[2] Majeng Motsisi, Mabeskraal, 3 February 2010.

problem – the failure of animals to reproduce and their propensity to experience frequent miscarriages.

In Setswana-speaking areas the polluting potential of women was generally called *mohato* – a term that also contained broader associations of supernatural threats. It was associated most strongly with impurities emanating from reproductive fluids and from death. Menstrual blood, the lochia and contact with a corpse were seen as especially dangerous. Women, in these conditions, were blamed not so much for spreading disease but for causing miscarriages and barrenness among livestock. They could also destroy the powers of the *muthi* that many stockowners applied to the cattle kraal in order to protect the animals from misfortune and to enhance their fertility. In QwaQwa, terms varied but *sesila* was particularly associated with dirt and impurity, while *sefifi* was a form of contamination associated with death.

The idea of *mohato* overlapped with that of *umkhondo* in Mpondoland (chapter 7). As in the case of *umkhondo*, it was applied to supernatural causes of ill health, but differed from witchcraft in that blame was not usually attributed to malevolent individuals. However, *mohato* has a narrower set of meanings than *umkhondo* – in that it is specifically associated with what anthropologists have called ritual pollution. *Mohato*, which Setswana-speaking colleagues suggest is a Sesotho word, may be a new term as we cannot find it in the anthropological literature on South Africa. It seems to express a hybrid concept that welds together older ideas about impurity with notions of infection. Before discussing this specific term, it is important to contextualise this gendered interpretation of ill health and infertility in animals that has been so prevalent in African societies.

In Mabeskraal, George Matlatsi explained how in 'the recent past' women had been brought before the tribal authorities for mingling with cattle, charged with causing abortions.[3] John Modisane from the same village proclaimed that he would call the police if he found a woman in his kraal and have her arraigned, not for trespass, but on the grounds that she was threatening the health and fertility of his cattle.[4] Abram Lebesa from QwaQwa stated that he would divorce his wife if she went into the kraal whilst menstruating as her presence would destroy the fertility of his animals and the protective powers of his *muthi*. The blood of women was essentially destructive.[5]

These reactions were quite extreme but the basic ideas were echoed in many testimonies from QwaQwa and North West Province. They drew on an ensemble of ideas that associated death, pollution and the agency of women as potential vectors of disease and misfortune. Anthropological literature shows that these were deeply set in southern African societies. In fact, historically, there seems to have been variation in the intensity of these beliefs and not all communities were as rigid as some of our informants suggested. Ethnographies

---

[3]   George Malatsi, Mabeskraal, 5 February 2010.
[4]   John Modisane, Mabeskraal, 2 February 2010.

dating back to the 1930s illustrate that restrictions against women handling cattle were not universal, especially in the northern parts of the country. H. Stayt observed that some Venda women did own cattle and there were fewer gender-driven prohibitions than in other parts of South Africa.[6] The Kriges also suggested that some Lovedu families allowed women into the kraal regardless of age.[7] Further south, regulations seemed to have been stricter at that time. Monica Hunter, researching in Mpondoland in the early 1930s, adopted the term 'ritual impurity' for *umlaza* in isiXhosa. This concept applied primarily to women, although men were in a state of *umlaza* after sex. Menstruating women could cause abortions if they entered the kraal or walked through cattle herds. If they drank milk whilst menstruating, the cattle would also become weak and miscarry. Menstruating women could intensify sickness in animals and negate the medicinal effects of potions and poultices. Women were also in a state of *umlaza* following a miscarriage or the death of a husband or child. They were a danger to sheep and goats, as well as cattle, but not to fowls or pigs, presumably because these were later introductions and seen as the province of women. To keep women in order, girls learnt if they disobeyed the taboos their periods would never cease.[8] By the 1970s, however, Hammond-Tooke thought that Pedi communities – where he was then researching – were more obsessed with pollution than the Nguni communities, adjacent to Mpondoland, where he had worked previously in the 1950s.[9]

In the 1960s Hermann Monnig described beliefs about *ditšhila*, which bore parallels with *umlaza*, among Pedi communities in the former Transvaal. He interpreted the term as dirt or impurity associated with reproductive cycles, birth and death. Those who came into contact with a 'contaminated' person could contract *makgoma,* a disease that caused people to wither and die. Women could also pass on *makgoma* to cattle if they came into contact with them in a state perceived as impure. *Ditšhila* was seen as an evil force that had to be treated and thwarted by medicines. The term *ditšhila* was still in use in QwaQwa in 2011 where informants applied it to very particular forms of ritual pollution (see below).[10]

In the 1970s, Harriet Ngubane discussed such contagion, or *umnyama*, in

---

[5]   Abram Lebesa, Ha-Sethunya, 2 March 2011.

[6]   H. Stayt, *The Bavenda* (London, Oxford University Press, 1931), 38.

[7]   E. J. Krige and J. D. Krige, *The Realm of a Rain-Queen: A Study of the Pattern of Lovedu Society* (London, Oxford University Press, 1943), 43.

[8]   Monica Hunter, *Reaction to Conquest: Effects of Contact with Europeans on the Pondo of South Africa* (London, International Institute of African Languages and Cultures, 1936), 46–47.

[9]   W. D. Hammond-Tooke: *Bhaca Society: A People of the Transkeian Uplands South Africa* (Cape Town, Oxford University Press, 1962), 21–24; *Boundaries and Belief: The Structure of a Sotho World-view* (Johannesburg, Witwatersrand University Press, 1981), 124–30; *Rituals and Medicine: Indigenous Healing in South Africa* (Johannesburg, Donker, 1989), 93–9; Mary Douglas, *Purity and Danger: An Analysis of Concepts of Pollution and Taboo* (London, Routledge, 2002), 155.

[10]   H. O. Monnig, *The Pedi* (Pretoria, van Schaik, 1967), 66–67.

isiZulu-speaking rural areas. Women, unlike their counterparts in Xhosa and Sotho-Tswana societies, played a big role in burials, handling the corpses of the dead which intensified their exposure to ritual pollution. Ngubane described these ideas in relation to the dangers symbolically posed by transitional or liminal socio-spiritual zones between life, death and ancestral status. Women both created life and buried the dead; their role spanned this world and that of the spirits and ancestors, transferring illness and evil from one realm to another. Given the value of cattle in African society, women had to be contained from animals as they could become the unwitting vectors of misfortune. Pollution could be removed by seclusion and ritual cleansing with the cessation of menses. Alternatively, women could be purified after a miscarriage or period of mourning by the passage of an allotted space of time, coupled with appropriate rituals.[11]

Traces and shadows, akin to the ideas of *umkhondo*, appear in other ethnographies and were often associated with death or women's ritual impurity. The Sepedi term *seriti*, for example, related to shadows and the shades of the dead. According to Monnig, the Pedi believed that if women disobeyed the taboos regarding the handling of cattle and respect for the kraal, they destroyed the *seriti* of the livestock and brought them harm.[12] In his work with Sepedi speakers in the lowveld in the 1990s, Isak Niehaus described *seriti* as an aura, or traces that we leave behind.[13] He differentiated between a whole person – an individual in the western understanding of personhood – and a 'dividual' or unbounded body whose fluids and spirits can leave the physical body, affecting people and the environment around them. *Seriti* was most perilous when associated with death and the traces of the departed. Niehaus's informants believed that a widow was possessed by the *seriti* of her dead husband throughout the one year of mourning, which ended with a ritual purification ceremony, as well as a sacrifice of livestock to the ancestors. The *seriti* was particularly strong in widows because of past sexual contact with their husbands, representing the transfer of a spiritual aura, as well as bodily fluids, between the two. The *seriti* of dead women was weaker than that of men hence widowers were less of a danger than widows. This had implications in relation to handling livestock and provides insights as to why many informants still regarded widows to be a particular danger to cattle, while widowers were not.

Berthold Pauw provided a synopsis of ethnographic ideas about Sotho rituals surrounding widowhood and drew in a Tswana comparison from Manamakgothe. When Pauw published his article in 1990, Manamakgothe was a Kgatla-Tswana settlement and part of Bophuthatswana, located near the Pilanesberg

---

[11] Harriet Ngubane, *Body and Mind in Zulu Medicine: An Ethnography of Health and Disease in Nyuswa-Zulu Thought and Practice* (London, Academic Press, 1977), 77–99.

[12] Monnig, *The Pedi*, 173–4, 66–67.

[13] I. Niehaus, 'Bodies, Heat and Taboos: Conceptualizing Modern Personhood in the South African Lowveld', *Ethnology*, 41, 3 (2002), 189–207.

Mountains, not far from Mabeskraal, one of our sites. In his words: 'people in certain ritually dangerous states can … have a negative effect on the fertility of cattle, the effectiveness of divination sets, medicines, and activities like beer brewing and pot making'. The presence of a widow 'has the effect of *go roma* or *go tlama* (binding) cows so that they do not calve'. Rituals pertaining to widowhood, he argued, survived the arrival of Christianity. Tswana widows had to identify themselves clearly in their year of mourning by wearing black clothing and carrying particular plants that were meant to make their footsteps safe. Widows scattered bits of *mogaga* bulb at crossroads and along roadsides to protect people and animals that passed after them. The year's mourning concluded with the slaughter of a goat and ritual washing with *mogaga* bulb mixed with aloes.[14] We discovered that *mogaga* is still an important ritual plant in the Mabeskraal/Pilanesberg area.

Mohato *in contemporary cattle-owning communities*

Many of these ideas were mentioned in our interviews, especially in North West Province and QwaQwa. In certain respects it is surprising and disturbing that they have survived so strongly. As noted, some of the communities in which we interviewed were relatively recent settlements, with populations from different ethnic and social backgrounds. Garankuwa and parts of QwaQwa held hundreds of thousands of people in peri-urban conditions. Many of our inform-ants had worked away from their villages and settlements, often in formal sector employment such as mines, factories, retailers and farms or, in the case of women, as domestic servants with white families. Moreover, livestock were being managed in a new spatial context. Kraals were often located on relatively small plots of land, adjacent to brick built square houses where it was very diffi-cult to separate women from livestock. Since Colin Murray's study of Lesotho in the 1970s, analyses of African families suggest that women's roles in rural and peri-urban communities have changed significantly, with many migrating and acting temporarily or permanently as heads of households.[15] High levels of unemployment amongst men have undermined their role as providers. Since the 1990s pensions have equalised the income streams for older men and women.

Despite the depth of social change, the majority of informants spoke of taboos against women entering the cattle kraal, handling cattle or passing

[14] Berthold Pauw, 'Widows and Ritual Danger in Sotho and Tswana Communities', *African Studies*, 49, 2 (1990), 75–99.

[15] Colin Murray, *Families Divided: The Impact of Migrant Labour in Lesotho* (Cambridge, Cambridge University Press, 1981); Belinda Bozzoli, *Women of Phokeng: Consciousness, Life Strategy and Migrancy in South Africa, 1900–1983* (London, James Currey, 1991); Deborah James, *Songs of the Women Migrants: Performance and Identity in South Africa* (Edinburgh, Edinburgh University Press, 1999).

between them on the roads. In some families women could not approach small stock either. It was usually the men who decided who could go into the kraal. These powerful assertions of male prerogative were not always realised in practice. As discussed later in this chapter, the taboos, although still enduring, had broken down in some households. Yet it is important to analyse them as a lingering expression of patriarchy rather than simply of concerns about ritual pollution. To our knowledge they have not been taken up in feminist writing in South Africa, yet remain a significant discourse and to a lesser extent a practice of gendered power in the lives of millions of South Africans. Even if women can to some degree negotiate their way around such restrictions, they act as recognition of asymmetrical relationships.

However, the nature and understanding of the taboos varied from household to household. Some claimed they enforced blanket taboos against all women, regardless of age. Others only banned women of childbearing age or women from outside the family. In North West Province the reasons were invariably to prevent abortions, although some informants stated women ruined the potency of medicinal herbs.[16] A number of testimonies clearly depicted a variety of standpoints. In Eva Sesoko's household, for example, 'women from the immediate family can go into the kraal to milk cows so long as they are not pregnant or menstruating. Outsiders are not allowed because they can cast spells on our cattle.' [17] She thus suggested a link between women and witchcraft. Patrick Sebeelo allowed women of all ages, including strangers, into the kraal so long as they were not menstruating, as only menstruation caused abortions.[18] Abraham Meno and Aaron Aobeng allowed pregnant women into their kraals, but not widows or menstruating women.[19] Andrew Mabaso believed that those who had miscarried should never go into a kraal 'as the cattle will replicate the women's misfortunes'.[20] A group of nine farmers from Shakung explained that there was no evidence that women did cause abortions – but it was tradition and 'common knowledge' that they did – 'the taboos were in place just in case'.[21]

Such statements emphasised the cultural idea that the kraal was very much a male space and some Tswana and Sotho informants expressed their adherence to the taboos from that perspective. Cattle were an important source of financial and cultural wealth and the kraal was symbolic of male ownership and status. Moses Tobosi claimed that many men in Mafikeng believed that women should neither go into the kraals, nor should they take a short-cut by walking through a herd of cattle, 'even if they are carrying heavy bundles of firewood

[16] Group Interview, Shakung, 19 January 2010.
[17] Eva Sesoko, Madidi, 23 Oct 2009.
[18] Patrick Sebeelo, Lokaleng, 3 Nov 2009.
[19] Abraham Meno and Aaron Aobeng, Mareetsane, 4 Nov 2009.
[20] Andrew Mabaso, Mafikeng, 5 Nov 2009.
[21] Group Interview, Shakung, 19 January 2010.

or buckets of water'.[22] Women effectively had to stand aside for men, represented here by their most important cultural asset – cattle. Tolomane Mohale, who ran a feedlot in Phuthaditjhaba, said he enforced these taboos even though he did not have a traditional kraal. He would not allow women to handle cattle as that would be bad for business and he would lose his customers because 'women lower the dignity of the kraal itself'.[23] July Chiloane allowed his wife to go into the kraal, but no other women of any age could enter this sacred space, 'otherwise the cows will abort'. Although he appeared to be a successful commercial farmer in Garankuwa and used biomedicines, he said he would never let a female vet or Animal Health Technician (AHT) treat his cattle in the kraal. 'Even if the animal is too sick to move, or in the throes of parturition', he said, 'I would rather that animal died than have a woman in the kraal. It is better to lose one cow than the rest of the herd.' Women were a danger to his animals and a threat to his assets. Chiloane asserted that women did not try to defy the taboos: 'Women know their place – they would never dare to challenge the views of men when it comes to managing cattle'.[24]

While these patriarchal, traditionalist and discriminatory attitudes were certainly articulated by men, the term *mohato* itself may be new and its reference points suggest that this remains a dynamic area of popular conceptualisation. Nketu Nkotswe, an English-speaking school teacher in Mabeskraal explained that *mohato* meant cross-infection both between humans and between humans and animals.[25] She did not associate *mohato* directly with germs; it seemed to be a more intangible form of contagion associated with women.[26] The situations in which infertility or disease could be spread by *mohato* varied between the testimonies. Several men regarded both menstruation and widowhood as a sickness. They saw women as essentially pathogenic.[27] Women could echo such thinking, though often in a less fundamental way. Eva Sesoko explained how *mohato* caused miscarriages. In her view it was pregnant women that made cattle abort. If a pregnant woman entered a kraal or walked amongst cattle on the road then the animals would miscarry. In contrast with Andrew Mabaso's testimony above, which suggested cattle could mirror the reproductive failure of women, Sesoko's ideas pointed to the inverse – a fertile woman would encourage sterility rather than fertility amongst cattle. Sesoko was concerned because many young women did not obey the taboos, especially if their families were not cattle owners. As a consequence, she had observed an exponential increase in abortions in cattle in recent years.[28]

[22] Moses Tobosi, Mafikeng, 6 Nov 2009.
[23] Tolomane Mohale, Phuthaditjhaba, 15 October 2010.
[24] July Chiloane, Garankuwa, 21 Oct 2009.
[25] Nketu Nkotswe, Mabeskraal, 1 February 2010.
[26] Nketu Nkotswe, Mabeskraal, 1 February 2010.
[27] George Malatsi, Mabeskraal, 5 February 2010.
[28] Eva Sesoko, Madidi, 23 Oct 2009.

Menstruation and the ability to bear children were enough to make women the conveyor-belts of *mohato*. Nketu Nkotswe suggested that menstrual blood was dangerous because of correlations with the blood of animals, slaughtered for ceremonies and purification rituals. Symbolically the blood of the butchered animal linked life with death and crossing these two spheres could be perilous, bringing disease and misfortune.[29] Hendrik Metswamere, a farmer and traditional healer from Mantsa, believed that it was not menstruation per se that was the issue, but the fact that all women, whether of childbearing age or post-menopausal, carried the spirits of the dead and the unborn. Women's 'footprints', a term used by Metswamere, bore the shades of the 'unliving' through the deaths of their husbands, children and even the babies that had never been conceived.[30] This posed a threat to society and had to be controlled by strict taboos. Mantsa was one of the most conservative communities we visited with homesteads more scattered than in peri-urban areas or betterment villages. The eight men interviewed there all declared that nobody allowed any women into the kraal or near their cattle, regardless of their age.

For many farmers in North West Province, widows were more dangerous than pregnant or menstruating women. A former official in the education department from Mafikeng explained that widowhood was both a state of sickness and healing. As the husband's body rotted in the ground, the blood of the widow altered too and this could be harmful to people and to animals. Widows were dangerous until they had undergone the purification ritual at the end of the mourning period and sacrificed a cow or goat to unite the husband's spirit with those of the ancestors. An unpurified widow who came into contact with animals could transfer the most virulent forms of *mohato* and sickness; death and infertility would follow. Hence he claimed that even educated families continued these traditions to protect their own health and that of their livestock.[31] Michael Mlangeni, from Mabopane near Pretoria, explained that widows were dangerous as they carried their husbands' spirit which could be intent on evil, seeking to bewitch the kraals of others by ensuring that the livestock could not reproduce. Widows were the instruments by which the dead could wreak revenge.[32] Similar ideas came through in interviews across the Setswana-speaking settlements that we visited.

Ideas about pollution and cattle disease in QwaQwa

Many informants in QwaQwa also depicted women as pathogenic. They placed particular emphasis on the dangers women posited to kraal *muthi* and, for that

[29] Nketu Nkotswe, Mabeskraal, 1 February 2010.
[30] Hendrik Metswamere, Mantsa, 12 February 2010.
[31] Anon, Mafikeng, 9 February 2010.
[32] Michael Mlangeni, Mabopane, 24 October 2009.

reason, believed it was important to ban all women from the kraal regardless of age. In this respect the practice of exclusion seemed to be more extreme than we encountered in Setswana-speaking areas or Mbotyi where owners allowed post-menopausal women into the kraal. Because stock theft was such a big issue in QwaQwa and many farmers used *muthi* in the hope of preventing it, there was a strong link between banning women and the security of the kraal. Daniel Khonkhe, for example, allowed his wife into the kraal but no other women. He thought the idea that women could cause abortions in cattle, or vice versa, was total nonsense. However, he believed that 'women from outside the household who enter the kraal may possess special *muthi* and bewitch my animals so they leave my kraal and follow her home'.[33] Women may not be violent stock thieves like some men, but in this conceptualisation, they could steal animals by more mysterious and sinister means. Other stockowners permitted their wives and daughters into the kraal because they thought that they could not harm the protective *muthi*, but affirmed that women from outside the family could destroy its potency.[34] The strength of kraal *muthi* was a significant issue for some inform-ants because they mentioned that thieves with more powerful medicines could become invisible. Their evidence for this was that their kraals had been treated, and that their dogs did not bark.

In QwaQwa in 2011 we encountered a number of specific words that people used to describe various types of pollution. These were more diverse than recorded either in the Tswana or in the Sotho ethnographies. Traditional healers articulated a more detailed and authoritative understanding as to why many Sotho people followed certain rituals. Rosalina Johanne, a healer, gave a partic-ularly complex analysis of 'dirt' which she associated with both menstruation and death. She used the word *ditšhila* (dirt) to describe women who were menstruating and could confer a range of misfortunes, including miscarriages and abortions in livestock:

> As it is not Sotho culture for a woman to announce her menstrual state, there are taboos against women entering the kraal until they have passed the menopause. I have no direct experience or evidence that menstruation causes abortions, but my ancestors have told me that is the case. My ancestors talk to me in my dreams. They tell me that this happened in the past and so there are taboos to protect the commu-nity and the traditions of the ancestors. The kraal is a sacred place .... It is the wealth of the household. For this reason, women should never enter kraals, nor should they pass amongst animals on the street.

Johanne also discussed *sesila* – another word used for widows in mourning.[35] The mourning period could last one month to one year depending on the

---

[33] Daniel Khonkhe, Ha-Sethunya, 13 October 2010.
[34] David Mphuthi, Phuthaditjhaba, 14 October 2010; Joseph Mapetla, Phuthaditjhaba, 15 October 2010; Ntseou Livestock Company, Phuthaditjhaba, 11 October 2010.
[35] Rosalina Johanne, Lusaka, 1 March 2011.

household. At the end of that time a cow was slaughtered and the widow was washed in a mixture of the cow's stomach contents, aloes and water. This ritual cleansing linked the living with the dead through the animal's sacrifice, while the smearing of chyme on the body represented a form of ancestral anoint- ment. In Mbotyi fresh stomach contents were smeared onto the body of a young woman who was being initiated as a *sangoma* during a ceremony that also involved the sacrifice of a cow.[36] *Sesila* itself was a danger to both livestock and humans. An unpurified widow who had sex with a man could transmit a disease called *mashwa*. So too could a woman who had had an abortion and not been purified. According to our translator Gavin Mohlakoana, many local people associated *mashwa* with HIV/AIDs, interpreting this relatively new disease according to their own understandings of dirt and attributing to women the responsibility for sexually transmitted diseases.

Another term that seemed to have a similar meaning to *sesila* was *sefifi*, asso- ciated with evil and bad luck that could exude from the mourning clothes. According to Abram Lebesa, women had to mourn their dead husbands for a year in his family and during that time they were in a state of *sefifi*: 'If a widow sees cows on the roadway she has to shout that she is *sefifi* so that the herders can shoo the animals out of the way.' Widowers could also be *sefifi*, but they were unlikely to cause disasters. The mourning period for men was short and widowers usually 'avoid the kraal not because they can harm the animals, but out of respect for the ancestors'. Widows and menstruating women could also ruin (*hosenya*) the protective *muthi* he buried in the kraal to ward off thieves and witches. Menstrual blood 'brings *seqetho* (bad luck) which causes abortions and infertility in cattle'. For this reason Lebesa never allowed any women to approach his kraal. In the hierarchy of perilous states though, '*sefifi* is the most dangerous of all to livestock, especially when it comes with widows. Women with *sefifi* can also transmit AIDS to men.'[37] April Nhlapo claimed that 'some women deliberately go into the kraal with the direct intent of causing abortions and ruining the cattle of their neighbours'. Men and women could be witches though men were more likely to be thieves; but women could destroy the wealth of a family by entering a kraal without the male owner's permission.[38]

Hammond-Tooke discussed the idea of *senyama* as a synonym to *sefifi* in rela- tion to Pedi culture.[39] In QwaQwa Johannes Motaung differentiated between the two terms. *Sefifi*, he said, referred to widows, *senyama* to menstruation. *Sefifi* was more dangerous as widows could cause all the cows in a kraal to abort, whereas a menstruating woman was only likely to induce a miscarriage in a single animal. He maintained that the ancestors enforced these beliefs because they wished to ensure the preservation of Sotho culture, and thus their own pre-

[36] *Sangoma*'s initiation/coming out after training, Mbotyi, 1 March 2009.
[37] Abram Lebesa, Ha-Sethunya, 2 March 2011.
[38] April Nhlapo, Lusaka, 28 February 2011.
[39] Hammond-Tooke, *Boundaries and Belief*, 124–30.

eminence within the Sotho mindset. It was important to respect the ancestors as your fortunes depended on them. [40] Maria and Anna Moneng associated *sefifi* with darkness rather than bad luck or evil spirits. The black mourning clothes covered not only the widow's body but also the shades of evil. A woman found on the streets in such garb after dark was likely to be beaten and could be killed to destroy the evil within. [41]

Johannes Mofokeng used the term *hlola* for *sefifi*. As in the case of the specific words, he suggested that there was some fluidity in the practices associated with mourning and cleansing. The black clothes themselves indicated a historical shift in ritual practices. Mofokeng explained how the period of mourning had declined for practical reasons – women had to go back to work. For this reason mourning lasted only a few weeks in many families, rather than months or a year. [42] What was most important was that widows performed the correct purification ritual at the end of the mourning period. Whatever fears individuals had about the wrath of the ancestors, Mofokeng's testimony revealed that people had adapted the beliefs and rituals to fit in with modern life. However, no-one in QwaQwa admitted to abandoning these traditions altogether.

Practices and challenges

In QwaQwa, as in North West Province, these beliefs were not supported with observations that women could bring misfortune. If asked about the evidence for women's responsibility for disease, informants responded with phrases like: this is our custom, or our tradition. In this respect attribution of animal disease to impurity differed from causation linked to the environment. It is striking that many informants were prepared to talk openly about this set of beliefs around pollution. However, it is possible that they did not wish to divulge their full opinions and offered generalised statements about custom because they were reluctant to question social norms.

Nevertheless, most informants seemed to express their views with conviction, and there were some who attributed particular events to the polluting influence of women. Michael Mlangeni associated miscarriages in his goats in 2007 to the fact that his pregnant niece had walked into the goat kraal. For him that was sufficient evidence that the taboos were necessary. [43] In Mareetsane there was general agreement amongst a group of five farmers that the reason why four cows had aborted recently was because a widow had been seen walking around the settlement, intermingling with the cattle. [44] Simon

[40] Johannes Motaung, Makwane-Tebang, 1 March 2011.
[41] Maria and Anna Moneng, Thebang, 3 March 2011.
[42] Johannes Mofokeng, Lusaka, 3 March 2011.
[43] Michael Mlangeni, Mabopane, 24 Oct 2009.
[44] Abraham Meno and Aaron Aobeng, Mareetsane, 4 Nov 2009.

Mathibedi from Mmakau stated that any news that a cow had aborted was attributed to a woman passing through the kraal or through a herd of cattle. No other causal evidence was needed.[45] Stockowners with these kinds of views did not regard infertility as a possible result of inherent physical problems, malnutrition or disease; infecundity could only be a consequence of the deliberate or inadvertent actions of women.

At a group meeting in Moiletswane, by contrast, there was a hint of a debate. Stephens Ndala stated that three of his cattle had miscarried lately and he was convinced that it was because a woman had passed through his herd. Piet Matjila was dubious and his scepticism was born out of the fact that his parents kept their cattle at a camp some distance from the village, where there were no women. Yet their cattle kept aborting for reasons that he could not explain. When questioned about the existence of diseases like brucellosis (contagious abortion), Matjila and his neighbours voiced their ignorance. They gave no indication that reproductive problems could be attributed to infections.[46]

This lack of knowledge about reproductive disorders probably helps to explain why these taboos are so strong in some communities. Although some farmers from throughout our field sites linked malnutrition or drought with miscarriages and difficult births, because the cows were too weak to deliver, there was little understanding that diseases such as brucellosis could also play a part in disrupting fertility. According to Johan Naude, the state veterinarian for QwaQwa, there were a number of reproductive problems in that area. With respect to brucellosis, the bacteria thrived in the placenta and uterine fluids which could contaminate the veld and infect grazing livestock. Although there was a vaccine, few used it. It was also a very difficult disease to explain to people because an infected animal might only abort once and then become pregnant afterwards. The problem was that the pregnant heifer remained an invisible carrier and could spread the disease during parturition.[47]

Naude also mentioned vibrosis and trichomoniasis, forms of venereal disease that affected bulls, which could be life-long carriers. Given the bulls and cows roamed together in the veld there was plenty of opportunity for cross-infection. These diseases made the cows sterile. Again there were vaccines, but even so the one for trichomoniasis was not always reliable.[48] None of the Sesotho-speaking informants mentioned buying any vaccines for reproductive disorders and these scientific names for diseases were unknown to them. Because farmers could not directly link the cause of abortions and infertility with the veld or visible environmental conditions, they were less likely to abandon explanations based on supernatural interpretations. Ideas surrounding reproductive failure remained

[45] Simon Mathibedi, Switch (Mmakau), 9 Nov 2009.
[46] Group Interview, Moiletswane Community Centre, 21 January 2010.
[47] Johan Naude, Bethlehem, 4 March 2011.
[48] Johan Naude, Bethlehem, 4 March 2011.

rooted in the way people imagined that communities, both living and dead, influenced their lives and their fortunes.

At times there was a sense of menace in the way livestock owners purportedly enforced the taboos against women. The onus lay with women to adhere to the restrictions. Stoki Taedi explained how traditional healers worked with stockowners to ensure compliance. Healers enjoyed status and respect in the community and, to some extent, they were feared. In her village of Mareetsane, the *sangoma*s banned women from the kraal, stating that menstruating women and unpurified widows caused abortions. Only post-menopausal women could enter the kraal at all. To enact these prohibitions, the healers buried 'stuff' (*muthi*) at the entrance of the kraals so if a menstruating woman ever entered the enclosure, she would never stop bleeding. As noted this penalty was mentioned in Hunter's discussion based on fieldwork in 1931–32 and perhaps had regional purchase across a long period.[49] Taedi affirmed that a woman would have to admit her transgression and seek help from the traditional healer who would know that she had broken the taboos. This threat, according to Taedi, was enough to keep women out of the kraals. Whatever doubts individuals might entertain about the likelihood of perpetual menstruation, they allegedly obeyed these orders, just in case they were true.[50] Even if individual women were not necessarily convinced about their responsibility, Taedi thought that the social stigma of being associated with wrong-doing and having to visit the *sangoma* for redemption was persuasive.

Although many women seemed to have internalised the taboos, some contested the restrictions by developing their own interpretations of pollution or by defying their fathers and husbands. The healer Rosalina Johanne, who felt it was important to maintain Sotho culture, had found a way of accommodating the traditions which she felt should be upheld. To stop abortions (and also in response to the reality of the lack of grazing lands) she kept one cow. Consequently, 'I have no kraal to enter and I can keep my cow tied up outside at night. I can milk my cow in the yard, menstruating or not'. In her analysis, it was the kraal rather than the animal that was sacred. Because she only had one animal she never had to worry about walking amongst her cattle so she could not bring about abortions in her own herd. Her cow was very healthy and reproduced annually. She sold her calves to prevent accumulating a small herd which she felt she might endanger by walking amongst them. They were a useful source of income.[51] In this case the supernatural justified a logical action, which was in tune with her financial needs and environmental constraints. But more generally we did not find that livestock owners invoked supernatural explanations for selling animals.

Maria and her daughter Anna Moneng, also from QwaQwa, had some

[49] Hunter, *Reaction to Conquest*, 46–47.
[50] Stoki Taedi, Mareetsane, 4 November 2009.
[51] Rosalina Johanne, Lusaka, 1 March 2011.

striking alternative views about pollution and the fertility of their animals. They lived without men, but with Anna's young children and their seven cows, five goats and three pigs. Maria went into the kraal as she was elderly, but Anna avoided it whilst menstruating as that could bring about abortions. At other times she entered the kraal and took the animals to the grazing lands. She also collected dung from the kraal and milked the cows in the enclosure. If she could not, she would ask one of her sons to perform these tasks. The Monengs believed that widows could make animals sterile but they did not cause abortions. They also held that men were equally dangerous after the death of their wives – an idea that did not come up in any other testimony. The Monengs did not think the length of mourning should be longer for women. In their view, regardless of gender, mourning should last two months for a child, six months for those in work, and 12 months for the unemployed. They had therefore adapted the mourning period to suit the practicalities of working life, at least to some degree. Workers did not have to remain in seclusion for six months as they had in the past. But they had to wear mourning clothes at work and stay apart from others, returning home before dark for their own safety. In addition, the Monengs banned adult men from the kraal. They believed that men could cause infertility in livestock if they had had sex in the last seven days. Since, as they put it, men could never be trusted to be honest about such matters, it was best to ban them altogether. They used the term *tsheloa* to describe the pollution sexually active men carried. The only male who could go into the kraal was an elderly man named Buti Malefane. Malefane was an experienced farmer who knew something about medicines, could diagnose diseases and tackle complicated births. They called on him if they had any problems with their animals because he could be trusted. It was unclear whether they trusted him because of his reputation as a healer or because they believed he was no longer sexually active.

No male informants referred to any taboos against men entering the kraal. However, Anna Moneng explained that when men found that their cows had not conceived, they would stop visiting their kraals if they had recently had sex. Men would never admit to this, according to Anna Moneng, 'because they think they are so important', but she felt that they did obey the laws of the ancestors.[52] The Monengs' interpretation of ritual impurity clearly had feminist elements and suggested that men might also entertain insecurities about the effects of their own presence on animals at certain times, especially if their livestock were in poor health or not reproducing. It is difficult to judge their testimony, which was not corroborated by other informants, but revealed how they modified and reinvented ideas about impurity. In their case they were explaining and rationalising their practice, where two women reared animals on their own in a changing working environment. They had found alternative arguments to justify their actions and explain infertility in livestock.

[52] Maria and Anna Moneng, Thebang, 3 March 2011.

Practicalities had an important bearing on the actual enactment of pollution taboos. In a minority of families women had long been allowed into the kraal because farmers needed their labour and in these cases they contested the idea that women undermined the fertility of livestock.[53] The veterinarian Johan Naude said that none of his female AHTs had experienced any problems entering a kraal to treat sick animals. He thought that these pollution taboos were 'talk rather than action' – an effort to verbally reinforce cultural traditions rather than non-negotiable practices.[54] Customs could be breached if female AHTs were likely to save animals. The analogy may be with respected women herbalists. Many of the healers interviewed in QwaQwa were women and male farmers said that they had no problem consulting a female *inyanga* if they wanted special *muthi* to treat diseases or doctor the kraal. The important thing was that she could be trusted to defend them from witchcraft and adversity. Women therefore presented different identities, and men also cast them with various personas according to particular situations or the needs of the moment. Women might cause abortions but they could also protect and heal.

In Mbotyi we also heard of diverse practices and it is clear that taboos did not operate uniformly. In the view of a few older men, *umkhondo* could, as mentioned in chapter 7, include the dangerous effects of menstruating women. Yet most informants did not explain it in this way. Women also disregarded such gendered strictures. Joselina Mtunyelwa, about 70 years old, recalled that she had herded cattle when a young girl because there were no boys in her homestead and she carried on when she married and her husband was away: she said there were no customs that stopped her from looking after cattle.[55] She accepted, however, that most women did not take these responsibilities.

Mamdenyeshwa Ndube, owner of a sizeable herd of 30, similarly herded livestock when she was young: 'if there was no boy at the homestead then the girls looked after the cattle.'[56] She went into the kraal, even when her husband was alive, as 'he liked girlfriends – he was away with his girlfriends'. In earlier times, she recalled, people said that women could not enter a kraal but 'since the kids have been schooling, I've been obliged to go in and to supervise the herding of cattle. People don't worry about it any longer: women just pass close to the kraal if the path is close – they walk on any path close to the kraal.' Our research assistant suggested that the ideas about pollution were still well known, and to the extent that they were followed in some homesteads, helped to create a certain collective identity in the village; a social ideal of kinds where people conceived themselves to be in tune with the ancestors.[57]

---

[53] Lott Motuang, Bollantlokwe, 10 November 2009; Tollo Josiah Mahlangu, Mabopane, 29 October 2009.
[54] Johan Naude, Bethlehem, 4 March 2011.
[55] Joselina Mtunyelwa, Mbotyi, 10 December 2011.
[56] Mamdenyeshwa Ndube, Mbotyi, 10 December 2011.
[57] Sonwabile Mkhanywa, Mbotyi, 10 December 2011.

Despite the testimonies from Mantsa, where all women were allegedly barred from the kraal, many Setswana speakers said they permitted post-menopausal women to go into the kraals, as they had become 'like men'.[58] Ethnographies recorded that older women lose certain of their gendered identities. In QwaQwa too, some men let elderly women in the kraal as the 'evil' lay in menstruation and not in the women themselves.[59] While the majority of live-stock owners in Mbotyi, especially those were larger herds, were men, a few elderly women interviewed became owners of and took control of livestock on the death of their husbands. Younger women could also purchase cattle. Spatial taboos did not seem to be a major issue for them.[60] Brucellosis and other repro-ductive diseases did not seem to be such a problem in Mbotyi which might help to explain why ideas about pollution were apparently not at the forefront of people's explanations for abortion. Stockowners did, however, have problems with dystocia, possibly because Brahman bulls had been brought in and the small, hybrid cows could not so easily cope with large foetuses.

Practices have clearly changed over a long period of time in some commu-nities but informants in North West Province and QwaQwa pointed to the 1980s and 1990s as a phase when respect for elders and gendered prohibitions declined. This coincided with widespread youth mobilisation and political protest in the rural areas.[61] According to Clansina Matlhatsi from Mabopane, women had started to go into the kraals since the 1980s; a change that she regarded as a positive step forward because many women were widows and could not afford herders.[62] Men also brought up this issue of widowhood. Josiah Letseka commented that women never used to go into the kraal, but now many did, and it was necessary that they learnt about cattle and animal health, in case they were widowed.[63] Zachariah Maubane, who taught in Mafikeng, but farmed near Warmbaths, attributed these changes to the fact that fewer men wanted to be herders, hence there was a labour shortage that effectively impelled widows and other women to take care of animals.[64]

A number of informants pointed to the 1980s as a time of dramatic trans-formations in the conceptualisation of farming and reproductive issues. Some men apparently feared the growing assertiveness of women, born out of the political struggle against apartheid in the 1980s and cemented by the general

[58] Don Modiane, Kraaipan, 11 November 2009.
[59] Chadiwick Mbongo, Lejwaneng, 12 October 2010.
[60] Joselina Mtunyelwa, Mbotyi, 10 December 2011; Mamdenyeshwa Ndube, Mbotyi, 10 December 2011; Thulebona Malindi, Mbotyi, 11 December 2011.
[61] Peter Delius, *A Lion Amongst the Cattle: Reconstruction and Resistance in the Northern Transvaal* (Oxford, James Currey, 1996); Ineke van Kessell, *Beyond our Wildest Dreams: The United Democratic Front and the Transformation of South Africa* (Charlottesville, University of Virginia Press, 2000).
[62] Clansina Matlhatsi, Mabopane, 29 October 2009.
[63] Josiah Letseka, Thethele (Mmakau), 13 November 2009.
[64] Zachariah Maubane, Mafikeng, 2 November 2009.

Photo 8.1 Woman with her cattle in Mabeskraal, NWP (KB, February 2010)

acquisition of the right to vote in 1994. According to Hendrik Metwamere in the conservative village of Mantsa, more of his cattle had aborted since 1994 despite the fact that he was a healer and knew how to use *muthi* to improve fertility. He attributed this to his impression that women had become far more politically and socially demanding and they had developed a greater sense of freedom: 'the government is ours; we do what we please.' Some women were starting to ignore the taboos and were now interacting with livestock in a new way.[65] Similarly in Fafung, Paulus Mmotsa complained, 'more and more of my cows are losing their calves. Last year I lost five calves due to miscarriages. Cows also have real problems giving birth. The calves get stuck and they die.' He attributed abortions and dystocia neither to disease nor malnutrition but to the fact that 'women are secretly going to the cattle post to sleep with the herders'. He made the explicit assumption that such things had never gone on before and, like Matlhatsi, he associated change with a decline in respect for the elders and the ancestors.[66] Both these narratives suggested that the end of apartheid psychologically created new opportunities for women.

For an earlier period, Hammond-Tooke associated some of the taboos with the possible incapacity of men to enforce the gendered patterns of authority that they aspired to wield.[67] There may be similar forces at work here with the menacing attribution of supernatural causation emphasising how impotent some men felt about the potential 'power' women could exert over their cattle,

[65] Hendrik Metswamere, Mantsa, 12 February 2010.
[66] Paulus Mmotsa, Fafung, 25 January 2010.
[67] Hammond-Tooke: *Boundaries and Belief*, 124–30; *Rituals and Medicine,* 93–99.

reinforced by the new social mobility available to women. This was perhaps a metaphor for deeper insecurities, surrounding men's inability to manage the homestead and their female relatives. In Mbotyi this was manifest in the diffi-culty that men had in controlling women's labour for agriculture (chapter 4). William Komane saw generational strife as disruptive in Mabeskraal. He told us that, since the 1980s, 'abortions, diseases and stock theft have all increased because the law has banned parents and elders from beating their children. Now there is no discipline and respect. Youngsters no longer respect their elders and the customs of their ancestors. They are abandoning their culture.' He thought this had repercussions beyond generating familial and inter-generational tensions – it created disharmony in society. Disharmony offended the living and the departed, exposing the community to sickness, *mohato,* crime and witchcraft.[68]

In QwaQwa three male members of the Ntseou Livestock Company lamented the fact that things were changing. They regarded the 1990s as a pivotal period in QwaQwa history in which women began to assert their rights. Once women had political rights they wanted entry to the kraal and they pushed for it. They claimed that Sotho women began to demand access to the cows so that they could milk them to feed their children. They wanted to save money for education rather than spending cash on unnecessary herders.[69] In fact, there had been a shortage of male labour due to migration to the cities for decades, but in their narratives the momentous political changes of the 1980s and 1990s were invoked to pinpoint social change. It is likely, as we noted above, that women were defying the taboos at much earlier stages, generating scepti-cism about the dangers of menstruation. Once some women began to mind and milk cattle, others demanded the same concessions.[70]

Maria Khoboko, a middle-aged woman from QwaQwa for instance, explained that she had never been allowed into the kraal in her youth and she believed then that menstruation did cause animals to abort or to become prey to thieves. However, by the 1990s, she had decided that this view was 'rubbish'. 'There is no evidence that women bring bad luck to animals. I persuaded my husband to let me into the kraal and I now help him with his sheep business.' Together they founded a wool cooperative with other local farmers in the village of Lejwaneng to try to improve the market for their crop.[71] Some men even embraced the changes – Daniel Khonkhe said that in his village of Ha-Sethunya there were a growing number of women looking after cattle, many of whom were divorced or widowed. In his view they have proved to be far

[68] William Komane, Mabeskraal, 4 February 2010.
[69] Ntseou Livestock Company, Phuthaditjhaba, 11 October 2010.
[70] Ntseou Livestock Company, Phuthaditjhaba, 11 October 2010; Ephraim Mafokeng, Phutha-ditjhaba, 15 October 2010; Thomas Mahlaba, Tseki, 15 October 2010; Jack Vezi, University of Free State, Phuthaditjhaba, 13 October 2010; Joseph Mapetla, Phuthaditjhaba, 15 October 2010.
[71] Maria Khoboko, Lejwaneng, 12 October 2010.

better at looking after animals than men. They are more attentive and solicitous to health needs. The problem with men, he stated, is that 'they are idle and drunk and will not do anything to help themselves'.[72] Although the majority of men interviewed held by the old taboos, at least verbally, political and social changes were undermining the idea that the kraal was exclusively a male domain. However, the withering away of customs and the ideas that under-pinned them raised new questions. If women did not bring about infertility in cattle – what did?

Despite these changes in some families, the above testimonies about *mohato* and related concepts have a strong and dynamic role and may be associated with reassertions of traditionalism in some areas. Respondents from the Ntseou Livestock Company accepted that women wanted their rights, but felt that this had been detrimental to the fate of livestock and the self-esteem of men. Predictably, they asserted that more cattle had aborted since the 1990s and stock theft had increased. They attributed a rise in larceny to the breakdown in *muthi* because women were working with animals, rather than to poverty, higher levels of crime or the failure of the police to maintain law and order.[73] Women's behaviour was a threat to the stability of society based on traditions in which everyone knew their role. Now there were many men who had no work, Khonkhe thought, because they had lost self-discipline and could no longer monopolise the management and care of animals. As a result they sensed a loss of identity and aimlessness in a changing world.[74]

Muthi *to protect the kraal and the cattle*

Throughout our interview sites a number of stockowners explained that they used *muthi* to protect the kraal and the cattle. Some invited traditional healers to their homesteads, while others used their own *muthi*, revealed to them in dreams or handed down through the generations. To varying degrees, individuals were protecting against witchcraft or *umkhondo*, counteracting the foot-steps of women, or attempting to enhance the fertility of their animals. The explanations for doctoring kraals could also reflect specific local problems. In QwaQwa, informants primarily aimed to prevent stock theft with protective *muthi*. In general we learnt about plants and remedies that were in common use but seldom about the more detailed mixtures used for difficult problems. Respondents from QwaQwa and North West Province were reluctant to talk about specific medical materials. As we noted in chapter 5, local knowledge is not freely dispersed and is sometimes guarded. By no means all people know the formula for the protective *muthi*. Another reason was the fear that if

---

[72] Daniel Khonkhe, Ha-Sethunya, 13 October 2010.
[73] Ntseou Livestock Company, Phuthaditjhaba, 11 October 2010.

someone knew a stockowner's recipe, they could make stronger *muthi* for malevolent purposes.

Writing about Mpondoland in the 1930s, Monica Hunter provided some details of the types of medicines people used to protect cattle and the kraal. Stockowners wanted to safeguard their cattle from a range of natural and supernatural phenomena deemed to be dangerous, such as lightening and attacks by witches' familiars like the *impendulu*. To defend an *umzi* (homestead) against sorcery and witchcraft, herbalists (*ixwhele*) used *inthelezi* medicines which they made into an infusion to sprinkle on the earth, or else they burnt the medicine in the huts and kraals to repel evil spirits. *Inthelezi* can refer to a number of different medicinal herbs. They also rubbed fat from pigs or sheep onto wooden pegs and inserted them into the ground between the huts. They believed that the treatment caught the *izilwana* (familiars), so they were unable to enter the kraal and they also protected the homestead against lightening. The Mpondo were also concerned about *umlaza* (pollution), and stockowners dosed their cattle with *intolwane* (*Elephantorrhiza elephantina*) in case they crossed the spoor of women. Some livestock owners burnt their own medicine in the kraal 'to make well the blood of the cattle' and to fortify them against disease and sorcery. A few incinerated tortoises in the kraal as they believed this enhanced the fertility of the herd.[75]

Hunter's account resonated with some of the practices we encountered in Mbotyi. Stockowners took a number of measures to treat or prevent *umkhondo*. They administered a mixture of bark and herbs called *umkhumiso* that differed from the plants used for diseases seen to have natural causes.[76] Some people burnt *umkhumiso* in the kraal whilst the animals were inside it, as a form of medical fumigation that protected the cattle and dispersed potential dangers. Lungelwa Mhlwazi dosed animals suffering from *umkhondo* with an infusion of powdered *intolwane* (*Elephantorrhiza elephantina*) or *imfingwana* (*Strangeria eriopus*).[77] This cycad (also called *mfingwane* or *imfingo*) grows in the forests and forest margins along the whole of the east coast of South Africa; its root and stem is widely used medicinally and the berries are poisonous. The *muthi*, she said, caused the fluid to drain from the joints and the sick animals recovered within a week. To keep the serpent away from the kraal, she sprinkled *imfingwana* on the ground to purify it.[78]

---

[74] Ntseou Livestock Company, Phuthaditjhaba, 11 October 2010; Daniel Khonkhe, Ha-Sethunya, 13 October 2010.

[75] Hunter, *Reaction to Conquest*, 65–66, 296–8.

[76] Sonwabile Mkhanywa, Mbotyi, 8 February 2009.

[77] Lungelwa Mhlwazi, Mbotyi, 19 February 2009; Jocelyn J. Fearon, 'Population Assessments of Priority Plant Species used by Local Communities in and Around Three Wild Coast Reserves, Eastern Cape, South Africa', unpublished MSc, Rhodes University (2010); www.plantzafrica.com/plantqrs/stangereriop.htm (accessed 30 May 2013).

[78] Lungelwa Mhlwazi, Mbotyi, 19 February 2009.

Myalezwa Matwana used a mixture of *imfingwana, amagamayo* and *isibande sehlati* (a species of long grass), burnt together in the kraal.[79] They were placed on top of hot embers and the cattle came together to imbibe the smoke. Another type of thatch grass *isiqungu* could be chewed to help protect medicines affected by women during their periods, and men when they had intercourse.[80] Delinkosi Soyipha mixed pig dung and crayfish eggs with water and gave it to the cows to drink.[81] These testimonies differed from those we heard in North West Province, because they suggested that it was sometimes possible to cure as well as prevent *umkhondo* – a disease or condition ascribed to supernatural influences. None of the Setswana-speaking respondents, who felt they had been victims of similar forces, had been able to treat their indisposed and dying cattle. *Umkhondo* seemed to be a general post hoc explanation of unidentified animal disease and frailty. This might help to explain its curability. *Umkhondo* affected cattle, sheep and goats, but not pigs. As in Ndlodu's testimony from QwaQwa it seems that porcines are less susceptible to supernatural forces than other livestock species, and pig fat is a component of some types of protective *muthi* (chapter 7).[82]

Protection against *umkhondo* ran parallel to and sometimes replicated those for the kraal more generally. In the latter case it was the space as a whole that was being safeguarded rather than a disease or symptom that was being cured. In Mbotyi, some stockowners still inserted pegs smeared with pig fat into the ground to ward off lightning because of the belief that pigs were more resistant to witchcraft. Lungelwa Mhlwazi, the specialist herbalist, treated kraals about twice a year in this way, which was more frequent than the annual doctoring described by Setswana- and Sesotho-speaking informants. She had a good knowledge of local flora and used a variety of herbs, including *intolwane, imfingwana, ndiyandiya, ingqubebe, umayime, maqweqwane* and *umahlogolosa*.[83] Mhlwazi told us a number of the specific plants used, though not all. Infusions of the bark and roots, scattered on the earth were designed to keep the maleficent serpents away; the roots of plants made a particularly potent infusion for counteracting supernatural forces. *Umbezo* (probably *Clutia pulchella*), produced a smell that

---

[79] Myalezwa Matwana, Mbotyi, 7 December 2011; Mandisa S. Ngwane, 'Socio-Cultural Factors that have Contributed to the Decrease of Plant Species in the Eastern Cape: A Case Study of the Etyeni Village, Tsolo, Tranksei', unpublished MEd, Rhodes University (1999), notes that *isibande* is a type of long grass used for thatching, for cleaning the mouth, for acne and for a musical instrument called *umrhube*. A. Kropf, *A Kafir-English Dictionary*, second edition, ed. R. Godfrey (Alice, Lovedale Mission Press, 1915), notes that it is used for matting for women during menstruation.

[80] Sonwabile Mkhanywa, Mbotyi, 7 December 2011.

[81] Delinkosi Soyipha, Mbotyi, 18 December 2011.

[82] Lungelwa Mhlwazi, Mbotyi, 19 February 2009.

[83] Fearon, 'Population assessments' mentions *ingqubebe* (*inkqubebe*) as commonly used in the Mkambati area, about 30–50 km north of Mbotyi along the coast, but we cannot identify the species yet. *Umayime* is probably *Clivia* species.

was said to repel both the serpent that brought *umkhondo* and also the *tokoloshe* (see chapter 7). For infertility she dosed the cattle with a decoction made from *imbizo*.[84] The word generally means a meeting, but was used locally to mean a plant medicine. Nongede Mkhanywa, the *sangoma*, and Myalezwa Matwana, the animal specialist, also gathered plants from the forest to make medicine that would counteract supernatural threats, either burning the herbs in the kraal or else burying them inside, at the entrance, in the centre and at the four corners.[85]

In the vicinity of Mbotyi a specialist healer claimed to treat broken bones with shadow medicine. Wellington Jikazi exorcised the earth by making four cuts in the soil with his knife where the shadow of the animal fell and then sprinkled *muthi* into the holes to heal the bones. He also gave them a drench to strengthen the bones. While Jikazi attributed some breakages to natural causes, evil spirits could push the person or the animal to the ground. He saw his treatment as driving away the supernatural causes as well as healing the wound.[86] Nongede Mkhanywa claimed that this form of healing was not general in Mpondoland but associated it with Sotho medical practices. Jikazi in fact came from East Griqualand, where there are more Sesotho speakers.[87] This was not unlike the use of *thobega/thobeha* in North West Province and QwaQwa, which people sometimes buried in the ground, at the spot where animals fell, to cure fractures. However, in those areas informants did not say that they associated breakages with supernatural causes (see chapter 5).

We were not able to garner as much information about the particular plants people used in Peddie to deal with witchcraft and pollution. One exception was a staunch Methodist and well-known local herbalist who brushed medicines on the walls and floor of the house to discourage these threats. He also applied a procedure called *ukubetelela* in which small sticks or wooden pegs with *umhla-belo* (medicine tied in bags), were driven into the ground with the back of an axe. They were placed in all the corners of the yard and in front of the door of every building of the homestead as well as at the entrance to the kraal, to protect the inhabitants and their livestock.[88]

In North West Province, informants discussed the sprinkling and burial of herbs rather than the use of wooden pegs to protect the kraal and the homestead. Although healers and many livestock owners said they treated the kraal every year, few were prepared to divulge the ingredients. Elphars Dinne, a healer from Mabeskraal, explained that he helped farmers to protect their property

---

[84] Lungelwa Mhlwazi, Mbotyi, 19 February 2009.

[85] Nongede Mkhanywa, Mbotyi, 25 February 2009; Myalezwa Matwana, Mbotyi, 26 February 2009.

[86] Wellington Jikazi, Mbotyi, 24 February 2009.

[87] Wellington Jikazi, Mbotyi, 24 February 2009; Nongede Mkhanywa, Mbotyi, 25 February 2009.

[88] Andrew Ainslie, 'Hybrid Veterinary Knowledges and Practices among Livestock Farmers in the Peddie Coastal Area of the Eastern Cape, South Africa' (unpublished report, 2009).

against misfortune. 'When farmers invite me to defend their kraals against infertility and abortions caused by menstruating women, widows or witchcraft', he
said, 'I grind up *mogaga* (*Drimia altissima*) and *sekaname* bulbs and mix them
with the roots of the *mokeye* tree'. He added these plants to water to create an
infusion that he sprinkled around the kraals every year. He grew *mogaga* and
sekaname in his garden but cycled to the mountains to collect fresh *mokeye*
whenever it was needed. Dinne asserted that many farmers in the village still
used this medicine and they asked him, or other healers, to prepare it for them
as they did not know how to mix it themselves. To forestall witchcraft he burnt
crushed *tshetlho* (a plant commonly used to treat retained placentas in North
West Province) and mixed the ash with goat's fat which he rubbed on whatever needed safeguarding. Some people made incisions in their skin or that of
their animals to fortify the protection. If animals were slow to reproduce he
made a mixture of herbs called *leswalo* which he placed on a special stone
(*legakwa*) in the middle of the kraal. The *leswalo* consisted of a plant called *malla
digangwa* and *tlhware* (python) fat. He believed that fumigating the cattle by
burning this *muthi* would boost their fertility.[89]

In QwaQwa, stockowners accepted that they were unable to protect animals
against theft in the veld because the rangelands were open to everyone. Farmers
just hoped that their herders could guard the animals effectively while grazing.
The kraal, by contrast, was a private, familial space which they believed they could
defend with special *muthi*. If an animal then went missing from the kraal, it was
because the thieves had visited an 'evil healer' who had provided them with
stronger *muthi*, which had enabled them to invade the cattle enclosure. When this
happened, the only recourse for the stockowners was to find a better *inyanga* who
could furnish them with more potent *muthi* to beat the thieves.[90] Sesotho speakers
not only doctored the kraal to try to stop theft, but also to protect cattle from evil
spirits and improve their fertility.[91] In QwaQwa most stockowners employed an
inyanga to make the medicine rather than preparing it themselves. As in the
Eastern Cape, healers sometimes treated the homestead too and sprayed medicine
directly onto the animals. None of the *inyanga*s we interviewed in QwaQwa
would reveal any of their herbal secrets. Some farmers described how the healers
buried the medicine at the corners of the kraal, at the entrance and at the milking
post located in the centre of the enclosure or just outside it. This resembled prac-

---

[89] Elphars Dinne, Mabeskraal, 4 February 2010.

[90] Daniel Khonkhe, Ha-Sethunya, 13 October 2010; Joseph Mapetla, Phuthaditjhaba, 15
October 2010.

[91] Ntsoeu Livestock Company, Phuthaditjhaba, 11 October 2010; Alan Malinga, University of
the Free State, 13 October 2010; Jack Vezi, University of the Free State, 13 October 2010; Jacob
Tshabalala, Phuthaditjhaba, 13 October 2010; Charlie Tshabalala, Phuthaditjhaba, 11 October
2010; Paulus Mosia, Ha-Sethunya, 13 October 2010; Sethunya Mathu, Ha-Sethunya, 13 October
2010; David Mphuthi, Phuthaditjhaba, 14 October 2010; Thomas Mahlaba, Tseki, 15 October
2010.

tices in Mbotyi.[92] Sometimes the ceremonies in QwaQwa could be more elaborate. As with other aspects of local knowledge, practices tended to be individualistic and there was no set formula for deflecting witchcraft and misfortune. Charlie Tshabalala said his healer dug a hole in the ground and buried something in the earth whilst praising the ancestors. He also rubbed *muthi* on the cows' tails for added protection against thieving.[93] Sometimes healers smeared a paste on the entrance of the kraal to ward off evil spirits and deter thieves.[94] Alan Malinga recounted how the healer sprinkled medicinal water around the kraal while delivering incantations to the ancestors and she buried *muthi* at the corners of the enclosure. The Malinga family also nominated a

> special dog that lives with the animals to protect them from evil. The healer has to make the dog strong so it can look after the cattle. She makes cuts in the skin and fills them with medicine. The dog stays in the kraal all the time and never goes out with the animals to the grazing lands. When the dog dies, we will choose a new dog to protect our kraal and our animals.[95]

Malinga did not know the precise reason for dedicating a dog to the kraal but it was possibly to ward off witches' familiars; Lungelwa Mhlwazi reported a similar strategy in Mbotyi.[96]

Thomas Mahlaba from QwaQwa had learnt to make his own medicines for supernatural protection. He burnt the dried bulb of *sheosheo* (*Athrixia fontana*) in the kraal whilst the cattle were inside. Inhalation of this smoke, and fumigation of the kraal, defended them from theft and also jackals and other predators.[97] Kenneth Mokoena also relied on family knowledge to doctor the kraal and mixed *bolele* (a type of algae) with soot and water which he sprinkled around the kraal and on the animals whilst invoking the ancestors. He did this at the start of summer and winter to ensure the fertility of his animals and to discourage witchcraft.[98] Solomon Mokotla used two different types of plants to overcome infertility. He took the leaves of the *podisa* plant, boiled them in water and sprinkled the infusion around the yard as 'it cleansed everything it touched'.[99] To ensure conception, he smeared *muthi* on the cows' vulvas which made them 'horny' so that they ran after the bulls. Mokotla prepared this aphrodisiac from the bulbs of the *kgapumpu* plant (*Eucomis bicolor*). He chopped up the bulb and mixed it with milk to make an unguent.

[92] Johannes Motaung, Makwane-Tebang, 1 March 2011; Michael Molibeli, Ha-Sethunya, 2 March 2011.
[93] Charlie Tshabalala, Phuthaditjhaba, 11 October 2010.
[94] Paulus Mosia, Ha-Sethunya, 13 October 2010.
[95] Alan Malinga, University of the Free State, 13 October 2010.
[96] Lungelwa Mhlwazi, Mbotyi, 19 February 2009.
[97] Thomas Mahlaba, Tseki, 15 October 2010.
[98] Kenneth Mokoena, Lusaka, 28 February 2011.
[99] Solomon Mokotla, Ha-Sethunya, 2 March 2011.

Photo 8.2 *Podisa.* Burnt in kraal and sprayed on cattle to protect them against witchcraft in QwaQwa (KB, October 2010)

By contrast, April Nhlapo thought plant materials were useless as the witches could make stronger *muthi* from similar ingredients. The most powerful medicine in his view was seawater, as witches would never enter the sea. The purity of the sea symbolised the antithesis of the impurity of the witch. Sea water was available at the local shop and people would bring it back if they travelled to Durban. Nhlapo regularly gave his cattle 500 ml to drink and also dosed himself.[100] Abram Lebesa believed you could combat a witch's *muthi* by using snakes:

> My grandfather taught me how to make a secret potion using plants and snake parts, which I bury in the four corners of the kraal, at the entrance and in the centre. I also sprinkle the cattle with this mixture. Then I bury the snake skin in a secret place in the kraal. Snakes are important as they protect animals from theft and witchcraft. The power of the snake will beat the crap out of any thief that invades the kraal. I do this every year and I also slaughter a cow for the ancestors. I have to please the ancestors so that my medicine will work. The blood of the cow brings the living and the dead together. The dead are happy and they make my *muthi* powerful.[101]

Potency was often discussed in this more metaphorical sense, rather than in relation to specific strength of medicines or concentration of the mix.

[100] April Nhlapo, Lusaka, 28 February 2011.
[101] Abram Lebesa, Ha-Sethunya, 2 March 2011.

Another ritual some Sotho stockowners performed, especially in the village of Ha-Sethunya, was the burial of bovine placentas in the cattle kraal. Farmers encouraged their cows to give birth in the kraal, but if that did not happen, they collected the afterbirths from the veld and brought them back to the enclosure. Burying the placenta was said to improve the fertility of the cows. Paulus Mosia also explained that it showed respect for cattle, which was important because cows had 'magical powers'. Cattle would look after the welfare of the household, if the owner cared for them physically, and ritualistically. Sheep and goat kraals were not sanctified spaces, as these animals lacked such powers, and their placentas were never buried there. Sethunya Mathu kept women out of his kraal specifically because he believed they could stop the interred placentas from conferring fertility.[102]

We also heard of women treating the space that they moved through in order to avoid spreading malevolent forces left behind in their footsteps. Rebecca Molekwa was a widow and herbalist from Mabeskraal who owned a large number of different types of livestock. To annul the *mohato* that widows in mourning imparted to humans and livestock, she mixed the bulb of *mogaga* with a second plant called *mosiama*. She and those she supplied sprinkled the mix as they walked through the fields, the kraals and along the roads that livestock might traverse.[103] Another healer from Mabeskraal, Elphars Dinne, corroborated the use of *mogaga,* which he mixed with other herbs to protect the kraal against menstruating women.[104] In QwaQwa women also purchased special herbs from the *inyanga*s to dispel pollution.[105]

Beliefs around the doctoring of the supernatural were contested. Some stockowners in QwaQwa explained why they were abandoning these traditions. Maria Khoboko, for example, stated that her family had ceased to doctor the kraal since they had become 'modernised'.[106] Gladice Mokoena had also come to dismiss kraal *muthi* on the basis that it was nonsense.[107] One could postulate that families who had made the greatest use of female labour in new roles related to livestock, through necessity or notions of progress, had decided to jettison the spiritual function of traditional medicines. In Sotho culture the application and efficacy of kraal *muthi* was closely tied to ideas of gender and the danger women allegedly posed to the potency of the medicine. If women like Mokoena and Khoboko could enter the kraal, then the purpose of applying *muthi* to the kraal became redundant because women would inevitably destroy its powers. Two male informants who stated they never used *muthi*, Konese

---

[102] Paulus Mosia, Ha-Sethunya, 13 October 2010; Sethunya Mathu, Ha-Sethunya, 13 October 2010.

[103] Rebecca Molekwa, Mabeskraal, 3 February 2010.

[104] Elphars Dinne, Mabeskraal, 4 February 2010.

[105] Johannes Motaung, Makwane-Tebang, 1 March 2011.

[106] Maria Khoboko, Lejwaneng, 12 October 2010.

[107] Gladice Mokoena, Ha-Sethunya, 13 October 2010.

Mofokeng and Ephraim Mofokeng (not related), also said that they allowed any women into their kraals, strengthening this assumption that the survival of this tradition was in part tied to the status of women in individual households.[108] On the other hand, the Monengs, who claimed men, rather than women, were the greatest danger to livestock, did treat the kraal with medicines, indicating once again how the adherence to rituals and medical beliefs could be highly adaptive and personalised.[109]

Conclusion

Ideas about *mohato*, and the rich vocabulary on pollution articulated by Sotho informants, remained a powerful element in the rhetoric of livestock management and tradition. These multi-dimensional concepts still flourished and it is striking that they have regionally specific forms. Although we could not observe this in detail, they seemed to impact strongly on gendered labour practices, as well as restrictions on women entering the kraals. To a degree these conceptualisations of supernatural forces work at a metaphorical level and are ways of bolstering patriarchy as much as interpreting diseases. In this sense such explanations of causation did not preclude biomedical treatments. As noted we found a greater flexibility about means of treatment than we did about interpretations of disease. Despite the legacy of patriarchal ideas, our interviews with women, and even some with men, indicated shifting gender relations in a number of families. Older women especially could become livestock owners and break the internal boundaries shaping space and gender relations. These interviews painted a picture of forceful women taking charge of their homesteads and developing a stronger role in the rural livestock economy.

[108] Konese Mofokeng, Phuthaditjhaba, 11 October 2009; Ephraim Mofokeng, Phuthaditjhaba, 15 October 2010.
[109] Maria and Anna Moneng, Thebang, 3 March 2011.

9
Conclusion

The dynamics of local knowledge

Understanding local knowledge has become a significant academic project amongst those interested in Africa and developing countries more generally.[1] We have explored a central body of African knowledge about livestock diseases. We have drawn on the small but expanding field of what others call ethnoveterinary research but we have attempted to move beyond this literature.[2] We have examined, first, changing patterns of local knowledge and the extent to which it has become hybridised. Second, we have analysed the relationship between local and scientific knowledge. Third, we have tried to understand an overarching range of ideas and practices and move beyond discussion of plant medicines, on the one hand, or witchcraft on the other.

Our informants were the descendants of communities that had been

[1]   Clifford Geertz, *Local Knowledge: Further Essays in Interpretive Anthropology* (New York, Basic Books, 1983); Laura Nader, 'The Three-Cornered Constellation: Magic, Science, and Religion Revisited' in Laura Nader (ed.), *Naked Science: Anthropological Inquiry Into Boundaries, Power, and Knowledge* (New York, Routledge, 1996); Richard Waller and Kathy Homewood, 'Elders and Experts: Contesting Veterinary Knowledge in a Pastoral Community' in Andrew Cunningham and Bridie Andrews (eds), *Western Medicine as Contested Knowledge* (Manchester, Manchester University Press, 1997), 69–93; F. Barth, 'An Anthropology of Knowledge,' *Current Anthropology*, 43, 1 (2002), 1–18.

[2]   Constance M. McCorkle, Evelyn Mathias and T.W. Schillhorn van Veen (eds), *Ethnoveterinary Research and Development* (London, Intermediate Technology Publications, 1996); Edward C. Green, 'Etiology in Human and Animal Ethnomedicine', *Agriculture and Human Values*, 15, 2 (1998), 127–31; P.J. Masika, W. van Averbeke and A. Sonardi, 'Use of Herbal Remedies by Small-scale Farmers to Treat Livestock Diseases in central Eastern Cape Province, South Africa', *Journal of the South African Veterinary Association*, 71, 2 (2000), 87–91.

colonised well over a century ago. They have lived in a state that has imposed practices rooted in scientific conceptualisations of disease causation and control. State veterinary institutions and policies have touched the lives of their ancestors and themselves. Yet their ideas and consciousness have not been completely colonised or transformed by such interactions. To a surprising degree, we have found that rural African livestock owners worked with older but dynamic understandings of disease and treatment that interacted vigorously with changing scientific ideas and conceptions of modernity. Overall we suggest relatively limited penetration of biomedical ideas about germs, or parasites such as ticks, in the explanations of disease. The dominant form of understanding rested in environmental and nutritional concerns.[3]

African farmers monitored ill health and closely observed the symptoms of disease in live animals. They also conducted post-mortems, often as part of the slaughtering process. In all but one research site, African communities had a wide variety of local names for diseases which did not always correlate with, or directly translate into, scientific terms. The most common cattle diseases at many of our field sites were tick-borne infections, particularly gallsickness (see chapter 3). The local names (*gala* in North West Province, *nyoko* in QwaQwa and *inyongo* in the Eastern Cape) sometimes encompassed a wider range of symptoms. Stockowners recognised many other key diseases such as heartwater – present in parts of North West Province and the Eastern Cape but absent from QwaQwa – and to a lesser extent redwater. Many Tswana informants had encountered blackquarter, for which they had multiple local names, and some farmers from QwaQwa said it was the most common and dangerous livestock disease there. Most were familiar with anthrax which stole lives in North West Province and the Eastern Cape but not in QwaQwa.

As we discussed in chapter 3, African farmers attached great weight to observations and interpretations of environmental conditions when explaining the causes of many animal diseases. In their view, the condition of the veld could foster good health, but equally it could bring about diseases like gallsickness and blackquarter. Changes in the consistency of the grasses and natural seasonal transformations in the pastures affected digestive functions and the overall health of livestock. Disease was seen as part of the natural cycle, although its intensity, frequency and impact could change in different years shaped by unpredictable natural events and by the arrival of new scourges. Environmental relationships were at the heart of many discussions of disease, and provided African communities with a context for the annual ebb and flow of particular ailments. Managing the nutrition of livestock remained central for interpreting and coping with disease.

With this perspective in mind, our findings also inform debates in environmental history. Research in this field has been focused on seeking an African environmentalism – for example manifested in the protection of forested areas

---

[3] Green, 'Etiology in Human and Animal Ethnomedicine'.

by 'guardians of the land'.[4] We were not looking specifically for a conservationist ethos, yet our findings did suggest the strength of environmental ideas and explanations. These were not essentially centred on the sanctity of the land or the preservation of nature. Rather respondents articulated ideas around the utility of natural resources.[5] For many of our informants, environmental forces were Janus faced. On the one hand the veld was the source of nutrition and vital medicinal herbs, praised for their curative properties. On the other hand the grasslands seemed to harbour the seeds of disease. Few spoke explicitly about the need to conserve resources, but their narratives included a concern about the loss of access to grazing and useful plants.

In the past, African and settler livestock owners used transhumance as a vital strategy to maximise nutrition and avoid diseases (see chapter 4). Transhumance was curtailed by state veterinary controls during the twentieth century and we found only remnants of such systems – largely in Mbotyi and QwaQwa. Aside from restrictions over movement, informants cited theft as the most common factor in the demise of these older systems of pastoralism. Transhumance is no longer a significant route to reducing morbidity and maintaining animal health although the transfer of animals to fresh pastures during the summer months still played a part in optimising nutritional resources.

Detailed knowledge of local environments, gained over many generations, has enabled African stock owners to identify plants which they regarded as effective treatments for diseases (see chapter 6).[6] A few plants stand out in some areas, such as aloes and *sekaname* in North West Province. In the Eastern Cape plant remedies are very diverse although two or three key species – particularly aloes and *Pittisporum* – recur in treatments for gallsickness. Stockowners greatly valued the laxative properties of plant remedies. Plants may also have acted as

---

[4]   J. M. Schoffeleers (ed.), *Guardians of the Land: Essays on Central African Territorial Cults* (Gweru, Mambo Press, 1979); Gregory Maddox, James Giblin and Isaria N. Kibambo (eds) *Custodians of the Land: Ecology and Culture in the History of Tanzania* (Athens OH, Ohio University Press, 1996); William Beinart and JoAnn McGregor (eds), *Social History and African Environments* (Oxford, James Currey, 2003); see especially introduction and article by Terence Ranger, 'Women and the Environment in African Religion: the Case of Zimbabwe', 72–86.

[5]   E. Kreike, 'Hidden Fruits: a Social Ecology of Fruit Trees in Namibia and Angola 1880s–1990s', in Beinart & McGregor (eds), *Social History and African Environments*, 27–42; M. L. Cocks, *Wild Resources and Cultural Practices in Rural and Urban Households in South Africa: Implications for Bio-Cultural Diversity Conservation* (Grahamstown, Rhodes University, 2006).

[6]   Masika, et al., 'Use of Herbal Remedies'; D. van der Merwe, G. E. Swan and C. J. Botha, 'Use of Ethnoveterinary Medicinal Plants by Setswana-speaking People in the Madikwe Area of the North West Province of South Africa', *Journal of the South African Veterinary Medical Association*, 72, 4 (2001), 189–96; D. van der Merwe, 'Use of Ethnoveterinary Medicinal Plants in Cattle by Setswana-speaking people in the Madikwe area of the North West Province', MSc Thesis, University of Pretoria 2000; A. P. Dold and M. L. Cocks, 'Traditional Veterinary Medicine in the Alice District of the Eastern Cape Province, South Africa', *South African Journal of Science*, 97, 9–10 (2001), 375–9; Cocks, *Wild Resources and Cultural Practices*.

analgesics and antiseptics. We noted about 60 plants in common use in the key research sites. Overall, curative strategies aimed to ensure that animals ate rapaciously and excreted well-formed and prolific dung.

We have illustrated how, especially in the early years after the political transition, farmers were encouraged to take over some former state functions such as dipping. More generally, stockowners' choices of treatment were becoming more individualised. A growing number saw biomedical treatments as quicker or more effective, but also believed that local plants were more suitable for certain conditions. Some informants, especially in QwaQwa, felt culturally committed to older practices as a statement about identity. Biomedical products were often used in a generalised way, to combat both named specific diseases and non-specific ailments. Terramycin LA, for example, has become something of a cure-all in the rural areas.

This shift towards biomedicines has significant implications for the further commodification of rural lives. It also potentially benefits major drug manufacturers such as Pfizer and Bayer. African livestock owners provided a growing market served by retail outlets selling veterinary medicines in the small towns. In general, the government veterinary service recommended recognised and branded drugs, and at stock days representatives of drug companies directly promoted these. However, these were relatively rare events and our evidence suggests that a number of medicines, particularly non-branded dip, were sold informally in the rural areas. While their advice clearly prioritised biomedicines, neither the state nor the drug companies heavily advertised or pushed them. Many interviewees would have used pharmaceutical products more regularly if they were cheaper and more easily available. While we did not research pharmaceutical companies, our impression is that the demand for biomedicines to some degree was coming from below, out of the process of local observation and experience that we have discussed.

Our evidence suggests that on the whole, plants used for healing named diseases and for protection against supernatural forces were different. There is some overlap, but particular combinations of plants were associated with specific conditions and it appears that these have been well-established for some time. Our older informants said they learnt them from their parents. Most of those we interviewed saw the key plant remedies as performing a natural rather than a supernatural function, and thought about their efficacy in relation to medicinal ends.

Scientific researchers have made attempts to isolate therapeutic chemical agents in a limited number of medicinal plants (see chapter 6).[7] They have recorded some positive indications but most experiments suggested that solutions in water released fewest of the active ingredients. Our informants used only water in which to mix their *muthi*s. Such research has, however, generally been

---

[7]   L. J. McGaw and J. N. Eloff, 'Ethnoveterinary use of Southern African Plants and Scientific Evaluation of their Medicinal Properties', *Journal of Ethnopharmacology*, 119, 3 (2008), 559–74.

done with one plant under laboratory conditions, and not in the field at particular seasons with a mix of plants in the same combinations that are deployed in local veterinary treatment. It is thus very difficult to judge the effectiveness of local plant medicines, especially because stockowners sometimes administered them along with biomedicines. Plant remedies might relieve symptoms while animals' natural immunities and resistance enabled them to recover. Some popular local treatments can be highly toxic, depending on the concentration or dose. *Sekaname* or *slangkop* can kill humans and livestock if taken in large quantities. Stockowners seldom administered standardised doses and the concentration of plant extracts in any particular mixture could vary greatly.

Overall our impression is of a gradual shift towards biomedicine, hastened by stockowners' incapacity to deal with the chronic problems of ticks and worms as well as some new scourges such as lumpy skin disease. While commitment to local knowledge is striking in some areas, informants also expressed uncertainty about its effectiveness. Transmission of knowledge is often restricted and we found a degree of secrecy about traditional remedies. Some respondents who wished to know more about medicinal herbs could not easily gain access to this information. For this reason, as well as the scarcity of key species in particular localities, the range of plant remedies available is probably declining. Many stockowners work with a fragmented and patchy understanding of both traditional remedies and biomedicines.

The limits of local knowledge

Literature on local knowledge has tended to affirm its value and to validate such alternative ways of knowing. In the later decades of the twentieth century, it was offered as a critique of top-down, coercive development strategies that were clearly failing in many contexts, and as a route to understanding why rural communities in so many poor countries rejected elements of scientifically based policies.[8] Our approach has determinedly taken local knowledge seriously and has examined its diverse content and practices. However, we also emphasise the limitations of local knowledge as a means of controlling diseases.[9]

In the veterinary sphere our research suggests the limits of disease conceptualisation. Scientists have identified a number of significant diseases that did not

[8]   Paul Richards, *Indigenous Agricultural Revolution: Ecology and Food Production in West Africa* (London, Unwin Hyman, 1985); James Ferguson, *The Anti-Politics Machine: 'Development', Depoliticization and Bureaucratic Power in Lesotho* (Cambridge University Press, Cambridge, 1990).

[9]   William Beinart, Karen Brown and Dan Gilfoyle, 'Experts and Expertise in Africa Reconsidered: Science and the Interpenetration of Knowledge', *African Affairs*, 108, 432 (2009), 413–33; Karen Brown, Andrew Ainslie and William Beinart, 'Animal Disease and the Limits of Local Knowledge: Dealing with Ticks and Tick-borne Diseases in South Africa', *Journal of the Royal Anthropological Institute*, 19, 2 (2013), 319–37.

appear in our interviews – for example, lungsickness, mastitis, brucellosis, trichomoniasis and vibrosis, whilst others such as *lamsiekte* or botulism only appeared occasionally in testimonies from North West Province. Stockowners recognised poisonous plants as a notable problem in parts of North West Province and QwaQwa; however, they were often familiar only with a single variety of toxic weed, rather the diversity of noxious flora identified by toxicologists. Disease vocabularies and understandings of causation can be very localised and bounded. Throughout our field sites we came across considerable uncertainty in diagnosis and hence also in treatment.

Many found it difficult to identify the most appropriate remedies for specific ailments and few were able to benefit from cooperative purchasing of drugs. Trust and word of mouth influenced approaches to biomedicines as well as local knowledge. While livestock owners were often enthusiastic about their own local medicine, they nonetheless evinced uneven perceptions of its efficacy. The general inability to find structures of local cooperation through which dipping could be continued, once state programmes collapsed, has resulted in much more individualistic approaches to tick control. Overall, informants complained that tick burdens had increased and they looked to the state to resurrect a more interventionist and coordinated strategy (see chapter 2). Veterinary scientists advocate a return to more tick-resistant cattle, such as Nguni breeds, but the schemes available were not attractive to the majority of smallholders that we interviewed.

Although environmental and nutritional ideas emerged as the most important explanations for diseases (see chapter 3), beliefs about the supernatural remained reasonably strong. Older ethnographies, written between the 1930s and 1950s, argued that African societies in southern Africa did not distinguish sharply between natural and supernatural causes. We can see the continued interweaving of these elements, but our interviews did indicate that stockowners made a distinction between the two and the combination of these ideas could be highly individualised (chapters 7 and 8).

Stockowners infrequently invoked witchcraft as a specific cause of death, unless an animal died suddenly or in mysterious circumstances (see chapter 7). Our findings are somewhat at odds with recent publications on witchcraft and human disease in South Africa that indicate a resurgence of occult ideas. Witchcraft has in part been associated with the massive human health and social crisis in the shape of HIV/AIDS.[10] This is not replicated in the veterinary sphere. There is widespread concern about animal health but witchcraft has not emerged – at least in our interviews – as a significant explanation.

A wider range of symptoms and deaths were attributed to a more ambient set of supernatural causes, expressed through concepts such as *umkhondo* in

[10] Adam Ashforth, 'An Epidemic of Witchcraft? The Implications of AIDS for the Post-apartheid State', *African Studies*, 61, 1 (2002), 121–43; Fraser McNeill, '"Condoms Cause AIDS": Poison, Prevention and Denial in Venda, South Africa', *African Affairs*, 108, 432 (2009), 353–70.

Mbotyi and *mohato* in North West Province. *Umkhondo*, precipitating weakness and sickness in animals, was explained as the infectious traces of malevolent agents or natural forces surrounding the kraal. *Mohato* was especially associated with pollution and the danger that women posed to livestock. We have not found this word in the ethnographies and it may be a new term for the range of taboos involving women's perceived capacity to trigger abortions in livestock. This concept, as well as the rich vocabulary on pollution articulated by Sesotho-speaking informants, might be a way of bolstering patriarchy as much as interpreting diseases. Farmers who invoked such explanations were not antipathetic to biomedical treatments. They could invest in the latest vaccines, anthelmintics for internal worms and pour-on acaricides to deal with ticks, at the same time as banning women from the kraal. Informants suggested that the use of plants would continue because biomedicines could not address such threats. Despite the legacy of patriarchal ideas, our interviews with women indicated shifting gender relations in a number of families. Older women especially could become livestock owners and break the internal boundaries shaping spatial and gender relations (see chapter 8).

In general, our interviews indicated that supernatural explanations were fading slowly but unevenly. These multi-dimensional ideas about infections and infertility were manifest in the different research sites, but clearly took regionally specific forms. Supernatural ideas remain dynamic, but our discussion resonates with Edward Green's overview of disease explanations. He contested the dominance of supernatural ideas in anthropological analyses of illnesses in Africa.[11] However, his idea of indigenous contagion theory is not entirely convincing in relation to our evidence. Few informants understood that some diseases could be transmitted from one animal to another.

In chapter five we quoted Caroline Serename from Kgabalatsane in North West Province. She lived in a substantial house with electricity and beautiful views across the valley towards the Magliesberg Mountains. She entertained us and her neighbours with conversation and Coca-Cola as we discussed both their ideas about livestock disease and their priorities for government assistance. Serename said: 'In the past there were no vaccines and animals were never identified as sick. Now animals are like children; they have to have injections and be cared for.'[12] Although not all informants were as committed to biomedicines as she was, we sensed a general but uneven shift towards the recognition of specific diseases and biomedical treatments. While there is no linear pattern of change, we suggest that the balance within local knowledge between more traditional and biomedical ideas are edging towards the latter.

---

[11] Green, 'Etiology in Human and Animal Ethnomedicine', quote from 129–30; Edward Green, *Indigenous Theories of Contagious Disease* (Walnut Creek CA, Alta Mira Press, 1999).

[12] Caroline Serename, Kgabalatsane, 13 November 2009.

Recommendations for policy and practice

Effective veterinary programmes should take account of dynamic local knowledge and work with the ideas held by hundreds of thousands of African livestock owners. This is particularly important in contemporary South Africa because much of the livestock is held on communal land. Moreover, the areas available to black farmers are expanding and many will continue to operate on a small scale. While the state no longer enforces highly centralised compulsory interventions, insensitivity to local practices can be counterproductive.

While we do not assert that all biomedicines are equally effective, available or affordable, we are also not arguing for medical relativism. It is one thing to understand local practices, and another to advocate them as a major strategy for dealing with livestock diseases. We think it is important to acknowledge the variability and limits of local knowledge and problems that might arise in its deployment in state-sponsored medical and development strategies.

Our recommendations (see appendix 1) grow out of a concern to facilitate improvements in the way that livestock owners manage the health of their cattle and gain access to knowledge about diseases and treatments. We are also suggesting potential strategies for more-effective state intervention and assistance. Support for individuals could involve the state, the private sector, NGOs and associations amongst stockowners themselves. It is important that veterinary policy is informed by research that takes into account African ideas and the constraints on African stockowners.

Appendix 1:
Recommendations

Role of the state

Research

At present, state veterinary policy operates largely through provincial governments. Expanding adequate research capacity at a provincial level, ideally in association with academic institutions, would be valuable. An example of effective research has been at the University of Fort Hare where Patrick Masika and his team are doing work that is highly relevant to state interventions and the recording of local medical knowledge. Tony Dold and Michelle Cocks at Rhodes have developed extensive records of local medicinal plants. The University of Pretoria and the Onderstepoort Veterinary Institute have sponsored phytochemical plant research drawing on local knowledge primarily from the northern provinces. We have used their publications extensively in our chapters. However, such researchers could communicate more across disciplines and make their work more easily available to government officers and smallholders. Such a strategy would require coordination between academics and local state officers. There is a major gap in research on the efficacy of administrative intervention.

Animal Health Technicians and interaction with communities

Veterinary surgeons at a district level are a fulcrum for policy implementation, regulation of diseases, and the circulation of information about treatment. A cluster of Animal Health Technicians (AHTs) serve in each district office and provide a critical interface with the community. The system of AHTs emerged in the 1990s as an alternative to the more-centralised implementation of policy through stock inspectors and dipping foremen. AHTs received better training than their predecessors so that they could be responsive to a much wider range of problems affecting the smallholder livestock sector.

This structure has considerable potential. AHTs in some areas are important in supervising dipping and helping with inoculation. Some also assist in the supply of biomedicines. Such activities, as well as stock days, bring them into regular contact with communities. Our experience in some areas, however, suggested that there were problems in maintaining motivation and successful outreach. Local trust and communications skills are important. There should be more opportunities for AHTs to share their experiences and discuss best practice. State officials may be able to identify where there has been locally successful outreach through the AHTs and find ways of replicating this. AHTs are often the only route for more isolated communities, which also have very limited knowledge of biomedicines, to gain access to more specialised drugs by purchasing them from AHTs. Although this opens opportunity for private benefit from public office, there is scope for discussion of this role by AHTs. They could also work with specialists in local knowledge to gather information on, for example, animal birthing practices and the preparation of plant remedies.

Stock Days

Stock days (also known as Information days) can be very effective as a route of communication and connection, although they do tend to be dominated by AHTs and representatives of pharmaceutical companies, and can be pedagogical in their style. The absence of common ground over naming of diseases is a strong manifestation of the divide between presenters and their audience (chapter 3). Stock days could be more hands-on occasions, focused around demonstrations and an exchange of knowledge, where local speakers can talk about their experiences and their methods. Our interviews suggest that people are open to learning more about biomedicines and biomedical explanations, but the transfer of knowledge would benefit from a more interactive context. The Eastern Cape Department of Agriculture has started an indigenous knowledge research group and this is one potential way of encouraging interaction. Stock days could also be a vehicle for disseminating written material in the local language.

Pamphlets

Fieldwork in North West Province, in particular, revealed a thirst for leaflets and printed material in the vernacular. Even if people were illiterate they found the written word empowering as they could use leaflets to ask for help about specific diseases and treatments at drug retail outlets. Leaflets could be a vehicle not simply for directives about appropriate drugs or therapeutics, but also for recording the diversity of local naming systems and plant treatments.

A series of basic pamphlets has been published in English and Afrikaans by J. A. Turton and others. They are designed to communicate elementary scientific understandings of common animal diseases.[1] A limited number of them, especially those dealing with anthrax and rabies, seem to have been translated into some African languages, but there has been no provision for many of the key diseases we encountered. We did not come across any of these leaflets in the

[1]   J. A. Turton: *Common and Important Diseases of Cattle*; *External Parasites of Cattle* (both Pretoria, Department of Agriculture, 2001) – the full range is available from www.nda.agric.za/publications.

field and our informants were not aware of them. Moreover, it would be valuable if the Department's pamphlets dealt more effectively with local concepts and names for diseases. Confusion over words offered opportunities for misdiagnosis and the inadequate prescription of drugs, which could be at best unnecessarily expensive for stock owners, and at worst harmful to their animals.

Onderstepoort Veterinary Institute has also issued a leaflet indicating when farmers should inoculate their animals for particular diseases. We distributed these to some farmers in North West Province; stockowners were particularly enthused by these worksheets and asked for them to be translated into the vernacular and to be more readily available. The willingness to accept inoculation as a prophylaxis demonstrated their openness to biomedicines, when affordable. Leaflets and vaccine programmes could easily be distributed at stock days and dipping days, and AHTs could take the time to explain the contents.

Social scientists and exchanges of knowledge

Specialists tend to dismiss local environmental or supernatural ideas about disease causation. Education should be a two way discussion and specialists need to find a way to interact with supernatural ideas. For example, the question of milking in Mbotyi is likely to be resolved only if someone can take on the debate about the baboons publicly, however strange such local ideas of causation may seem to outsiders (chapter 7). Problems surrounding dystocia and abortions, which are a major concern of farmers in North West Province, need to be addressed by interactive education on brucellosis and other reproductive diseases and disorders, if herds are to increase and notions of ritual pollution are to be further eroded.

There is a case for incorporation of social scientists into provincial veterinary departments, which employ largely scientifically trained staff at various levels. Social scientists could focus directly on questions of communication and liaison with communities. An analogy could be made with the National Parks Service, and the provincial environmental departments, which have specifically employed social officers to engage in community relations. In a number of places in the country there are university research groups and NGOs that are concerned with questions of local knowledge and community organisation and management.

Subsidies and management of dipping and inoculation

In earlier years the state took responsibility for the infrastructure of rural veterinary services, in particular dip tanks and the associated record keeping. Dipping fees or other systems of local taxation partly covered the cost. The state also provided vaccines and distributed them through suitable storage facilities at the veterinary offices. Dipping in tanks is no longer a complete solution to the control of ticks and tick-borne diseases (chapter 2). It is now recognised that a combination of different strategies is necessary for this complex area of disease control. Many of our informants were enthusiastic about more regular access to dipping tanks as the prohibitive cost of pour-on acaricides undermined their ability to manage tick numbers. The problem is that the dipping infrastructure is eroded or absent and dipping committees are given responsibility for maintenance of tanks and procuring acaricides. The committees have some presence in the Eastern Cape and there are similar, more informal, organisations in QwaQwa, but there appear to be no such structures in North West Province where farmers bemoaned the loss of the dipping tank. Emptying and refilling the tank is time consuming and difficult to do by hand. Pumps are ideal but hard to maintain.

In the light of past difficulties in levying specialist taxes for dipping operations, there is a strong argument that the costs of maintaining dips should come out of central or provincial government coffers. The counter-argument, which was attractive to the ANC government in its early years in office, was that dipping benefited only a restricted number of people so livestock owners should take responsibility for themselves. However, there are so many families in South Africa with some livestock that this provision can be considered a general benefit. Localising responsibility for the maintenance of tanks and supply of dip has not generally been successful (chapter 2).

Nguni Cattle Schemes

Indigenous Nguni cattle are being promoted as a complementary strategy to address tick-borne infections because they appear to have developed resistance over centuries.[2] The University of Fort Hare Nguni Project, initiated in 1998, has bred herds of this strain and distributed animals in surrounding communities. The North West Province Provincial Government has encouraged farmers to join in its Nguni Cattle Scheme set up in 2007 (chapter 2). Advocates of pure-bred Nguni see their re-dispersal in the African areas as a major route forward in developing disease control. The Bonsmara and Brahman breeds also share some of these characteristics and have the advantage of being larger. These are exciting initiatives but the present projects are small and restricted. Relatively few African livestock owners have the capital to invest in pedigree bulls, neither could they control reproduction in communal areas. This problem has always undermined livestock improvement programmes. Some, but not all, scientists think that because Nguni are small, they are prone to dystocia from copulating with larger bulls. At a grassroots level, we found that the clamour was for a return to state-run dipping rather than efforts to transform their herds. When we raised this in our report-back sessions, farmers in Mbotyi as well as in Fafung and Bollantlokwe in North West Province felt that Ngunis were too small and the state schemes too restrictive.[3]

[2] Arthur Spickett, factsheet 'Integrated Tick Management,' received 22 February 2011.
[3] Meeting with farmers in Bollantlokwe, 27 November 2012; Fafung, Thursday 29 November; Mbotyi, 5 December 2012.

Census and information gathering

In our interviews with senior members of the Eastern Cape veterinary service, the question of a provincial livestock census was raised as an important element for strategic interventions and planning of veterinary services. Dr Ivan Lwanga-Iga at Dohne Agricultural Research Station felt that in recent years the state had not put sufficient resources into data collection and this undermined the possibility for planning. He sensed that many diseases were under-reported and mortality rates from specific diseases were unknown.[4]

The possibilities of finding the resources for a full count of animals on a specific day are remote and our interviews suggest that the historical suspicion of livestock censuses, associated with forced culling and other government interventions, remain. However, the old techniques for gathering statistics through inoculation and dipping days still provide potential for systematic collection of data on livestock numbers and even on incidence of diseases. State vaccination campaigns covering different animal diseases receive quite wide public support and, if space were made available for interaction and information gathering over the period of a census year, then a fuller numerical and epidemiological picture could emerge without significant extra expense.

Facilitation of private sector, NGO and individual initiatives

There are a number of areas where individuals rather than the state alone could take responsibility. All of these potential improvements are bound up with central problems of labour and costs.

Herding, labour and surveillance of animals

One central question concerns the herding of livestock. For the most part, as we have mentioned, the pastures in the African-occupied communal areas are still relatively open with few effective fences. In North West Province and QwaQwa, most owners organised herding for their animals, but this was less common in the Eastern Cape, especially where the animals were left out at night in pastures. The absence of daily care is more conducive to straying, theft and a lack of attention to health – for example the early manifestations of disease or high tick burdens. In Mbotyi most owners leave their animals out in the pasturelands during the summer months and this is probably one reason why these cattle appear to have higher tick burdens than those in North West Province. Patrick Masika at Fort Hare, who has done most of his research in the Amatola Basin in the Eastern Cape, thought that there was diminishing emphasis on keeping animals clean of ticks.[5]

---

[5]   Patrick Masika, University of Fort Hare, 23 February 2011.

In earlier decades boys and youths were responsible for herding cattle. As an increasing proportion of boys went to school, herding duties within the family have generally fallen on the older men and sometimes women. Some employed herders, but this is perceived as an expensive option, especially as there is relatively little direct cash income from the herds and flocks. In North West Province, widows in particular engaged herders because they felt they lacked experience in managing livestock. However, they expressed uneasiness in leaving animals in the care of those outside the family, who could be involved in stock theft. As in the case of women's agricultural labour, wage work has not fully replaced family labour (chapter 4).

We have no direct recommendation as to how such problems can be overcome. But the difficulty of finding labour for the livestock economy, in a context where youth employment is so high, is striking. The work is isolating and young people did not speak about the camaraderie of herding which was so evident in the fond reminiscences of older men. There is a crisis of kinds in the communal rural areas around the rejection of agricultural work amongst young people, and at the same time, the wages offered are also insufficient to attract them to it. This issue is little addressed in thinking about development in the rural areas.

Maintaining kraals

Many of the kraals and areas around the homesteads are piled with dung and are a breeding ground for worms and infections such as mastitis. In Mbotyi, even some of the better and bigger herds lived in conditions which were highly conducive to worm reproduction and infestation. Ever since the veterinary services were established in the 1870s, officers have fulminated against the condition of the kraals in South Africa and agricultural officers wept about the failure to use manure on the fields. We are joining this age-old chorus!

Litter

The amount of litter in gardens and in the veld, especially in villages in North West Province, is also damaging to animal health. Gardens and grasslands around settlements are strewn with plastic bags, bottles and food wrappers (chapter 3). Providing municipal services in rural and peri-urban areas remains an enormous challenge in South Africa. In Kwakwatsi (the township near Koppies) groups of volunteers collected rubbish. But the local state is essential for sustained refuse management even in the rural areas.

Practices such as more effective herding, as well as keeping animals and kraals cleaner, and removing litter from pastures, would all make a difference to animal health. Recommendations on this front are necessarily sensitive and problematic, because they are so deeply bound up with the availability and control of family labour, as well as attitudes to animals. However, they could also form part of the debate at stock days.

Livestock sales and turnover

For many years, state departments of agriculture advocated a higher turnover of African cattle as one of the keys to rural development. The idea was that this would both bring in additional income to rural communities and diminish the number of livestock on the veld. Their call was for fewer and better cattle. Officials perceived that sales would help to resolve the environmental crisis about overgrazing. The state encouraged smallholders to sell off a proportion of their herds every year in order to keep overall numbers at a lower and more stable level.

We found that most stockowners were still reluctant to sell cattle on a regular basis, though this applied less to goats, where there was a strong demand for ceremonies and sacrifices. Access to markets also has a bearing on the willingness of stockowners to sell animals. In the case of Mbotyi, however, it is difficult to see how stockowners might increase external sales. First, the livestock are too strongly affected by disease to be of much interest to those selling through retail outlets such as supermarkets and butcheries. Generally stockowners try to sell their animals when they are too old and the meat too tough to be of much interest to retailers. Second, as discussed in chapter 3, the price of cattle in Mbotyi tends to be higher than in the heartlands of the Eastern Cape or Kwa-Zulu Natal so there is relatively little incentive for those in the village to sell outside. For purchasers the costs of transport from such a remote area is an inhibiting factor. In effect meat supplies are provided at a reasonably high level by informal slaughtering and distribution through ceremonies and events.

In peri-urban areas such as Garankuwa in North West Province, where a lower proportion of the population own livestock and where supermarkets and retail outlets are more accessible, there are probably fewer opportunities for smallholders to sell their livestock despite the bigger market for packaged meat. Furthermore, Tswana smallholders also felt disadvantaged compared with commercial producers because they were unable to access abattoirs on the grounds, they claimed, that their meat was likely to be infested with tapeworms, or the animals had not received a full schedule of vaccines. Consequently they were forced to sell their livestock at auctions, for a lower return. Long-advocated solutions like increasing auctions within the African rural areas are unlikely to provide a resolution in themselves.

Costs of medication

The question arises as to whether the state should supply free or subsidised medication. In the Eastern Cape, the dip is in theory provided gratis by the Department of Agriculture as are some key inoculations, such as anthrax that can affect people, and blackquarter, which is included in a joint vaccine. In North West Province, the administration of anthrax vaccination is patchy and in some villages, such as Mantsa, stockowners complained that no AHT or vet had inoculated their livestock for several years, and human as well as animal deaths from anthrax were on the rise. Protecting people from zoonotic diseases such as anthrax, brucellosis and rabies, should be the function of the state and here there is an argument for free public animal vaccination campaigns to protect people as well as livestock. While not everyone is a livestock owner, everyone, especially in poorer communities, is susceptible to these serious infections. Disease protection is a universal benefit.

In the case of non-zoonotic diseases, there is an argument for individual stockowners to be responsible for medication because they benefit directly. However, poorer rural livestock owners, especially those with smaller herds, pay a disproportionately high cost for their medicines and have very limited access to adequate storage facilities. Their costs could be significantly reduced by cooperative purchasing systems, or the provision of key broad-spectrum drugs such as Terramycin (oxytetracycline) and ivermectin, which are very widely used, at closer to their cost price. This is probably not an area for state involvement as it could provide an opportunity for corruption, but it is an area where the private sector suppliers, such as the big pharmaceutical companies, could consider distribution at a fairer price, and the production of generic drugs. At present rural retail outlets charge a high mark-up. We have illustrated some differentials in prices at various centres (chapter 2).

Some AHTs already play an informal role in the commercial distribution of medicines and, although this presents the potential for private gain by state employees, it may be that they are in the best position to provide drugs for specific diseases. They can sometimes buy in bulk and they are qualified to administer biomedical products. We saw this informal supply system operating in connection with lumpy skin disease. An AHT bought the vaccine in an urban centre where it was cheaper, and arranged with the community to inject their livestock for payment. While it may be unwise in some respects for AHTs to become involved in private deals, the state doesn't appear to be discouraging such transactions. An alternative would be cooperative purchasing by communities themselves, but when we raised this possibility people suggested that there was not sufficient trust or they operated too individualistically.

Transmission of knowledge

One of our most striking findings concerned the constraints on transmission of knowledge within communities around plant remedies (chapter 5). Those with knowledge sometimes felt reluctant to share this with others. They were constrained by lack of trust, concerns about witchcraft or the desire to preserve the commercial value of their practices. In fact, most of the plant remedies we discuss in this book are not secret. They have been recorded and published in a diverse range of texts, mainly academic. However, such materials are not generally available to a wider audience in the African rural areas. There is a strong case for disseminating information about the relevant plants, the plant combinations and the dosages used in combatting particular diseases – which are less generally recorded.

Promotion of and experimentation with local medicines

Scientists and experts from the Department of Agriculture may be uneasy about dissemination of local knowledge that has not been subject to systematic research or laboratory testing, and where issues such as dosing remain opaque. However, laboratory experiments have shown that some of the local plants in common use can have therapeutic properties, and they are frequently deployed. For example, Viola Maphosa, who researched *Elephantorrhiza elephantina* at Fort Hare, confirmed its value as an anthelmintic and advocated dissemination of this evidence to African rural communities. Broadcasting such findings might encourage stock owners to test plants and develop best practice. Poisonous plants are especially problematic, for example *sekaname,* used widely in North West Province. This plant is unlikely to gain endorsement from scientists because of uncertainty about dosage, but more information about it would be helpful to the many that use it. By validating local knowledge, cooperation between specialist officers and African smallholder farmers, could be expanded. Kobus Eloff, a phyto-chemist at the University of Pretoria, suggested experiments with easily available and relatively cheap solvents such as acetone, which might be more effective than water in releasing medicinal properties from plants. These could be made available to local farmers with instructions about drying, grinding and mixing.[6] When we raised this idea of using acetone at the report-back in Fafung, stockowners were uneasy and reluctant to experiment with new ways of extracting chemical properties.[7] People connected plant medicines with tradition and local identity.

Over-exploitation of plants and cultivation of medicinal species

Those researching plants used in traditional medicines have sounded alarms at the over-exploitation of valuable species, and a number of recommendations have been posited. On the one hand they advocate regulation of plant collection which is difficult to enforce; on the other they propose expanded cultivation of medicinal flora. There is clearly great potential in the latter strategy because it could provide both additional supply as well as a source of income for African rural communities.[8] A number of nurseries are already engaged in commercial production of species widely used in local medicine. Roodeplaat Agricultural Research Station is experimenting with mass cultivation of popular species. Specialists often prefer wild plants, but as we have noted, some species are already grown in gardens, such as aloes (chapter 6).

---

[6] Discussion with Prof Kobus Eloff, Pretoria, 25 February 2010.
[7] Report-back to farmers in Fafung, 29 November 2012.
[8] K. F. Wiersum, A. P. Dold, M. Husselman and M. L. Cocks, 'Cultivation of Medicinal Plants as a Tool for Biodiversity Conservation and Poverty Alleviation in the Amatola Region, South Africa', in R. J. Bogers, L. E. Craker and D. Lange (eds), *Medicinal and Aromatic Plants* (2006).

African Ideas about Diseases and Conditions Associated with the Environment

English	Setswana	Sesotho	isiXhosa (Mbotyi)	isiXhosa (E. Cape)	African ideas	Scientific explanations
Abortion / miscarriage	Pholotso	Folotsa	Ukupile	Ukuphunza	Associated with drought or malnutrition, women entering the cattle kraal or walking amongst cattle. Not generally associated with diseases.	Miscarriages can be due to diseases such as brucellosis, vibrosis and trichomoniasis, controllable by vaccination. Malnutrition and genetics can also play a part.
Anthrax	Lebete (spleen); Kwatsi		Mbende (spleen)	Ubende; nyamamakhwenkwe	Sudden death with bloody emissions from the orifices. Associated with grasses, dirty blood, witchcraft.	Death occurs from 2–72 hours after incubation. Fever, laboured respiration, bloody diarrhoea, swellings of the tongue and throat. Animals found dead with a black tarry discharge exuded from the nose, mouth and anus. Caused by bacteria bacillus anthracis. Controllable by vaccination. Zoonotic.
Blackquarter	Serotswana; Leotwana; Ramokutwane; Letsogwane; Sephatlho	Serotswana	IsiDiya	IsiDiya	Identified mainly by blackened and stinking muscles at post-mortem. Associated with grasses and malnutrition.	Only affects younger animals under 3 yrs. Often healthy animals are infected and mortality can be 100 per cent. Animals have severe swellings

English	Setswana	Sesotho	isiXhosa (Mbotyi)	isiXhosa (E. Cape)	African ideas	Scientific explanations
						and lameness especially in one hindquarter. Death within 24 hours of onset of symptoms. Often found dead in kraal. At post-mortem muscles appear spongy, dark and bloody and smell rancid. Caused by a bacterium Clostridium chauveoei or clostridium novyi. Controllable by vaccination.
Blood disease	Bolowetse ra madi		Iphela igazi		Associated with dirty blood or weakness leading to a diminished appetite and libido; maybe associated with blackquarter and anthrax.	No scientific classification.
Footrot	Tlhakwana	Mokaka			Found in muddy kraals; some associate it with bont tick lesions between digits of goats.	Hooves swell and the digits become separated. Painful; animals become lame. Caused by bacteria – fusobacterium necrophorum. Found in faeces, urine and water.
Gallsickness	Gala	Nyoko	Inyongo	Inyongo	Post-mortems reveal enlarged and split	Animals develop swollen gallbladders. Symptoms include

English name	Afrikaans	Zulu	Symptoms	Cause and control
			gallbladder which can yellow other organs. Symptoms of constipation, dry nose, lagging behind the herd. Cause usually lies in the seasonal change of grasses. Not generally seen as tick-borne.	fever; jaundice; refusal to graze; constipation; dry nose. Caused by bacteria of the genus Anaplasma. Blue ticks Boophilus decoloratus are the primary vectors. Controlled by dipping and vaccines.
Gifblaar (Afrikaans – toxic plant)	Mohau		Toxic plant that is most dangerous in spring before the rains come. Rapid death.	Cardiotoxicosis. Varied indistinct symptoms – salivation, breathing difficulties, slimy faeces, bellowing, staggering. Latent period of 24 hours during which the toxin is absorbed into the bloodstream from the rumen. Cattle often drop dead after exercise or drinking water.
Heartwater	Heartwater	Amanzana Umkhondo	Post-mortems reveal fluid around the heart and thorax. Animals are weak. Often regarded as a new disease. Some in NWP associate it with bont ticks.	Symptoms include fever, nervous symptoms, diarrhoea, death. Caused by Cowdria ruminantium, spread by bont tick (Amblyomma hebraeum). Controlled by dipping and antibiotics.
Horsesickness Sterf		Isimoliya	Environmental causes including worms in the veld.	Viral disease, spread by Culicoides (midges). Symptoms include fever, sweating, lethargy,

English	Setswana	Sesotho	isiXhosa (Mbotyi)	isiXhosa (E. Cape)	African ideas	Scientific explanations
						coughing, breathing difficulties, swellings of the eyes and head. Controlled by vaccination.
Botulism / Lamsiekte	Mokokomalo; Magetla (shoulder)				Symptoms of lameness, poor appetite and constipation. Attributed to poor grasses but not to bones or toxins.	Toxins paralyse the muscles of the body, including the tongue and tail. Caused by a very strong toxin (Clostridium botulinum) produced by bacteria growing in rotting carcases. Cattle suffering from a phosphorous deficiency are most likely to gnaw bones and ingest the toxin. Controllable by vaccination. Zoonotic.
Lumpy skin disease	Bolowetse ba letlano; Sekgwakgwa	Skin en vel	Umhlaza	Ungqakaqha	Lesions on the hide. In NWP and QwaQwa seen as a new disease and associated it with midges as the AHT had told them that. Mpondoland seen as general category of skin problems.	Caused by a virus. Spread by midges. Nasal discharge followed by nodules that appear all over the body from 1 to 4 cm in size. Nodules may disappear or else the skin dies off leaving open sores that can become infected by maggots and fail to heal easily. Controllable by vaccination.
Lung disease			Ihashe		Post-mortems reveal rotten lungs, stuck to thorax.	No scientific classification.

Disease	Local names	Local description	Veterinary description
Mastitis		Not specifically associated with infection, but tick damage.	Bacterial infections cause inflammation of the teats and discharge from mammary glands, occludes the udders. Control by antibiotics.
Myasis		Mpondoland associated with worms that creep into the brain.	Fly strike: flies lay their larvae in wounds caused by tick bites and other lesions. Can go septic.
Orf	Semmee; Dikakana; Selongwane	Lesions on the mouths of goats, caused by browsing on thorny vegetation	Pustules form around the lips and muzzle of sheep and goats. Caused by a parapoxvirus. Spread by grazing/browsing. Controlled by antibiotics.
Redwater (babesiosis)	Omo khibidu; Motlapologo khibidu; Mbendeni; Ihlwili; Amanzabomvu	Post-mortems reveal a swollen liver. Symptoms include red diarrhoea. Linked to dirty blood and seasonal changes to the grasslands.	Symptoms include fever, loss of appetite, red urine, anaemia, jaundice, death. Caused by Babesia parasites; spread by ticks such as blue tick (Boophilus decoloratus). Control by dipping, vaccines, antibiotics.
Retained placenta		Common problem throughout field sites but no explanations of causation.	Placenta fails to drop after birth. Causes can be malnutrition, disease or genetic defects.
Scab	Lepalo; Lekgwakgwa Ibuwa	Sores on a sheep; associated with biting	Scaly, crusty lesions on woolly areas of sheep and goats. Mites

English	Setswana	Sesotho	isiXhosa (Mbotyi)	isiXhosa (E. Cape)	African ideas	Scientific explanations
					insects	infect the ears. In serious cases anaemia and emaciation follow. Caused by Acari mites.
Slangkop (toxic plant)	Sekaname				Not generally seen as toxic, but an important medicinal plant. Some in NWP drench cattle with diluted slangkop as a form of inoculation.	Cardiotoxicosis. Indicators include apathy, drooping head, grinding teeth, rapid or irregular heart rate, rumen failure. Potentially rapid death from cardiac arrest.
Three-day stiff sickness			Nonkhwanyana		Associated with stiffness in limbs but not with a viral infection.	Virus spread by midges. Stiffness in limbs, animals lie down a lot. Often a nasal discharge and copious dribbling. Not usually fatal. Controllable by vaccine.
Tulp (toxic plant)	Teledimo	Tele			Seasonal poisoning associated with first rains and green grasses. Animals seem dizzy and suddenly drop dead. Some drench cattle with diluted tulp as a form of inoculation.	Cardiotoxicosis. Indicators include apathy, drooping head, grinding teeth, rapid or irregular heart rate, rumen failure. Potentially rapid death from cardiac arrest.
Worms (internal parasites)	Ndlozi		Izilo; Iskelem		Associated with grasslands and climatic conditions, especially	Causes diarrhoea, weight loss, anaemia.

| Worms (poisonous creatures in the veld) | Mohamba ka ndlwana; Sebokwana sa tlhaga | heat and humidity. Not linked to dung in Mpondoland.

NWP idea that worms are toxic and disappear with the burning off of dew. | Not recognised scientifically. |

Appendix 3:
African Ideas about Supernatural Causation

Setswana	Sesotho	isiXhosa (Mbotyi)	African Ideas
Mohato			Cross-infection – namely from women to cattle. Causes miscarriages and misfortune for the kraal.
	Bodisa		Women spoil kraal muthi
	Ditshila		Dirt associated with menstruation.
	Hlola		Bad luck associated with mourning
	Hosenya		'To ruin' – women ruin the cattle and protective muthi
	Senyama		Bad luck that women transmit to cattle in kraals
	Sequetho		Bad luck from menstruation
	Sesila		Dirt associated with death
	Tsheloa		Infertility conveyed by men who have recently had sex.
		Ichanti	Serpent associated with misfortune and disease, especially when its tracks are crossed.
		Imfene	Baboon familiar to which the difficulties of milking are ascribed.
		Umkhondo	Supernatural danger threatening livestock. Associated with muddy kraals, traces in the dew and menstruating women. Causes weakness and range of symptoms.

Appendix 4:
Plants and Diseases

These are plants that informants mentioned in their interviews. Most of these appear in the text. This is not a comprehensive list of all the plants that researchers have identified for these regions. We have tried to find the botanical name where possible. Sources for botanical names include John Mitchell Watt and Maria Gerdina Breyer-Brandwijk, *The Medicinal and Poisonous Plants of Southern and Eastern Africa* (Edinburgh, Livingstone, 1962); R. A. Paroz, *A List of Sotho Plant Names* (Maseru, Basutoland Scientific Association, 1962); R. B. Bhat and T. V. Jacobs, 'Traditional Herbal Medicine in Transkei' *Journal of Ethnopharmacology*, 48, 1 (1995), 7–12; Ben–Erik van Wyk , *Medicinal Plants of South Africa* (Pretoria, Briza, 1997); Ben–Erik van Wyk and Nigel Gericke, *People's Plants: A Guide to Useful Plants of Southern Africa* (Pretoria, Briza, 2007); A. P. Dold. and L. M. Cocks, 'Traditional Medicine in the Alice District of the Eastern Cape Province, South Africa', *South African Journal of Science*, 97, 9–10 (2001), 375–9; D. van der Merwe, G. E. Swan and C. J. Botha, 'Use of Ethnoveterinary Medicinal Plants by Setswana-speaking People in the Madikwe Area of the North West Province of South Africa', *Journal of the South African Veterinary Association*, 72, 4 (2001), 189–96'; PlantZAfrica website: www.plantzafrica.com.

Plants used in North West Province for natural and supernatural diseases

Plant – local name	English/Afrikaans/ Latin name where known	Usage	Method of usage	Where used
Kgophane/Mokgopa	Aloe spp.	Treatment for gallsickness and constipation; anthelmintic; purify the blood, make the blood bitter to repel ticks; poultice; tonic for all species of livestock; fertility enhancer;	Infusion made with water as a drench; crush leaves and mix with salt or ash to make a lick; grilled to release the sap to make a poultice for sores	Ubiquitous use in NWP

273

Plant – local name	English/Afrikaans/Latin name where known	Usage	Method of usage	Where used
Lengana	Wormwood; Artemeisa Afra	post-parturition cleanser; treatment of and prevention for anthrax		
		Coughs in goats (+humans)	Feed the leaves or make an infusion	Shakung
Lekatse		Retained placenta	Ground roots added to ash from thatch roofs and salt and water	
Lepate (also known as tshetlho)	Dicerocaryum eriocarpum/seneciodies	Retained placenta; venereal disease in bulls	Infusion	Ubiquitous use in NWP
Lethole	Hydnora johannis	Cleanse the blood; prevent reproductive diseases	Ground into a powder	Shakung
Makananguane	Dicerocaryum eriocarpum/seneciodies	Retained placenta; post-parturition cleanser; cure for gallsickness and blackquarter; disinfectant	Infusion made with water	Name for lepate in Mabeskraal
Makgonatsotlhe	Sphedamnocarpus pruriens	Retained placenta; eyewash; blackquarter	Infusion of roots	Magogoe
Malla digangwa		Fertility enhancer	Mixed with python fat and burnt in kraal	Mabeskraal

Manyana	Dagga, marijuana; Cannabis sativa	Preventative for horsesickness	Infusion made with coffee, salt and tobacco; added to water trough	Mantsa
Mathubadifala	Boophane disticha?	Fertility enhancer; venereal disease in bulls; post-parturition cleanser; blood cleanser	Crushed leaves mixed with crushed sekaname bulb	Bollantlokwe, Slagboom
Mogaga	Drimia altissima	Ritual cleansing; protects cattle from widows and menstruating women	Flakes of bulb scattered on earth or whole bulbs thrown into kraal; mixed with mosiama to make infusion	Mabeskraal
Mogato		Ulcers and abscesses, especially on udders	Ointment from leaves and goats milk	Mantsa
Mokeye	Acacia gerrardi	Protect the kraal from widows and menstruating women	Pieces of root mixed with mogaga and sekaname bulb	Mabeskraal
Mongololo or mongollo	Pouzilzia mixta?	Retained placenta and dystocia; cure for gallsickness, diarrhoea and poor appetite	Infusion made from roots; bark mixed with salt as a lick	Mabeskraal
Monna Maledu	Hypoxis hemerocallidea	Cleanse the blood; treat sores; diarrhoea	Ash from grilled plant dries up sores	Shakung; Sutelong
Moralla	Gardenia volkensii	Prevents anthrax	Burn bark in kraal	Mabeskraal
Moretlwa	Grewia flava	Prevents blackquarter	Twigs covered with pig fat and smeared on cattle	Kgomo Kgomo

Plant – local name	English/Afrikaans/ Latin name where known	Usage	Method of usage	Where used
Morototshwetsheue		*Laxative; cure for gallsickness; prevents anthrax; blood cleanser*	*Infusion from the roots; mixed with sekaname and mosalashopeng for anthrax*	*Sutelong; Kgomo Kgomo*
Mosalashopeng	*Withania somifera*	*Cure for anthrax*	*Mixed with sekaname and mosalashopeng.*	*Sutelong, Kgomo Kgomo*
Moshitluane		*Blindness; stiff joints*	*Decoction made from roots*	*Shakung*
Mosiama		*Mixed with mogaga for ritual cleansing*	*Crushed with pestle and mortar to make powder*	*Mabeskraal*
Mositsane	*Elephantorrhiza elephantina*	*Laxative; gallsickness*	*Crushed, added to water and dried.*	*Mafikeng Mabeskraal*
Motsha	*Acacia niolotica*	*Fractures*	*Used to make splints*	
Nata		*Tick repellent – makes blood biter*	*Twigs from the tree placed in feeding trough*	*Mabeskraal*
Phate ya ngaka	*Hermannia depressa?*	*Protection from sekaname poisoning*	*Decoction of roots and sekaname bulb*	
Rramburo	*Scilla natalensis?*	*Retained placenta*	*Soak the bulb in water to make slimy infusion*	*Shakung*

Setswana	Scientific name	Uses	Preparation	Locality
Sebete	*Senna italica*	Constipation; diarrhoea; gall-sickness; blood purifier; cure and preventative for anthrax	Decoction made from roots; sometimes roots are ground up and salt added	*Mabeskraal*
Sebete bete		Gallsickness; retained placenta; abortions	Creeper added to water	*Mafikeng; Mantsa*
Sekaname	*Drimia spp.*	Cleanse the blood; preventative and cure for anthrax; tick repellent – makes blood bitter; heals sores; treat blackquarter and redwater; constipation; anthelmintic; retained placenta; kraal medicine	Flakes of bulb steeped in water as a drink or disinfectant; ground up and placed in trough; mixed with porcupine or kudu dung for blackquarter	*Used throughout NWP*
Sekhalo	*Ziziphus zeyheri*	Diarrhoea in calves from too much milk	Infusion made from roots	*Ramatlabama*
Serokolo	*Carissa bispinosa*	Prevents cross-infection from humans to animals (mohato)	Infusion	*Mabeskraal*
Seswagadi	*Jatropha zeyheri*	Retained placenta	Decoction made from roots	*Mabeskraal*
Thobega	*Seddera sufruticosa*	Speeds up healing of fractures; placed on fractures and buried in ground	Burnt to make ash	*Mabeskraal*

Plants used in QwaQwa for natural and supernatural diseases

Plant – local name	English/Afrikaans/ Latin name where known	Usage	Method of usage
Bolele	*Algae spp.*	*Protect the kraal*	*Sprinkled around kraal*
Hloenya / Loenya	*Dicoma anomala*	*Improve fertility; gallsickness; tonic*	*Mixed with salt, water and mohaladitue for fertility; made into a poultice to rub onto vagina; mixed with mositsane for gallsickness*
Kgamane	*Rumex spp.*	*Gallsickness*	*Decoction from roots*
Kgapumpu	*Eucomis bicolor*	*Infertility drug*	*Infusion from leaves*
Kgware	*Pelargonium caffrum*	*Gallsickness*	*Decoction from crushed bulb*
Lebejana	*Asclepias spp.*	*Tick repellent; wounds; laxative for gallsickness*	*Infusion; may be mixed with aloes, or salt and vinegar for gallsickness*
Lekhala	*Aloe spp.*	*Gallsickness*	*Decoction from leaves*
Leshokgwa	*Nasturtium officinale / Xysmalobium undulatum*	*Gallsickness*	*Decoction made from ground bulbs*
Letuetlene		*Diarrhoea from green grasses*	*Decoction from root*

Moduane	Willow; Salix spp.	Anthelmintic	Scatter leaves in kraal
Mofifi	Rhamnus prinoides	Gallsickness; acaricide; appetite enhancer; footrot	Boil the leaves to make an a spray; drink for gallsickness
Mohaladitue	Zantedeschia albomaculata	Improve fertility; retained placenta	Infusion
Mohalakane	Aloe spp.	Gallsickness; constipation; emetic to remove plastics	Infusion; sometimes traditional beer is added
Mosisidi	Salvia spp.	Gallsickness	Decoction made from roots
Mositsane	Elephantorhizza elephantina	Diarrhoea; worms; retained placenta; treat gallsickness and blackquarter	Decoction made from bulb as laxative; ground up and added to salt as a lick
Mothimolo	Asclepias decipiens	Constipation	Infusion
Phate ya ngaka	Hermannia depressa?	Blackquarter, constipation	Decoction made from leaves
Podisa		Protection from witchcraft	Crushed leaves added to water to make a spray
Poho tshehla	Phytolocca spp. / Xysmalosium undulatum	Gallsickness; constipation	Decoction made from bulb; may add potassium permanganate
Qobo	Gunnera perpensa	Retained placenta	Decoction made from bulbs

Plant – local name	English/Afrikaans/ Latin name where known	Usage	Method of usage
Selepe	Hermannia geniculata	Protection from witchcraft	Crushed bulb made into a drink and mixed with other plants
Setimamollo		Blackquarter	Ash smeared on wire inserted in dewlap
Sheosheo	Athrixia fontana	Protect the kraal	Roots and bulb burnt in the kraal to protect cattle
Thobeha	Seddera sufruticosa	Fractures	Ash placed on the ground where animal fell; bulb shavings placed on wound
Tlwele	Calpurnia sericea	Diarrhoea	Infusion made from leaves; other plants may be added such as motoloana and mosokelo

Plants used in the Eastern Cape for natural and supernatural diseases

Plant – local name	English/Afrikaans/ Latin name where known	Usage	Method of usage	Where used
Amagamayo		Umkhondo	Burnt in kraal	Mbotyi
Balsemkopiva	Bulbine frutescens?	Gallsickness		Kat River
Ikahala	Aloe spp.	Gallsickness	Infusion	Kat River
Imfinguana	Cycad; Strangeria eriopus	Umkhondo	Root made into an infusion	Mbotyi
Imbizo		Infertility	Decoction or burnt in kraal	Mbotyi
Impitchi	Peach tree; Prunus persica	Tick wounds; myasis	Poultice	Mbotyi
Injalamba	Ipomoea spp.	Gallsickness	Creeper stem mixed with other plants to make an infusion	Mbotyi
Intoluane	Elephantorrhiza elephantina	Sheep dewormer; umkhondo; protect homesteads	Powdered root made into an infusion	Mbotyi
Intongana	Drimia robusta?	Redwater	Mixed with other plants to make an infusion	Mbotyi
Isibande sehlati	Long grass	Umkhondo	Mixed with other plants and burnt in kraal	Mbotyi

Plant – local name	English/Afrikaans/Latin name where known	Usage	Method of usage	Where used
Isihlungu	Acokanthera	Snakebite	Infusion	Mbotyi
Isifthi	Baphia racemosa	Gallsickness	Infusion	Peddie
Isiqungu	Thatch grass	Counteract effects of pollution on muthi	Chewed	Mbotyi
Mathunga	Haemanthus albiflos	Gallsickness	Infusion	Peddie
Mhlolo	Grewia lasiocarpa	Retained placenta	Mixed with other plants to make an infusion	Mbotyi
Nieshout (umthathi)	Sneezewood; ptaeroxylon obliquum	Gallsickness; three-day sickness	Mixed with other plants to make a decoction	Kat River
Ubuhlungu	Acokanthera	Gallsickness	Mixed with other plants to make an infusion for gallsickness	Kat River
Uluimi lenkomo	Gasteria spp.?	Redwater	Mixed with other plants to make an infusion	Mbotyi
Umbezo	Clutia pulchella	Umkhondo; witches' familiars		Mbotyi
Umayime	Clivia spp.	Protect homesteads	Mixed with other plants	Mbotyi
Umkhazi	Typha capensis	Retained placenta	Mixed with other plants to make an infusion	Mbotyi

Umkhomakhoma	Fern; Nephrodium athamanticum	Retained placenta		Mbotyi
Umkhumiso		Umkhondo	Mix of bark and leaves	Mbotyi
Umkhwenkhwe	Cheesewood; pittisporum viridiflorum	Gallsickness	Bark mixed with other plants to make an infusion	Mbotyi; Kat River
Umlung'mabele	Knobwood; zanthoxylon capense	Gallsickness	Leaf mixed with other plants to make an infusion	Mbotyi; Kat River
Umsintsi	Coastal coral tree; erythryna caffra	Retained placenta	Mixed with other plants to make an infusion	Mbotyi
Umzane	White ironwood; vepris lance-olata	Gallsickness	Leaf mixed with other plants to make an infusion	Mbotyi
Uqangazana	Clerodendrum glabrum	Redwater; sores; tick damage	Leaf mixed with other plants to make an infusion	Mbotyi
Uvendle	Pelargonium reniforme	Gallsickness	Mixed with aloes	Masakhane

283

Appendix 5:
Non-Plant Remedies

North West Province

Substance used	Disease or condition	Method of usage	Region
Alwyn (crystallised aloes)	Constipation	Substitute for aloe plants	Shakung
Blousal (washing powder)	Retained placenta; cure blindness	Mixed with water	Madidi; Mareetsane
Coffee	Boost appetite; tonic	Mixed with salt and licks	Mabeskraal; Lokaleng
Coca-Cola	Constipation; gallsickness; remove wrappers from stomachs; retained placenta	May be mixed with vinegar to make stronger laxative	Sutelong; Mabeskraal
Cow hide	Lumpy skin disease; warts	Put strips of cow hide around neck – cure	Sutelong; Mantsa
Dettol	Lumpy skin	Inject into the neck	Mabeskraal; Magogwe; Dithakong; Mafikeng
Diesel	Lumpy skin; fly strike	Use as an ointment	Mabeskraal
Fish oil	Laxative; gallsickness		Ramatlabama
Heated rods	Diarrhoea; stiffness	Heated rods placed on skin to create burnt stripes; burns under the tail for diarrhoea	Sutelong; Madidi
Jeyes Fluid	Dip to kill ticks; disinfect kraal	Sprinkled/smeared on cattle	Widespread use

Kudu dung	Protect cattle from blackquarter; mohau poisoning	Mixed with water; sometimes *sekaname* and sorghum beer are added	Sutelong; Slagboom; Kgomo Kgomo
Methylated spirits	Pain killer; diarrhoea		Shakung
Millipedes	Blindness	Dried and crushed and placed in eye	Widespread use
Mowa	Retained placenta	Ash scrapped off inside thatched roofs	Ramatlabama; Mantsa
Paraffin	Disinfect wounds; topical application against ticks; tick repellent	Smeared on wounds; added to salt as a lick	Shakung; Mabeskraal
Petrol	Disinfect wounds from ticks and fly strike; teat injuries	Smeared on lesions	Mabeskraal; Lokaleng
Pig fat	Sores	Ointment	Kgomo Kgomo; Mabeskraal
Potassium permanganate	Tonic; laxative; gallsickness; fractures; cure-all for chickens	Added to water trough	Widespread use
Salt	Laxative; anthelmintic; appetite enhancer; prevent blood diseases, anthrax, blackquarter	Blocks of salt placed in kraal as licks	Widespread use
Sorghum beer	Laxative; retained placenta; treatment for blackquarter and mohau poisoning	Drink; mixed with vinegar for blackquarter	Widespread use
Tar	Prevents blackquarter	Inserted into dewlap	Died out by 1980s in many villages but still used in Kgomo Kgomo and Mabeskraal
White sugar	Treat blindness	Sprinkled into the eye	Mabeskraal; Shakung
Wire	Prevent anthrax and blackquarter	Inserted into the dewlap	Old practice that was widespread but now dying out

Non-Plant Treatment and Preventatives – QwaQwa

Substance used	Disease or condition	Method of usage	Region
Cow hide	Lumpy skin disease; warts	Put strips of cow hide around neck – cure	
Fish oil	Laxative; gallsickness		
Jeyes Fluid	Dip to kill ticks	Sprinkled/smeared on cattle	
Madubula (tar–acid disinfectant)	Blackquarter	Mixed with water to make a drink	
Millipedes	Blindness	Dried and crushed and placed in eye	
Motor oil	Footrot; lesions from tick bites and scab	Soak sacks with oil at entrance to kraal to cure footrot; smear on wounds for bites	
Potassium permanganate	Tonic; laxative; gallsickness	Added to water trough or mixed with plants like mositsane	
Salt	Laxative; anthelmintic; appetite enhancer; cure blackquarter	Blocks of salt placed in kraal as licks	
Seawater	Protection from witchcraft	Drink	
Soot	Cleanse the reproductive tract of cows	Mixed with water; drink	
Wire	Prevent anthrax and blackquarter	Inserted into dewlap; smeared with ash from burnt setimamolla shrub	

Non-Plant Treatment and Preventatives – Eastern Cape

Substance used	Disease or condition	Method of usage	Region
Cowhide	Warts		Kat River
Cuttle fish or squid shell	Eye problems		Mbotyi
Epsom Salts	Gallsickness	Mixed with water to make a laxative	Kat River
Jeyes Fluid	Sores		Mbotyi
Madubula (tar–acid disinfectant)	Wounds		Mbotyi
Motor oil	Tick control	Mixed with dip	Mbotyi
Pig fat	Protect the kraal	Smeared on pegs	Mbotyi
Pig dung	Umkhondo	Mixed with crayfish eggs	Mbotyi
Salt water	Gallsickness; internal parasites	Drink	Peddie
Umxhube (mixture of maize & sorghum)	Gallsickness	Mixed with other plants	Kat River

Select Bibliography

Ainslie, Andrew (ed.). *Cattle Ownership and Production in the Communal Areas of the Eastern Cape, South Africa.* Programme for Land and Agrarian Studies, Research Report 10 (Cape Town: University of Western Cape, 2002).

Ainslie, Andrew. 'Farming Cattle, Cultivating Relationships: Cattle Ownership and Cultural Politics in Peddie District, Eastern Cape', *Social Dynamics*, 31, 1 (2005), 129–56.

Ainslie, Andrew. 'Keeping Cattle? The Politics of Value in the Communal Areas of the Eastern Cape Province, South Africa'. Unpublished PhD, University College, London, 2005.

Ashforth, Adam. *Witchcraft, Violence and Democracy in South Africa* (Chicago: University of Chicago Press, 2005).

Barth, F. 'An Anthropology of Knowledge', *Current Anthropology*, 43, 1 (2002), 1–18.

Beinart, William. 'African History and Environmental History', *African Affairs*, 99, 395 (2000), 269–302.

Beinart, William. 'Environmental Origins of the Pondoland Revolt', in Stephen Dovers, Ruth Edgecome and Bill Guest (eds), *South Africa's Environmental History: Cases and Comparisons* (Cape Town: David Philip, 2002), 76–89.

Beinart, William. 'Ethnic Particularism, Worker Consciousness and Nationalism: the Experience of a South African Migrant, 1930–1960', in S. Marks and S. Trapido (eds), *The Politics of Race, Class and Nationalism in Twentieth Century South Africa* (London: Longman, 1987), 286–309.

Beinart, William. 'Soil Erosion, Conservationism and Ideas about Development: a Southern African Exploration', *Journal of Southern African Studies*, 11, 1 (1984), 52–83.

Beinart, William. *The Political Economy of Pondoland, 1860–1930* (Cambridge: Cambridge University Press, 1982).

Beinart, William. *The Rise of Conservation in South Africa: Settlers, Livestock and the Environment, 1770–1950* (Oxford: Oxford University Press, 2003).

Beinart, William. 'Transhumance, Animal Diseases and Environment in the Cape, South Africa', *South African Historical Journal*, 58, 1 (2007), 17–41.

Beinart, William. 'Transkeian Smallholders and Agrarian Reform', *Journal of Contemporary African Studies*, 11, 2 (1992), 178–99.

Beinart, William. *Twentieth Century South Africa* (Oxford: Oxford University

Press, 2001).

Beinart, William. 'Vets, Viruses and Environmentalism: the Cape in the 1870s and 1880s', *Paideuma*, 43 (1997), 227–52.

Beinart, William, Brown, Karen and Gilfoyle, Daniel. 'Experts and Expertise in Colonial Africa Reconsidered: Colonial Science and the Interpenetration of Knowledge', *African Affairs*, 108, 432 (2009), 413–33.

Beinart, William and Wotshela, Luvuyo. *Prickly Pear: the Social History of a Plant in the Eastern Cape* (Johannesburg: Wits University Press, 2011).

Brown, Karen. 'Poisonous Plants, Pastoral Knowledge and Perceptions of Environmental Change in South Africa, c. 1880–1940', *Environment and History*, 13, 3 (2007), 307–32.

Brown, Karen. 'Tropical Medicine and Animal Diseases: Onderstepoort and the Development of Veterinary Science in South Africa 1908–1950', *Journal of Southern African Studies*, 31, 3 (2005), 513–29.

Brown, Karen. 'Veterinary Entomology, Colonial Science and the Challenge of Tick-borne Diseases in South Africa during the late Nineteenth and early Twentieth Centuries', *Parassitologia,* 50, 3–4 (2008), 305–19.

Bundy, Colin. '"We Don't Want Your Rain, We Won't Dip": Popular Opposition, Collaboration and Social Control in the Anti-Dipping Movement 1908–16,' in W. Beinart and C. Bundy, *Hidden Struggles in Rural South Africa* (London: James Currey, 1987), 191–221.

Cocks, M. L. *Wild Resources and Cultural Practices in Rural and Urban Households in South Africa: Implications for Bio-Cultural Diversity Conservation* (Grahamstown: Rhodes University, 2006).

Coetzer, J. A. W. and Tustin, R. C. (eds), *Infectious Diseases of Livestock* (Oxford: Oxford University Press, 2004).

Digby, Anne. *Diversity and Division in Medicine: Health Care in South Africa from the 1800s* (Oxford: Peter Lang, 2006).

Dold. A. P. and Cocks, M. L. 'Traditional Medicine in the Alice District of the Eastern Cape Province, South Africa', *South African Journal of Science*, 97, 9–10 (2001), 375–9.

Dold, A. P. and Cocks, M. L. *Voices from the Forest: Celebrating Nature and Culture in Xhosaland* (Auckland Park: Jacana, 2012).

Ernst, Waltraud (ed.), *Plural Medicine, Tradition and Modernity, 1800–2000* (London: Routledge, 2002).

Gehring, Ronette. 'Veterinary Drug Supply to Subsistence and Emerging Farming Communities in the Madikwe District, North West Province, South Africa'. MMedVet (Pharmacology), University of Pretoria, 2001.

Gilfoyle, Daniel. 'Anthrax Vaccination in South Africa: Economics, Experiment and the Mass Vaccination of Animals, c. 1900–1945', *Medical History*, 50, 4 (2006), 465–90.

Gilfoyle, Daniel. 'Veterinary Research and the African Rinderpest Epizootic: The Cape Colony 1896–1898', *Journal of Southern African Studies*, 29, 1

(2003), 133–54.

Gilfoyle, Daniel. 'Veterinary Science and Public Policy at the Cape 1877–1910'. Unpublished DPhil thesis, University of Oxford, 2002.

Green, Edward. *Indigenous Theories of Contagious Disease* (Walnut Creek, CA: Alta Mira Press, 1999).

Hajdu, Flora. *Local Worlds: Rural Livelihood Strategies in Eastern Cape, South Africa* (Linkoping: Linkoping University, 2006).

Hammond-Tooke, W. D. *Bhaca Society: A People of the Transkeian Uplands South Africa* (Cape Town: Oxford University Press, 1962).

Hammond-Tooke, W. D. *Boundaries and Belief: The Structure of a Sotho Worldview* (Johannesburg: Witwatersrand University Press, 1981).

Hammond-Tooke, W. D. *Rituals and Medicine: Indigenous Healing in South Africa* (Johannesburg: Donker, 1989).

Hlatshwayo, M. and Mbati, P. A. 'A Survey of Tick Control Methods used by Resource-poor Farmers in the Qwa-Qwa Area of the Eastern Free State Province, South Africa', *Onderstepoort Journal of Veterinary Research*, 72, 3 (2005), 245–9.

Hunter, Monica. *Reaction to Conquest: Effects of Contact with Europeans on the Pondo of South Africa* (London: International Institute of African Languages and Cultures, 1936).

Jenjezwa, Vimbai. 'Stock Farmers and the State: A Case Study of Animal Healthcare Practices in Hertzog, Eastern Cape Province, South Africa'. Unpublished MA, University of Fort Hare, 2010.

Kellerman, T. S., Coetzer, J. A. W., Naude, T. W., Botha, C. J. *Plant Poisonings and Mycotoxicoses of Livestock in Southern Africa* (Oxford: Oxford University Press, 2005).

Kepe, Thembela. 'Medicinal Plants and Rural Livelihoods in Pondoland South Africa: Towards an Understanding of Resource Value', *International Journal of Biodiversity Science and Management*, 3, 3 (2007), 170–83.

Le Vaillant, Francois. *Travels into the Interior Parts of Africa by Way of the Cape of Good Hope in the Years 1780, 1781, 1782, 1783, 1784 and 1785* (London: G. G. & J. Robinson, 1790).

Maphosa, Viola. 'Determination and Validation of Plants used by Resource-limited Farmers in the Ethno-veterinary Control of Gastro-intestinal Parasites of Goats in the Eastern Cape Province, South Africa'. Unpublished PhD thesis, University of Fort Hare, 2009.

Masika, P. J., Sonandi, A. and van Averbeke, W. 'Perceived Causes, Diagnosis and Treatment of Babesiosis and Anaplasmosis in Cattle by Livestock Farmers in Communal Areas of the Central Eastern Cape Province, South Africa', *Journal of the South African Veterinary Association*, 68, 2 (1997), 40–4.

Masika, P. J., Sonandi, A. and van Averbeke, W. 'Tick Control by Small-Scale Farmers in the Central Eastern Cape Province, South Africa', *Journal of the South African Veterinary Association*, 68, 2 (1997), 45–8.

Masika, P. J., van Averbeke, W. and Sonandi, A. 'Use of Herbal Remedies by Small-scale Farmers to Treat Livestock Diseases in Central Eastern Cape Province, South Africa', *Journal of the South African Veterinary Association*, 71, 2 (2000), 87–91.

McGaw, L. J. and Eloff, J. N. 'Ethnoveterinary Use of Southern African Plants and Scientific Evaluation of their Medicinal Properties', *Journal of Ethnopharmacology*, 119,3 (2008), 559–74.

Mekonnen, S., Bryson, N. R., Fourie, L. J., Peter, R. J., Spickett, A. M., Taylor, R. J., Strydom T. and Horak, I. G. , 'Acaricide Resistance Profiles of Single- and Multi-host Ticks from Communal and Commercial Farming Areas in the Eastern Cape and North-West Provinces of South Africa', *Onderstepoort Journal of Veterinary Research,* 69, 2 (2002), 99–105.

Moyo, B. 'Determination and Validation of Ethno-Veterinary Practices used as Alternatives in Controlling Cattle Ticks by Resource-limited Farmers in the Eastern Province, South Africa'. Unpublished MSc dissertation in Animal Science, Fort Hare University, 2008.

Moyo B. and Masika, P. J. 'Tick Control Methods used by Resource-limited Farmers and the Effect of Ticks on Cattle in Rural Areas of the Eastern Cape', *Tropical Animal Health and Production*, 41, 4 (2009), 517–23.

Ngubane, Harriet. *Body and Mind in Zulu Medicine: An Ethnography of Health and Disease in Nyuswa-Zulu Thought and Practice* (London: Academic Press, 1977).

Niehaus, Isak. *Witchcraft, Power and Politics: Exploring the Occult in the South African Lowveld* (Cape Town: David Philip, 2001).

Pauw, Berthold. 'Widows and Ritual Danger in Sotho and Tswana Communities', *African Studies,* 49, 2 (1990), 75–99.

Sekokotla, Malesela. 'Assessing Implementation of Veterinary Extension on Control of Cattle Parasites in Moretele District North West Province'. Unpublished MSc (Veterinary Sciences), University of Pretoria 2004.

Smith, Andrew. *Contribution to South African Materia Medica*, third edn (Grahamstown: Cory Library, Rhodes University, 2011). First published by Lovedale Press, 1895.

Soyelu, O. T. and Masika, P. J. 'Traditional Remedies used for the Treatment of Cattle Wounds and Myiasis in Amatola Basin, Eastern Cape Province, South Africa', *Onderstepoort Journal of Veterinary Research*, 76, 4 (2009), 393–7.

Spickett, A. M., van der Merwe, D. and Matthee, O. 'The Effect of Orally Administered *Aloe Marlothii* Leaves on *Boophilus Decoloratus* Tick Burdens on Cattle', *Experimental and Applied Acarology,* 41 (2007), 139–46.

Tamarkin, Mordechai. *Volk and Flock: Ecology, Identity and Politics among Cape Afrikaners in the Late Nineteenth Century* (Pretoria: University of South Africa Press, 2009).

Van der Merwe, D. 'Use of Ethnoveterinary Medicinal Plants in Cattle by Setswana-speaking people in the Madikwe area of the North West Province'. Unpublished MSc Thesis, University of Pretoria 2000.

Van der Merwe, D., Swan, G. E. and Botha, C. J. 'Use of Ethnoveterinary Medicinal Plants by Setswana-speaking People in the Madikwe Area of the North West Province of South Africa', *Journal of the South African Veterinary Association*, 72, 4 (2001), 189–96.

Van Wyk, Ben-Erik. *Medicinal Plants of South Africa* (Pretoria: Briza,1997).

Van Wyk, Ben-Erik and Gericke, Nigel. *People's Plants: A Guide to Useful Plants of Southern Africa* (Pretoria: Briza, 2007).

Watt, John Mitchell and Breyer-Brandwijk, Maria Gerdina. *The Medicinal and Poisonous Plants of Southern and Eastern Africa* (Edinburgh: Livingstone, 1932 and 1962).

Index